AF361761

Cantilever-Based Sensors

Cantilever-Based Sensors

Editors

Bruno Tiribilli
Paolo Paoletti
Joao Mouro

MDPI • Basel • Beijing • Wuhan • Barcelona • Belgrade • Manchester • Tokyo • Cluj • Tianjin

Editors
Bruno Tiribilli
Institute of Complex Systems
CNR - National Research
Council
Sesto Fiorentino (Firenze)
Italy

Paolo Paoletti
Department of Mechanical,
Materials and Aerospace
Engineering
School of Engineering,
University of Liverpool
Liverpool
United Kingdom

Joao Mouro
Institute of Complex Systems
CNR - National Research
Council
Sesto Fiorentino (Firenze)
Italy

Editorial Office
MDPI
St. Alban-Anlage 66
4052 Basel, Switzerland

This is a reprint of articles from the Special Issue published online in the open access journal *Sensors* (ISSN 1424-8220) (available at: www.mdpi.com/journal/sensors/special_issues/cantilever_based_sensors).

For citation purposes, cite each article independently as indicated on the article page online and as indicated below:

LastName, A.A.; LastName, B.B.; LastName, C.C. Article Title. *Journal Name* **Year**, *Volume Number*, Page Range.

ISBN 978-3-0365-1652-3 (Hbk)
ISBN 978-3-0365-1651-6 (PDF)

Contents

Preface to "Cantilever-Based Sensors"

Microcantilevers are typically rectangular-shaped beams, approximately 100 μm 200 microns long, 20 μm 40 microns wide, and 0.5 μm 1 microns thick, and are made of silicon or silicon nitride. Their mechanical response is often described as a very soft spring. The static deformation of a cantilever allows for detection of the smallest forces with unprecedented sensitivity, whereas the resonance frequency of its dynamic response can be used to measure extremely small masses or fluid properties. Cantilever-based sensors have received considerable interest in the last few decades as they offer an unparalleled opportunity for the development of highly sensitive biophysical and chemical sensors, employed in a very wide spectrum of applications. These sensors have been widely utilized in electronics, automotive and aerospace systems, biophysics, environmental monitoring, and medical diagnosis sectors, among others. Their working principle is often based on the interaction between a micrometric cantilever and its surrounding medium, where the mechanical device responds to changes in some environmental property, such as temperature, pressure, flow, density, viscosity, or the presence of some analytes of interest. The resonance frequency response of a microcantilever can be used to measure extremely small masses or fluid properties.

An interesting and widely used application of micromechanical sensors is as gas sensors. The authors in [1] developed a fully coupled finite-element model for the frequency response of a cantilever-based photoacoustic gas sensor. In this work, the amplitude of vibration of a microcantilever was used as a transducer for the pressure wave generated by the gas molecules to be detected. The model was validated with experimental data and used to optimize and miniaturize previously reported sensor designs.

The authors in [2] developed a technique for imprinting particles directly on a triangular-shaped cantilever and used the resonant response of the cantilever to determine the adsorbed mass on the surface. The number of particles on the surface was counted for further validation of the results by inspecting the cantilever using scanning electron microscopy (SEM).

The review paper presented in [3] discusses the scanning and actuation techniques for cantilever-based fiber optic endoscopic scanners. These micromechanical components contribute to further miniaturization of endoscopes, resulting in the ability to image previously inaccessible small-caliber luminal organs, enabling the early detection of lesions and other abnormalities in these tissues.

Piezo-resistive silicon cantilevers are promising candidates in performing high-speed surface roughness measurements due to their low mass, low probing force, and high signal linearity. The authors of [4] investigated the trackability of two different cantilever microprobes by building a theoretical dynamic model, by measuring their resonant response, and by performing experiments on reference surfaces.

The review in [5] consists of a comprehensive analysis of the rich dynamical response of a microcantilever oscillating in a viscous fluid and how this response can be used for mass and rheology sensing. This paper also takes a look at future challenges in the field, such as practical implementations of sensors, the interaction of devices with non-Newtonian viscoelastic fluid, or the expected progress in the microfabrication of new geometries.

All of the previous contributions demonstrate how new cantilever-based sensors are continuously investigated for the development of innovative applications. However, this Special Issue also collects several contributions about atomic force microscopy (AFM), which is an extraordinary technique that reached maturity many years ago. High-quality, commercially available instruments are largely employed in a variety of fields providing impressive quality images. Nevertheless, many researchers continue to study new solutions to improve the application of this technique and to push its limits.

The recent advancements in microcantilever-based force sensors for AFM applications together with the recent advancements in using high-speed AFM for the study of biological samples were reviewed in [6].

The authors in [7] proposed a new theoretical model for use in contact resonance atomic force microscopy. This model captures the tip-sample interaction by imposing appropriate boundary conditions in the Euler–Bernoulli beam equation. The predictive capabilities of the model were verified with finite-element analysis.

Tracking of biological and physiological processes on the nanoscale is fundamental in nanomedicine. AFM techniques are typically used in this area, but these are severely limited when non-transparent fluids, such as blood, are used due to the limitations caused by the optical readout. The authors in [8] proposed a new solution where the optical readout is replaced by piezo-resistors embedded in the cantilever, allowing them to present images of the hemostatic process of blood coagulation in a turbid liquid.

AFM techniques can also be used to the measure structural and nanomechanical properties of tissues or cell–cell forces to elucidate elusive phenomena in physiology and pathology. In [9], the authors used AFM to capture the topographical details of stromal collagen fibrils in porcine cornea and to calculate their elastic modulus as a function of progressive amylase treatment. The response of the ultrastructure of the cornea to chemicals can be important in understanding a number of ocular disorders. In [10], the authors detected the small forces involved in cell–cell interaction using a cell that adheres to the substrate and a second cell attached to the cantilever free end. The importance of accurately controlling the experimental condition is outlined by the authors, as shown by the influence of the experimental conditions on the shape of the cell–cell force curves.

Biomedical applications will benefit from the development of a new Scanning Probe Microscopy (SPM) configuration, as proposed in [11]. This paper reported a series of design solutions that significantly increase the stability, precision, and usability of a modified Lateral Molecular Force Microscope. The full automation of this new SPM configuration with force sensitivity and stability tailored for biological samples offers a particularly attractive instrumentation for future biomedical applications.

Aknowledgments

The editors express their gratitude to all authors for their enthusiastic and interesting contributions and to all reviewers for their invaluable work in making sure that the published papers are rigorous and of interest to the community. Finally, we also direct our appreciation to all members of the {Sensors} editorial office for their kind support and readiness during the creation of this Special Issue.

BIBLIOGRAPHY

1 - Zhou, S. Fully Coupled Model for Frequency Response Simulation of Miniaturized Cantilever-Based Photoacoustic Gas Sensors. Sensors 2019, 19(21), 4772; https://doi.org/10.3390/s19214772.

2 - Nyang'au, W.; Setiono, A.; Schmidt, A.; Bosse, H.; Peiner, E. Sampling and Mass Detection of a Countable Number of Microparticles Using on-Cantilever Imprinting. Sensors 2020, 20(9), 2508;

https://doi.org/10.3390/s20092508.

3 - Kaur, M.; Lane, P.; Menon, C. Scanning and Actuation Techniques for Cantilever-Based Fiber Optic Endoscopic Scanners—A Review. Sensors 2021, 21(1), 251; https://doi.org/10.3390/s21010251.

4 - Xu, M.; Li, Z.; Fahrbach, M.; Peiner, E.; Brand, U. Investigating the Trackability of Silicon Microprobes in High-Speed Surface Measurements. Sensors 2021, 21(5), 1557; https://doi.org/10.3390/s21051557.

5 - Mouro, J.; Pinto, R.; Paoletti, P.; Tiribilli, B. Microcantilever: Dynamical Response for Mass Sensing and Fluid Characterization. Sensors 2021, 21(1), 115; https://doi.org/10.3390/s21010115.

6 - Alunda, B.; Lee, Y. Review: Cantilever-Based Sensors for High Speed Atomic Force Microscopy. Sensors 2020, 20(17), 4784; https://doi.org/10.3390/s20174784.

7 - Jaquez-Moreno, T.; Aureli, M.; Tung, R. Contact Resonance Atomic Force Microscopy Using Long, Massive Tips. Sensors 2019, 19(22), 4990; https://doi.org/10.3390/s19224990.

8 - Leitner, M.; Seferovic, H.; Stainer, S.; Buchroithner, B.; Schwalb, C.; Deutschinger, A.; Ebner, A. Atomic Force Microscopy Imaging in Turbid Liquids: A Promising Tool in Nanomedicine. Sensors 2020, 20(13), 3715; https://doi.org/10.3390/s20133715.

9 Kazaili, A.; Abdul-Amir Al-Hindy, H.; Madine, J.; Akhtar, R. Nano-Scale Stiffness and Collagen Fibril Deterioration: Probing the Cornea Following Enzymatic Degradation Using Peakforce-QNM AFM. Sensors 2021, 21(5), 1629; https://doi.org/10.3390/s21051629.

10 - Oropesa-Nuñez, R.; Mescola, A.; Vassalli, M.; Canale, C. Impact of Experimental Parameters on Cell–Cell Force Spectroscopy Signature. Sensors 2021, 21(4), 1069; https://doi.org/10.3390/s21041069.

11 - Szeremeta, W.; Harniman, R.; Bermingham, C.; Antognozzi, M. Towards a Fully Automated Scanning Probe Microscope for Biomedical Applications. Sensors 2021, 21(9), 3027; https://doi.org/10.3390/s21093027.

Bruno Tiribilli, Paolo Paoletti, and Joao Mouro
Editors

Article

Fully Coupled Model for Frequency Response Simulation of Miniaturized Cantilever-Based Photoacoustic Gas Sensors

Sheng Zhou [1,2]

[1] Department of Physics and Astronomy, VU University Amsterdam, 1081 HV Amsterdam, The Netherlands; s.zhou@vu.nl

[2] LaserLaB Amsterdam, VU University Amsterdam, 1081 HV Amsterdam, The Netherlands

Received: 10 October 2019; Accepted: 1 November 2019 ; Published: 2 November 2019

Abstract: To support the development of miniaturized photoacoustic gas sensors, a fully coupled finite element model for a frequency response simulation of cantilever-based photoacoustic gas sensors is introduced in this paper. The model covers the whole photoacoustic process from radiation absorption to pressure transducer vibration, and considers viscous damping loss. After validation with experimental data, the model was further applied to evaluate the possibility of further optimization and miniaturization of a previously reported sensor design.

Keywords: photoacoustic spectroscopy; viscous damping; sensor miniaturization; COMSOL

---

1. Introduction

Photoacoustic (PA) spectroscopy has been recognized as a sensitive trace gas detection technique [1]. In a typical PA system, a modulated light beam with a proper wavelength is sent into a gas sample cell where target gas molecules are excited by the beam. A pressure wave is produced by the periodic thermal expansion of the gas sample results from collisional relaxation of excited gas molecules and is detected by various pressure transducers. The detected pressure wave amplitude is used to evaluate the concentration of target gas in the sample. Commonly, the acoustic resonance of the gas column in the cell is used to improve the PA signal and a commercial microphone mounted on the gas cell is used to detect the pressure wave. Finite element methods (FEM) have been successfully implemented to evaluate the eigen frequencies, modes, and frequency response of PA cells with complex geometries for cell design optimization, as analytical solutions are restricted to simple cell geometries [2–4]. Furthermore, it has been demonstrated that FEM can even quantitatively simulate the PA signal [5]. Most FEM models, however, do not include the pressure transducer in the model. This is feasible when commercial microphones, which normally have a flat frequency response under 20 KHz, were used as the pressure transducer. Those microphones can be regarded as part of the rigid wall of the PA cell when the PA signal frequency is much lower than 20 KHz. However, with the development of a new branch of PA systems that utilize the mechanical resonance of the pressure transducers for PA signal improvement, it becomes essential to include the pressure transducer in the FEM model and consider the acoustic-vibration coupling to simulate the PA system properly [6,7]. Additionally, with the continuous development of miniaturized PA sensors, it is important to include the viscous damping in the FEM model. In fact, viscous damping may play a dominant role on PA signal reduction when the viscous penetration depth is at the same order of the gas cell dimensions [8,9].

Recently, we introduced a series of cantilever-based miniaturized PA sensors to the field in which customized micro-cantilevers were used as the pressure transducer and glass tubes with an inner diameter as small as 0.6 mm was used as gas cells [10–12]. To guide the development of such gas sensors towards further optimization and miniaturization, a fully coupled sensor model based on Comsol Multiphysics is proposed in this paper. The model covers the whole photoacoustic process from radiation absorption to pressure wave generation, and then to transducer vibration. Viscous damping was included as an energy loss mechanism. This model is applied to a PA sensor reported in [12], showing that it can match the experimental data on the sensor frequency response quantitatively well. Simulations of how different sensor parameters were going to influence the PA signal are further given, suggesting the possibilities and limitations of miniaturization and signal enhancement for cantilever-based PA sensors.

2. Experimental Setup

The simulation model is based on an experimental setup as schematically shown in Figure 1, which was described in detail in [12]. The PA sensor is shown on the right part of the figure. It mainly consists of an one end sealed transparent glass tube as the cell and a cantilever beam hanging a micromirror over the cell inlet as the pressure transducer. The micromirror was made from a piece of square glass plate (width = 300 µm, thickness = 30 µm), and was made transparent at the center and reflective elsewhere. Two fibres were aligned towards the micromirror, of which the excitation fiber transmits the excitation laser beam into the cell through the transparent region of the micromirror, while the readout fiber points towards the reflective region so that an interferometer (OP1550, OPTICS11) it is connected to can read the micromirror deflection signal. When the wavelength of the excitation laser (Oclaro, TL5000VCJ) is modulated around one of the target gas' absorption line (1530.37 nm for Acetylene in this case) by a current source, gas molecules inside the cell are heated up periodically and hence generate a pressure wave that vibrates the micromirror. A lock in amplifier (SRS865, SRS) connected to the interferometer was used to extract the micromirror vibration amplitude signal at 2nd harmonica of the excitation frequency. All components of the PA sensor were enclosed in a gas permeation tube because the sensor was designed to be immersed in transformer oil for dissolved C_2H_2 detection in our experiments. A photo of the inner structure of the sensor part in the setup was included in the figure to indicate the scale (optical fibres in the figure have a diameter of 0.125 mm). To be able to compare the real sensor performance with the simulation model described hereafter, the sensor's frequency response was collected when it was immersed in an oil standard sample (True North) with a certificated dissolved C_2H_2 concentration ($10(1 \pm 10\%)$ ppm). During data collection, the central excitation wavelength was fixed at the C_2H_2 absorption line and the excitation frequency was swept around half the resonance frequency of the transducer.

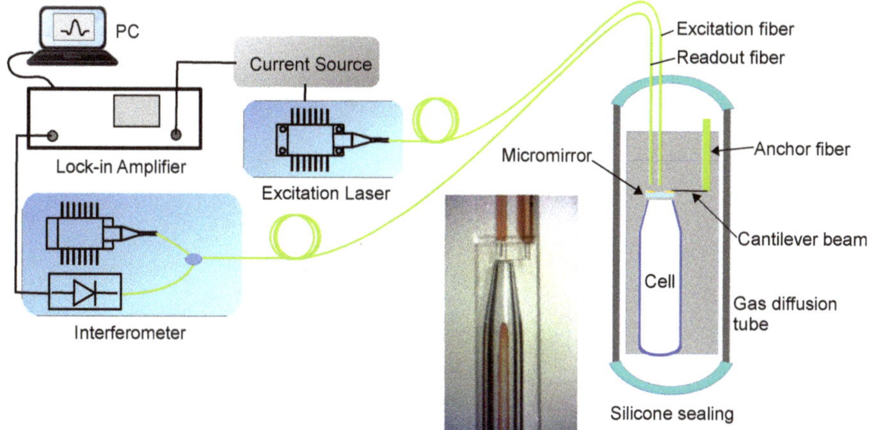

Figure 1. Experimental setup.

3. Modeling of the PA Sensor

A sketch of the PA sensor modeled in Comsol Multiphysics is shown in Figure 2a. To reduce the computational cost, a 2D axisymmetric model was used. The cantilever pressure transducer was simplified with a round silica glass micromirror (diameter of 300 µm, thickness of 30 µm) combined with a spring foundation constraint that was applied to it. Ideally, the spring constant of the spring foundation should be set to the same as that of the cantilever beam of the pressure transducer. However, because the mechanical properties of the cantilever beam were unknown in our experiments, the spring constant was chosen to ensure that the resonance frequency value resolved by the model overlaps with that of the experiment's result. The permeation tube of the sensor was also not included in the model for simplicity. Key parameters of the experimental setup that were directly implemented in the model are listed in Table 1.

Table 1. Parameters used for the model.

Parameter	Value	Description
P	24	Excitation laser power (mW)
t1	30	Micromirror thickness (µm)
r1	300	Cell main body inner radius (µm)
l1	13	Cell length (mm)
r2	133	Cell inlet inner radius (µm)
t2	50	Cell wall thickness (µm)
g	15	Cell inlet to micromirror gap size (µm)
l2	1.5	Cell inlet section length (mm)

In order to get a correct assessment of the damping loss around the micromirror and the cell inner wall, and to include the radiation absorption process directly into the model, the gas in the gas cell and around the cell inlet was modeled with a thermoviscous interface by simply using the material properties of standard air defined in the Comsol library. Because of the viscous property and much higher thermal conductivity of the cell walls, they were treated as no-slip and isothermal boundaries. A pressure acoustics layer was used to truncate the computational domain. The cell wall in the pressure acoustic domain was set as a sound hard boundary. The outer spherical perimeter of the pressure acoustic domain was set as spherical wave radiation condition so that acoustic wave experiences little reflection on the perimeter. Other parameters, such as radius of domains and maximum mesh element size, were chosen to ensure the convergence of the simulation results (e.g., parameter independent frequency response, and relative error estimations smaller than 0.005), as well as reasonable computational time.

The energy transfer from the excitation laser beam to the gas molecules in the cell was modeled using a heat source condition in the air domain inside the cell, with heat power density calculated below.

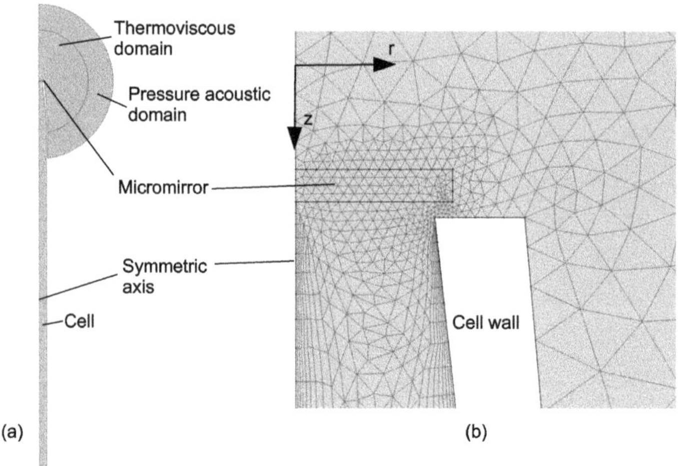

Figure 2. Model of the PA (photoacoustic) sensor (**a**) and mesh at the micromirror region (**b**).

Assuming that the transparent window at the center of the micromirror does not affect the excitation laser beam propagation, the beam divergence from the excitation fiber tip can be treated as a Gaussian beam with an initial beam diameter (full $1/e^2$ width), $2w_0$, equal to the mode-field diameter of the fiber (10.5 μm). The beam radius at a position z (set $z = 0$ at the excitation fiber tip surface) can be described as:

$$w = w_0\sqrt{1 + (\frac{\lambda z}{\pi w_0^2})^2} \tag{1}$$

with λ being the wavelength. The light intensity of the beam can be then described as:

$$I(r,z) = I_0 \exp(-\frac{2r^2}{w^2}) \tag{2}$$

$$I_0 = kP \exp(\frac{2}{\pi w^2}) \tag{3}$$

where I_0 represents the on-axis light intensity, P is the power of the excitation laser, and r is defined as the distance to the beam axis. k is the transmission coefficient of the excitation laser beam from the laser to the cell, which considers the coupling loss between fiber connectors and power loss due to micromirror reflection, absorption and scattering. The heat power density at any location in the domain, which defines the heat source input in the model, is finally calculated as [6]:

$$H(r,z) = \alpha I(r,z) \tag{4}$$

where α represent the absorption coefficient of the target gas in the cell. According to the calibration result detailed in [12], a dissolved C_2H_2 concentration of 10 ppm in the transformer oil corresponds to a gas phase C_2H_2 concentration of 8 ppm in the gas cell. This leads to an absorption coefficient of $9.26 \times 10^{-4}\,\text{m}^{-1}$, calculated at atmosphere pressure and room temperature, based on the HITRAN database [13].

The mesh around the micromirror region is presented in Figure 2b. To properly resolve the excitation laser beam, boundary layer mesh elements were applied to the region around the cell axis, with a first layer thickness of $w_0/2$. Similarly, to include the viscous damping properly, boundary layer mesh elements were applied to the air close to the cell wall with a first layer thickness of $dvisc/5$ and edge mesh elements were applied to the cell inlet to a micromirror gap region with a maximum element size of $min(dvisc/3, gap/3)$, where $dvisc$ is the viscous penetration depth [14], defined as:

$$dvisc = \sqrt{\frac{\mu}{\pi\rho f}} \tag{5}$$

with μ as the dynamic viscosity, ρ as the static density, and f as the PA signal frequency.

4. Results

The model was run to evaluate the micromirror vibration amplitude as a function of the heat power density modulation frequency. The simulated pressure profile at the resonance frequency is included in Figure A1. The simulated vibration amplitude of the micromirror central point around the first Eigen mode is compared to that measured in the experiment, as shown in Figure 3. From the frequency response curves, one can calculate the quality factor of the system [15]. The quality factor was calculated to be 10 for both experimental and simulation results and is independent from the transmission coefficient value set in the model. This indicates that the model can simulate the viscous damping that occurred in the sensor very well. Moreover, when k was set to 0.75, the simulated frequency response of the sensor almost overlapped with the experimental one. When k was varied between 0.5 and 1, the simulated frequency response was at the same order as that measured with the experiment. This result shows that the proposed model could simulate the frequency response of real PA gas sensor qualitatively well.

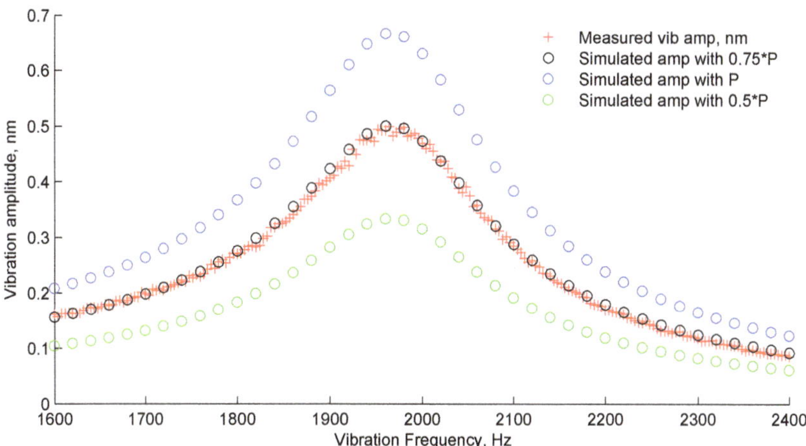

Figure 3. Experimental and simulated sensor frequency response.

It is well known that, distinguished from conventional absorption spectroscopy, the signal in photoacoustic spectroscopy is independent from the cell length and becomes higher when the cell radius is reduced [16]. With increasing interests developed around photonic chip based gas sensors [17–20], it would be interesting to evaluate the miniaturization potential of cantilever-based PA sensors, as a guidance for the possible photonic integration of photoacoustic gas sensors in the future. Parametric sweeps based on the above model were carried out for this purpose. Single parameters including cell length, cell inner radius, cell inlet to micromirror gap size, and cantilever spring constant were swept around their designed values, and the sensor frequency response was simulated and collected. Three parameters, including resonance frequency, micromirror vibration amplitude at resonance, and quality factor, were further extracted from the simulated frequency response and plotted in Figure 4.

To calculate the quality factors, simulated frequency response curves were fitted to a displacement amplitude function of forced oscillation [15]:

$$A = \frac{b}{\sqrt{(f^2 - f_0^2)^2 + (rf)^2}} \tag{6}$$

where f_0, r, b are the fitting parameters. The fitting reached a relative residual smaller than 10% for all points. The quality factor is then calculated by:

$$Q = \frac{f_0}{r} \tag{7}$$

As seen in Figure 4, the oscillation behavior of the frequency response while sweeping the cell length suggests some complex coupling between the gas volume and the cantilever. Assuming an acoustic velocity of 340 m/s, the acoustic wavelength (λ) at 1940 Hz is about 176 mm, which is approximately two times the period of the oscillations. However, according to this simulation, the cell length at around 25 mm leads to the highest PA signal, even though it is much shorter than $\lambda/4$. There is a proper hypothesis for this behavior.

Firstly, due to the diameter shrinkage at the cell outlet region and/or the influence of the air gap between the micromirror and the cell outlet, a cell with a length of 25 mm acts equivalent to a $\lambda/4$ long tube resonator with one end closed, which has a fundamental acoustic resonance of around 1940 Hz.

Secondly, when increasing the cell length from 25 mm, every increment of cell length by $\lambda/2$ creates a new match of the cantilever mechanical resonance with a higher acoustic resonance mode of the tube resonator, which leads to a new vibrational amplitude peak.

This hypothesis is partially verified by the pressure profile data captured while the cell length was set to 25 mm, 105 mm, and 185 mm respectively, as shown in Figure A2. Pressure standing waves are clearly visible at those cell lengths.

While sweeping the spring constant, the vibration amplitude at resonance ($f = f_0$) doesn't follow the inverse of frequency ($1/f$) accordingly as defined by Equation (6). This could be explained by a match between the acoustic resonance of the air volume and the mechanical resonance of cantilever, when the spring constant is around 3 N/m.

Furthermore, increasing the cell radius could increase the PA signal most significantly, which shows that the cell diameter in the current sensor design is not optimal. The model predicts that the PA signal increased by two times if the cell radius was changed from 0.3 mm to 0.8 mm, while other parameters are fixed.

The modeling results for gap size sweep indicates that when the gap size was reduced from 15 μm to 5 μm, the PA signal doubled, which is likely due to better coupling between the pressure wave and the pressure transducer. A further reduction of the gap size reduced the PA signal a lot, most likely due to the so-called "breathing effect" of the narrow gap region [21].

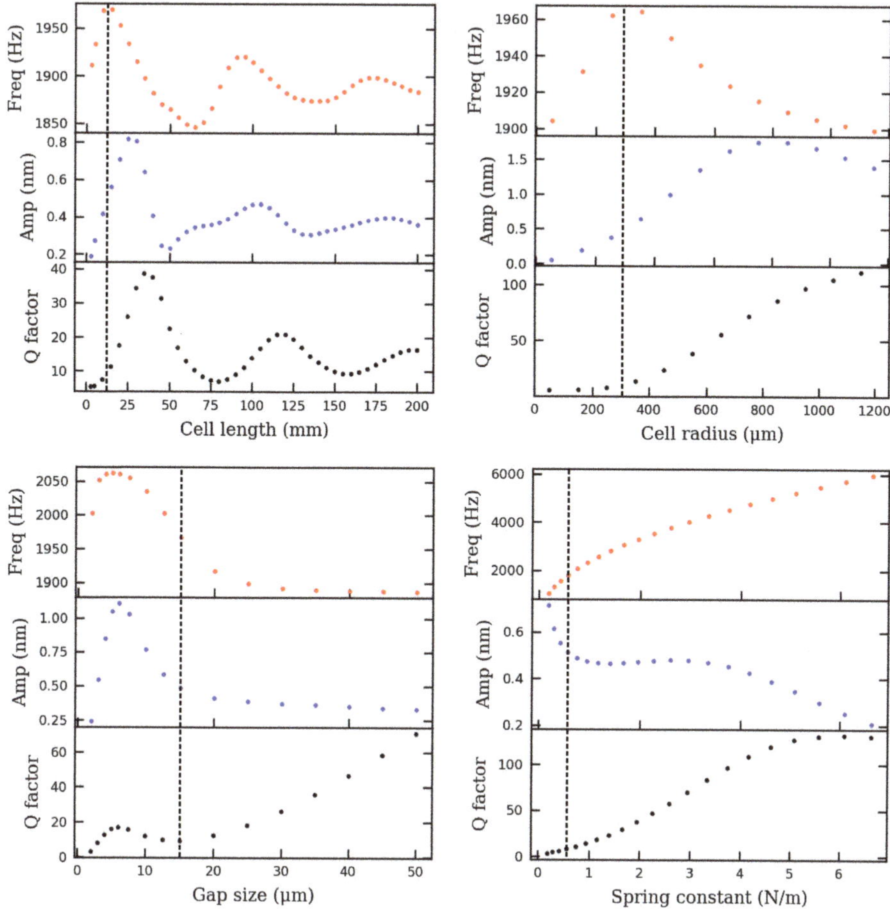

Figure 4. Influences of geometrical parameters to resonance frequency, cantilever vibration amplitude, and quality factor of the PA sensor. Dashed lines indicate original sensor parameters.

5. Conclusions and Discussion

In conclusion, a fully coupled cantilever-based PA sensor model was introduced in this paper. It was applied to a miniaturized PA sensor reported in [12], showing that it could match experimental results quantitatively well.

The model was further applied to investigate how different sensor parameters were going to influence the sensor performance. According to the parametric sweep results, it seems that the signal to noise ratio of the sensor reported in [12] could be further improved by reducing the gap size, considering environmental acoustic noise as the dominant noise source [11]. Moreover, it seems that matching the acoustic and mechanical resonances of the PA cell and the pressure transducer could be an effective signal enhancement method that is worth further investigation. Furthermore, as seen from the results, the cell radius and the cell length were already limiting the sensing performance of the current sensor design. Further miniaturization is expected to increase the influence of viscous damping on the cell wall and hence deteriorate the gas sensing performance even further. Based on the modeling results, a semi photonic integration approach, where all components other than the gas cell are integrated into a photonic circuit, might combine the high sensitivity of photoacoustic spectroscopy with the mass production potential of photonic chips in future.

Funding: This work was supported by European Research Council (ERC) (Grant No. 615170) and LASERLAB-EUROPE (Grant No. 654148).

Conflicts of Interest: The author declares no conflict of interest.

Appendix A. Simulated Pressure Profile at Resonance Frequencies

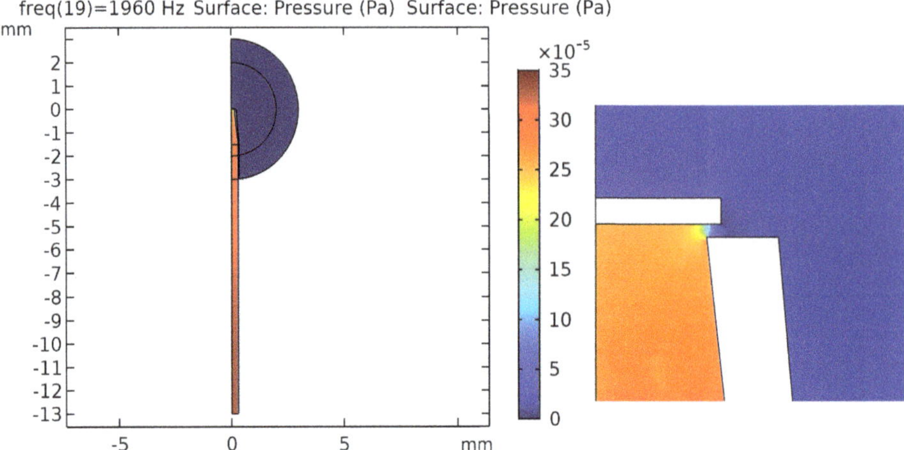

Figure A1. The left figure shows the simulated pressure profile of the sensor cell at resonance frequency. The right figure shows an enlarged view of the cell gap region.

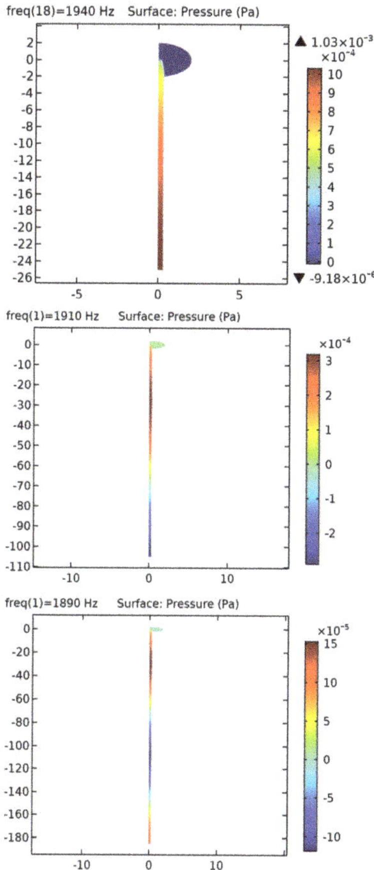

Figure A2. Pressure profile in the cell at resonance frequencies when the cell length is set to 25 mm, 105 mm, and 185 mm respectively. The x axis of each graph is manually scaled for visibility.

References

1. Sigrist, M.W. Trace gas monitoring by laser photoacoustic spectroscopy and related techniques (plenary). *Rev. Sci. Instrum.* **2003**, *74*, 486–490. [CrossRef]
2. Baumann, B.; Wolff, M.; Kost, B.; Groninga, H. Finite element calculation of photoacoustic signals. *Appl. Opt.* **2007**, *46*, 1120–1125. [CrossRef]
3. Haouari, R.; Rochus, V.; Lagae, L.; Rottenberg, X. Topology Optimization of an Acoustical Cell for Gaseous Photoacoustic Spectroscopy using COMSOL® Multiphysics. In Proceedings of the COMSOL Conference, Rotherdam, UK, 16 November 2017; pp. 1–6.
4. Duggen, L.; Frese, R.; Willatzen, M. FEM analysis of cylindrical resonant photoacoustic cells. In *Journal of Physics: Conference Series*; IOP Publishing: Bristol, UK, 2010; Volume 214; p. 012036.
5. Parvitte, B.; Risser, C.; Vallon, R.; Zéninari, V. Quantitative simulation of photoacoustic signals using finite element modelling software. *Appl. Phys. B* **2013**, *111*, 383–389. [CrossRef]
6. Firebaugh, S.L.; Roignant, F.; Terray, E.A. Modeling the response of photoacoustic gas sensors. In Proceedings of the COMSOL Conference, Boston, MA, USA, 8–10 October 2009.

7. Ma, Y.F.; Tong, Y.; He, Y.; Long, J.H.; Yu, X. Quartz-enhanced photoacoustic spectroscopy sensor with a small-gap quartz tuning fork. *Sensors* **2018**, *18*, 2047. [CrossRef] [PubMed]

8. Parvitte, B.; Risser, C.; Vallon, R.; Zéninari, V. Modelization of Photoacoustic Trace Gases Sensors. In Proceedings of the 2012 COMSOL Conference, Milan, Italy, 10–12 Ocotber 2012.

9. Glière, A.; Rouxel, J.; Parvitte, B.; Boutami, S.; Zéninari, V. A coupled model for the simulation of miniaturized and integrated photoacoustic gas detector. *Int. J. Thermophys.* **2013**, *34*, 2119–2135. [CrossRef]

10. Zhou, S.; Slaman, M.; Iannuzzi, D. Demonstration of a highly sensitive photoacoustic spectrometer based on a miniaturized all-optical detecting sensor. *Opt. Express* **2017**, *25*, 17541–17548. [CrossRef] [PubMed]

11. Zhou, S.; Iannuzzi, D. A fiber-tip photoacoustic sensor for in situ trace gas detection. *Rev. Sci. Instrum.* **2019**, *90*, 023102. [CrossRef] [PubMed]

12. Zhou, S.; Iannuzzi, D. Immersion photoacoustic spectrometer (iPAS) for arcing fault detection in power transformers. *Opt. Lett.* **2019**, *44*, 3741–3744. [CrossRef] [PubMed]

13. Rothman, L.S.; Gordon, I.E.; Babikov, Y.; Barbe, A.; Benner, D.C.; Bernath, P.F.; Birk, M.; Bizzocchi, L.; Boudon, V.; Brown, L.R.; et al. The HITRAN2012 molecular spectroscopic database. *J. Quant. Spectrosc. Radiat. Transf.* **2013**, *130*, 4–50. [CrossRef]

14. Blackstock, D.T. *Fundamentals of Physical Acoustics*; ASA: Monroe, MI, USA, 2001.

15. Moloney, M.J.; Hatten, D.L. Acoustic quality factor and energy losses in cylindrical pipes. *Am. J. Phys.* **2001**, *69*, 311–314. [CrossRef]

16. Firebaugh, S.L.; Jensen, K.F.; Schmidt, M.A. Miniaturization and integration of photoacoustic detection. *J. Appl. Phys.* **2002**, *92*, 1555–1563. [CrossRef]

17. Robinson, J.T.; Chen, L.; Lipson, M. On-chip gas detection in silicon optical microcavities. *Opt. Express* **2008**, *16*, 4296–4301. [CrossRef] [PubMed]

18. Yebo, N.A.; Bogaerts, W.; Hens, Z.; Baets, R. On-chip arrayed waveguide grating interrogated silicon-on-insulator microring resonator-based gas sensor. *IEEE Photonics Technol. Lett.* **2011**, *23*, 1505–1507. [CrossRef]

19. Xie, Z.; Cao, K.; Zhao, Y.; Bai, L.; Gu, H.; Xu, H.; Gu, Z.Z. An optical nose chip based on mesoporous colloidal photonic crystal beads. *Adv. Mater.* **2014**, *26*, 2413–2418. [CrossRef] [PubMed]

20. Tombez, L.; Zhang, E.; Orcutt, J.; Kamlapurkar, S.; Green, W. Methane absorption spectroscopy on a silicon photonic chip. *Optica* **2017**, *4*, 1322–1325. [CrossRef]

21. Koskinen, V.; Fonsen, J.; Kauppinen, J.; Kauppinen, I. Extremely sensitive trace gas analysis with modern photoacoustic spectroscopy. *Vib. Spectrosc.* **2006**, *42*, 239–242. [CrossRef]

Article

Sampling and Mass Detection of a Countable Number of Microparticles Using on-Cantilever Imprinting

Wilson Ombati Nyang'au [1,2,*], **Andi Setiono** [1,3], **Angelika Schmidt** [1], **Harald Bosse** [4] and **Erwin Peiner** [1]

[1]   Institute of Semiconductor Technology (IHT) and Laboratory for Emerging Nanometrology (LENA), Technische Universität Braunschweig, D38106 Braunschweig, Germany

[2]   Department of Metrology, Kenya Bureau of Standards (KEBS), 00200 Nairobi, Kenya

[3]   Research Center for Physics, Indonesian Institute of Sciences (LIPI), Kawasan Puspiptek Serpong, Tangerang Selatan 15314, Indonesia

[4]   Precision Engineering Division, Physikalisch-Technische Bundesanstalt (PTB), 38116 Braunschweig, Germany

*   Correspondence: wilombat@tu-braunschweig.de

Received: 20 March 2020; Accepted: 24 April 2020; Published: 28 April 2020

Abstract: Liquid-borne particles sampling and cantilever-based mass detection are widely applied in many industrial and scientific fields e.g., in the detection of physical, chemical, and biological particles, and disease diagnostics, etc. Microscopic analysis of particles-adsorbed cantilever-samples can provide a good basis for measurement comparison. However, when a particles-laden droplet on a solid surface is vaporized, a cluster-ring deposit is often yielded which makes particles counting difficult or impractical. Nevertheless, in this study, we present an approach, i.e., on-cantilever particles imprinting, which effectively defies such odds to sample and deposit countable single particles on a sensing surface. Initially, we designed and fabricated a triangular microcantilever sensor whose mass m_0, total beam-length L, and clamped-end beam-width w are equivalent to that of a rectangular/normal cantilever but with a higher resonant frequency (271 kHz), enhanced sensitivity (0.13 Hz/pg), and quality factor (~3000). To imprint particles on these cantilever sensors, various calibrated stainless steel dispensing tips were utilized to pioneer this study by dipping and retracting each tip from a small particle-laden droplet (resting on a hydrophobic n-type silicon substrate), followed by tip-sensor-contact (at a target point on the sensing area) to detach the solution (from the tip) and adsorb the particles, and ultimately determine the particles mass concentration. Upon imprinting/adsorbing the particles on the sensor, resonant frequency response measurements were made to determine the mass (or number of particles). A minimum detectable mass of ~0.05 pg was demonstrated. To further validate and compare such results, cantilever samples (containing adsorbed particles) were imaged by scanning electron microscopy (SEM) to determine the number of particles through counting (from which, the lowest count of about 11 magnetic polystyrene particles was obtained). The practicality of particle counting was essentially due to monolayer particle arrangement on the sensing surface. Moreover, in this work, the main measurement process influences are also explicitly examined.

Keywords: piezoresistive microcantilever mass sensor; resonant frequency; dispensing tip; droplet; particle sampling; adsorption; PMMA; magnetic polystyrene particles

---

1. Introduction

The need and demand for a cost-effective and reliable fluid-based particles sampling and counting technique is of an inestimable significance, e.g., in biomedicine, to detect physical/chemical and biological particles, and diagnosis of diseases [1], etc. Conveyance and gravimetric detection of these particles with suspended microchannel resonators [2,3] has recently been achieved with negligible damping and high mass sensitivity. The sensors (e.g., microcantilevers) can also, for instance, be dipped into a solution containing particles or analytes of interest such as the feline coronavirus to adsorb and detect them [4–6]; but, low mechanical quality factor Q and randomized particles adsorption are the inevitable outcomes. Alternatively, liquid-borne media can most conveniently be transferred onto a sensing surface through droplet dispensing coupled with solvent evaporation. Nonetheless, a ring-cluster of particles (also called coffee-ring effect) is often observed at the edges of a dried liquid droplet [7–10]. This is a typical phenomenon that is manifested, for instance, after the evaporation of impure water droplets on a solid surface, deposition of DNA/RNA microarrays with functional and particle coatings [11], disease diagnostics and drug discovery [12], lithography patterning [13], particle and biomolecule separation and concentration [11].

The coffee-ring phenomenon is majorly caused by the pinning of a contact line of the drop edges to the substrate, and the radial outward-flow from the center (of the droplet) of carrier liquid during evaporation, which eventually transports the suspended particles to the rim [14]. Moreover, the particles should adhere to the substrate surface and the evaporation rate be high near the edge of the droplet. Consequently, the solvent that is lost to the ambient atmosphere (through evaporation at the rim of the droplet) is primarily compensated by the fluid flow (accompanied with the solutes/particles) from the center of the droplet.

The particle ring deposits have, however, been eliminated or suppressed by various techniques. For instance, Yuinker et al. (2011) used ellipsoidal-shaped or a mixture of both spherical and a small number of ellipsoidal suspended particles [15] to suppress the cluster-ring effect. Elsewhere, the ring phenomenon has been managed and suppressed by controlling and optimizing of drop temperature [16], using surfactants [17], and tuning the particle concentration and droplet size [18], etc. It should be noted, however, that in cases where determination of particle concentration (or number of particles) is necessary, the cluster-ring deposits (see Figure 1) make particles counting extremely difficult or even impractical. The latter is quite explicit particularly if the adsorbed particles form non-uniform multilayers on the solid surface. By tuning the particle concentration, conventional liquid dispensing [19,20] can be utilized to deposit and realize a relatively small particle concentration [18]. This is, however, a pressure-driven process, and the dispensing tips are often inevitably clogged [20]. With dip-pen nanolithography [21,22], an atomic force microscope (AFM) tip (used as a pen) is dipped into a desired molecular ink; and then the sampled ink (coated on the apex of the atomically sharp tip) is transferred directly onto the substrate (from the tip/meniscus to the meniscus/surface interface). But this is a serial process characterized with low throughput. Moreover, limited substrates and inks can be used with this method. Additionally, the expensive and fragile micro/nano-sized AFM tips deployed in this scanning-probe-based direct-writing method limits the versatility of the technique. Similarly, using a polymer stamp, i.e., poly(dimethylsiloxane) (PDMS), with a predesigned pattern, micro-contact-based printing [23] can be applied to pattern self-assembled monolayers (SAMs) and deliver numerable particles onto substrate surfaces. This approach is however difficult to integrate with resonant mass sensors.

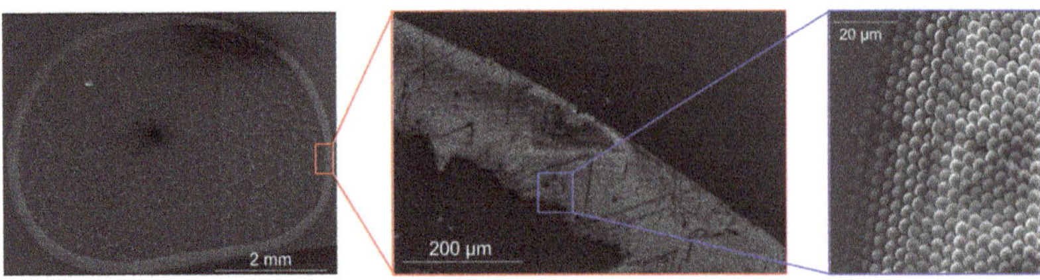

Figure 1. Typical SEM image of a cluster-ring deposit of polymethylmethacrylate (PMMA) particles arising from droplet dispensing on an *n*-type (100) silicon-substrate surface.

Consequently, an easy-to-implement practical approach for assembling a monolayer or uniformly multi-layered particles on a sensing surface is of desirable interest. In this study, a particle-imprint method is presented as a flexible and versatile approach for delivering small countable amounts of particle samples on solid surfaces. Unlike droplet dispensing, this method does not require dispensing air pressure nor complicated equipment. Thus, making it a cost-efficient alternative technique for sampling and depositing particles on sensing surfaces.

The particle-imprint method (in this work) involves dipping a dispensing tip into an arbitrary sized particles-laden droplet followed by substrate/sensor contact to deposit the particles. With this method, a specified tip can be dipped and retracted from a small droplet containing assorted monodispersed microparticles (μPs). Afterwards, the particles-laden tip immediately contacts the cantilever sensing area for a defined time duration (to deposit the particles). During this tip/sensor contact period, we have shown that some particle solution detaches from the tip and adheres onto a hydrophilic sensing surface (i.e., silicon bulk substrates and microcantilevers). In congruence with our initial hypothesis, monolayer particles arrangement on these sensors and substrates has also been realized.

In the subsequent sections, we therefore present details on how assorted particle samples, i.e., PMMA and magnetic polystyrene particles, were localized and deposited onto in-house fabricated microcantilever sensors utilizing our present approach and setup. PMMA particles (in a functionalized state, i.e., if the particle surface is bound with surfactant molecules) are most widely used as biomedical materials due to their biocompatibility. Recently, these particles have increasingly been applied as drug carriers (e.g., antibiotics), fillers for cosmetic and dental surgery, and for vaccine formulation [24] and colon cancer treatment [25]. Moreover, their application in colloidal lithography has recently been demonstrated [26]. Similarly, functionalized magnetic polystyrene particles have found manifold applications in various fields. Of interest is their use in separating biomolecules (e.g., antibodies, proteins, and nucleic acids etc.) [27], and separating and sorting of cells [28]. They are also utilized as tracers for magnetic particle imaging (MPI) [29], agents for diagnostics and targeted destruction of cancer tumors through local delivery of heat (hyperthermia), and for image contrast enhancement of diseased tissues [30] and targeted drug delivery [31]. In addition, they have notably been used in environmental pollution mitigation to remove oil from waste water [32]. In our study though, we have primarily utilized unfunctionalized forms of PMMA and magnetic polystyrene particles (as discussed in the subsequent section) and determined their mass concentration on the sensing surface. Further work to simultaneously determine the magnetic moment from the mass of magnetic polystyrene particles, utilizing a (modified) measurement setup, is intended. This will help to characterize and effectively render the use of these particles in magnetic resonance imaging (MRI) and MPI applications. In the current study, the particles samples (deposited by particle-imprint method) and adsorbed on in-house fabricated microcantilever mass sensors were quantified based on in-plane resonant frequency f_0 response measurements and vividly compared with particle counts from scanning electron microscope (SEM) images.

2. Materials and Sensor Fabrication

2.1. Particle Samples

In this study, we used magnetic polystyrene particles (micromer®-M, hereafter denoted as MPS) from micromod Partikeltechnologie GmbH, Rostock, Germany; and polymethylmethacrylate (PMMA) from Sigma-Aldrich Inc., St. Louis, Missouri, USA. The nominal particle diameters (assuming spherical shape) and densities were about 2 μm and 1.1 gcm^{-3} (MPS μPs); 2 μm and 1.18 gcm^{-3} (PMMA μPs). Experimental samples were prepared by tuning the particle concentration by diluting the original solution with deionized water to realize various concentration levels ranging from approximately 0.01 mg/mL to 2 mg/mL.

2.2. Calibration of Dispensing Tips

Prior to use, our stainless-steel dispensing tips were calibrated (at Physikalisch-Technische Bundesanstalt (PTB)) to determine their geometrical dimensions and shapes. This was performed using X-ray computed tomography (xCT) and optical reference measurements utilizing a coordinate measuring machine (CMM). The inner and outer diameters of the stainless-steel tip (from Nordson EFD Inc., East Providence, RI, USA), shown in Figure 2, were about 0.117 mm ± 10 µm and 0.236 mm ± 1 µm, respectively. The tips were cylindrical; and all the diametric measurements were taken at intervals of about 0.5 mm by fitting a circle to the determined surface. The assigned uncertainty values were computed with about 95 % confidence interval ($k = 2$). The main uncertainty factors considered included the repeatability of the measured diameters and the maximum permissible sphere distance error (for inner diameter) and maximum length measurement error (for outer diameter).

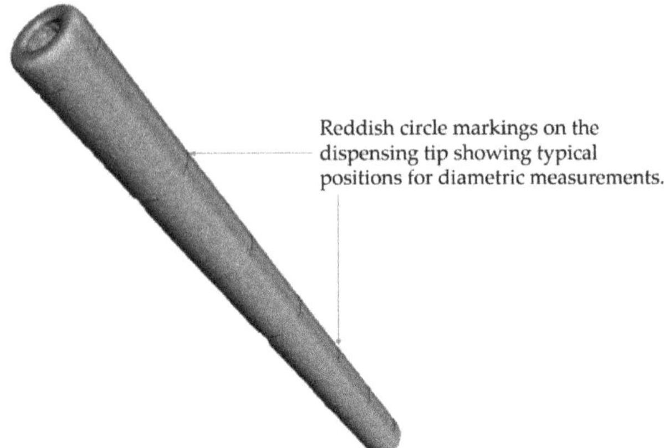

Figure 2. X-ray computed tomography (xCT) image showing a 3D rendering of the surface of capillary of a stainless-steel dispensing tip.

2.3. Cantilever Sensor Design and Fabrication

In Table 1, we show the cantilever geometric dimensions and the simulated characteristics by finite-element modeling (FEM) using Comsol Multiphysics 4.4b. The free-end configurations of these microcantilevers (as depicted in Figure 3) were either rectangular or triangular, and the thickness t of all the sensors are essentially fixed (i.e., $t = 15$ µm). The triangular free-end of first type of triangular cantilever (TCant1, cf: Figure 3b) is equilateral-shaped (with sides = 700 µm, and length L_2 = 606 µm), and it is positioned at $L_1 = 394$ µm from the fixed-end of the cantilever (with beam width w = 170 µm). The length of the rectangular segment $L_1 = 394$ µm $\approx 2/5L$, where L is the total cantilever length of TCant1 i.e., $L_1 + L_2$; which corresponds to the length $L = 1000$ µm of the regular/rectangular cantilever (RCant1), as depicted in Figure 3a. The two sensors have different cantilever masses, i.e., $m_0 = 9.76$ µg (TCant1) and $m_0 = 5.94$ µg (RCant1). Similarly, their flexural fundamental resonant frequencies (and sensitivities) also differ significantly with $f_0 \approx 185.0$ kHz ($\rightarrow \sim 0.04$ Hz/pg) and $f_0 \approx$ 220 kHz ($\rightarrow \sim 0.07$ Hz/pg) corresponding to TCant1 and RCant1 sensors, respectively. In the current study, we have designed a second type of triangular cantilever (TCant2, cf: Figure 3c) with an isosceles triangular-free-end whose base b (just like TCant1 sensor) is positioned at $L_1 = 394$ µm (from fixed-end) but its magnitude is twice the clamping beam width (i.e., $b = 2w$). The total cantilever length L (i.e., from the fixed-end to apex/free-end) for TCant2 sensor was nonetheless the same as TCant1 and RCant1 sensors (i.e., $L = 1000$ µm). Additionally, by tuning the base-width b of the triangular-free end of TCant2 sensor (to $b = 2w$), the cantilever mass ($m_0 = 5.94$ µg) is rendered equivalent to the mass of the regular cantilever (RCant1) sensor (having same total length L and fixed-beam width

w). Although RCant1 and TCant2 sensors have the same L and m_0, a larger resonant frequency and mechanical quality factor was, however, observed for TCant2 ($f_0 \approx 271$ kHz, $Q \sim 3000 \pm 150$) than RCant1 ($f_0 \approx 220$ kHz, $Q \sim 2000 \pm 200$), as shown in Table 1. This is supposedly due to their geometrical (shape) differences. Moreover, the isosceles (triangular fee-end) in TCant2 could possibly cause little damping and lesser effective cantilever mass compared to the rectangular counterpart. Nevertheless, further work is required to unravel the factors behind the observed differences. For cantilevers of equal length and similar free-end geometries but different triangular base-widths ($\rightarrow$ different m_0 values) i.e., TCant2 and TCant1 sensors, their fundamental in-plane resonant frequencies, mechanical quality factors, and sensitivities also differ accordingly (as shown in Table 1). TCant1 cantilever ($m_0 = 9.76$ µg) yields a smaller $f_0 \approx 185$ kHz ($\rightarrow \sim 0.04$ Hz/pg) compared to $f_0 \approx 271$ kHz ($\rightarrow \sim 0.09$ Hz/pg) from TCant2 ($m_0 = 5.94$ µg). Moreover, the latter offers an enhanced mass sensitivity and improved quality factor. It should be noted that all our sensors were designed to be excited in the in-plane bending mode of vibration. The main material parameters used in cantilever FEM simulations in Comsol Multiphysics were density (2.33 gcm^{-3}), Young's modulus (170 GPa) and Poisson ratio (0.28) for silicon, and volume (i.e., geometrical dimensions of the sensor). All our cantilevers are piezoresistive and work in a dynamic mode.

Table 1. Cantilevers design parameters and simulated characteristics.

Symbol	Definition	Cantilever Sensor Design/Type		
		Rectangular (RCant1)	Triangular (TCant1)	Triangular (TCant2)
L_1	Rectangular step-length (µm)	-	394	394
L_2	Triangular step-length (µm)	-	606	606
L	Cantilever total length (µm)	1000	1000	1000
w	Width (fixed-end) (µm)	170	170	170
t	Thickness (µm)	15	15	15
b	Triangular-free-end: base (µm)	-	700	340
m_0	Cantilever mass (µg)	5.94	9.76	5.94
f_0	Resonant frequency (kHz)	220	185	271
S_m	Mass sensitivity (Hz/pg)	0.07	0.04	0.09
Q	Quality factor [i]	2000 ± 200	1800 ± 200	3000 ± 150

(i) Measured mechanical quality factor.

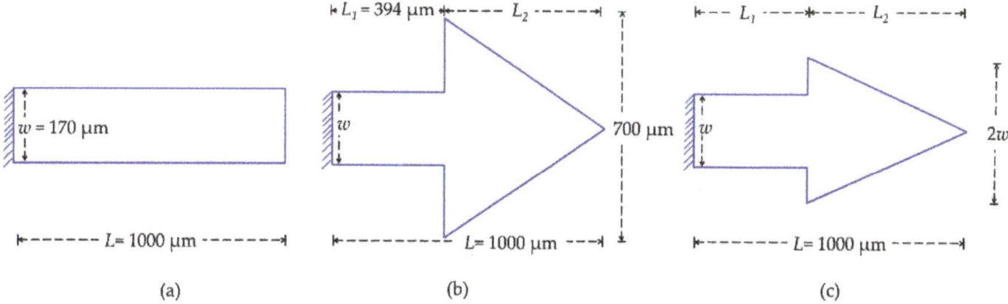

Figure 3. Schematic designs of (**a**) Rectangular (regular) cantilever (RCant1), and triangular cantilevers (**b**) first type (TCant1) and (**c**) second type (TCant2). The shape of the triangular free-end of TCant1 is equilateral (with the base $b = 700$ µm) whereas that of TCant2 is isosceles (with $b = 2w = 340$ µm).

To manufacture these cantilevers, n-type (100) silicon wafers (Siegert Wafer GmbH, Aachen, Germany) were utilized as base material through a bulk-micromachining fabrication process [33,34]. Initially, these wafers were diced from the silicon wafers into $\sim 30 \times 30$ mm^2 substrates. Prior to use, the bulk silicon substrates were thoroughly cleaned to remove from their surfaces any organic contaminants (e.g., dust particles, lubricants, grease, silica gel, etc.), ionic contaminants (mostly from inorganic compounds), and silicon dust or metallic debris, etc. To clean them, each substrate was boiled (for ~5 min) in a 1:1 oxidant mixture solution of sulfuric acid (H_2SO_4, 96 %) and hydrogen peroxide (H_2O_2, 30 %) contained in a quartz glass beaker. The substrate was then immersed in a water bath for about 5 min before rinsing it with deionized water and blow drying with nitrogen gas. Such a cleaning process is oxidative; and it therefore yields a hydrophilic surface.

The main cantilever fabrication steps comprised of:

- Substrate preparation (i.e., pre-cleaning),
- Thermal oxidation,
- Photolithography,
- Dopant diffusions: n^+ and p (1100 °C); p^+ (1200 °C),
- Contact holes formation,
- Metallization: Cr: 300 Å / Au: 3000 Å,
- Membrane formation, in which the backside was etched in a cryogenic inductive-coupled plasma reactive ion etching (Cryo-ICP-RIE) process using SF_6/O_2,
- Structuring and free-release of cantilever through a second Cryo-ICP-RIE process (SF_6/O_2).

Basically, each pre-cleaned bulk silicon substrate (~30 × 30 mm^2) was initially thermally oxidized (in a furnace at a temperature $T \approx 1100$ °C for about 100 min). Subsequently, the oxidized sample was cleaned by sonication (using acetone as cleaning agent), then mounted and spin-coated with a positive photoresist (AZ 5214, Shipley) prior to patterning (by photolithography using MJB4 mask aligner from SÜSS MicroTec, Garching, Germany). Patterning was essentially useful in defining various microstructural features of interest ranging from n^+, p, and p^+ doping sites, contact holes, metallization, membrane, and the cantilever, respectively. Each of these features was realized using a specific mask design.

In patterning the metal-line connections, for instance, metallization was done by first depositing a 30-nm-chromium layer (which serves as an adhesive layer) and secondly, 300-nm-gold layer through an electron-beam deposition process. Subsequently, a lift-off process (in acetone) was carefully undertaken to remove the photoresist from the samples. Since this is a critical step in the fabrication process, a thorough microscopic inspection of the metallized samples was performed to assess the continuity of the connection lines. Hereafter, the sample was thermally oxidized and patterned for backside etching to define the membrane or thickness of the cantilever. Therefore, Cryo-ICP-RIE etching was carried out at a temperature of approximately −80 °C for approximately 56 min. Lastly, the cantilevers were patterned and freely released through a cryogenic etching process (at −95 C for about 15 min). This was the stage at which the desired free-end cantilever configuration (i.e., rectangular, or triangular) was realized, as depicted in Figure 4.

Figure 4. SEM images of silicon-based piezoresistive cantilever sensors after fabrication and basic cleaning processes, with (**a**,**b**) depicting TCant1 and TCant2 triangular cantilever sensors.

Usually, after fabrication, the cantilever surface was still covered with the photoresist. To remove it, the sensor was soaked and cleaned in acetone and thoroughly water-rinsed before drying in a gentle stream of nitrogen gas. Alternatively, oxygen plasma cleaning was performed. Both cleaning methods can reliably be used to clear the photoresist from the sensors.

3. Particle-Imprinting Process

3.1. Particle Sampling and Tip Coating

In Figure 5, we schematically illustrate the sampling and depositing of particles onto a silicon bulk substrate or cantilever-based sensor. Basically, two main steps were involved:

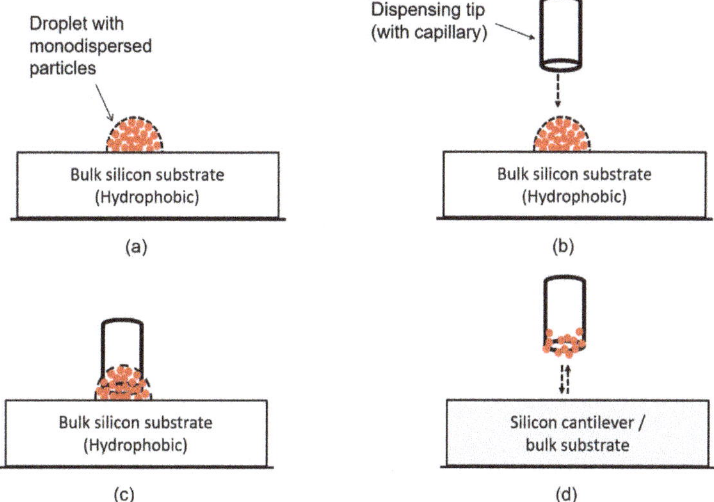

Figure 5. Schematic illustration of the particle sampling and deposition by particle-imprinting process. (**a**) Arbitrary sized droplet containing monodispersed particles on a hydrophobic silicon surface. (**b**) The dispensing stainless-steel tip is positioned above the droplet and then moved (using a 3D micro-positioning system) into the droplet. (**c**) The tip is dipped into the particles-laden droplet to coat it with the particles solution. (**d**) After retraction from the droplet, the particles-coated tip is moved onto the sensing surface to contact and deposit the particles thereon before it is retracted therefrom.

Firstly, a particle solution was initially prepared by homogenously mixing (by sonication) the selected monodispersed particle solution. Then, an arbitrary-sized small drop (see Figure 5a) was deposited on a pre-cleaned hydrophobic silicon surface (~10 ×10 mm^2) under ambient conditions. In this case, the small droplet served as a particle reservoir. The particles, which are suspended in a fluid medium e.g. in a droplet, exhibit a random motion.

A dispensing stainless-steel tip (D-tip) was then moved (Figure 5b) and dipped into the small droplet (Figure 5c). Consequently, as illustrated in Figure 5d, the tip surface (apex) is coated (or inked) with the particles, i.e., a thin (liquid) film would form or adhere on the tip surface upon retraction from the droplet. The particle solution adheres on the surface of the tip courtesy of capillary action.

It should be noted that the tip apex and the sensing surface were carefully aligned in the same horizontal plane. Therefore, initially, a contact between the D-tip apex (without particle solution) and the surface was established, tested, and optically inspected using an USB camera (Mz-902, Oowl Tech Ltd., Hong Kong, China). The point of contact (i.e., z_0) was carefully noted and the tip was then moved (by means of a micro-positioning system as depicted in Figure 6) into the particle reservoir (i.e., to point z_0 in the drop) for particle coating (for a defined time period).

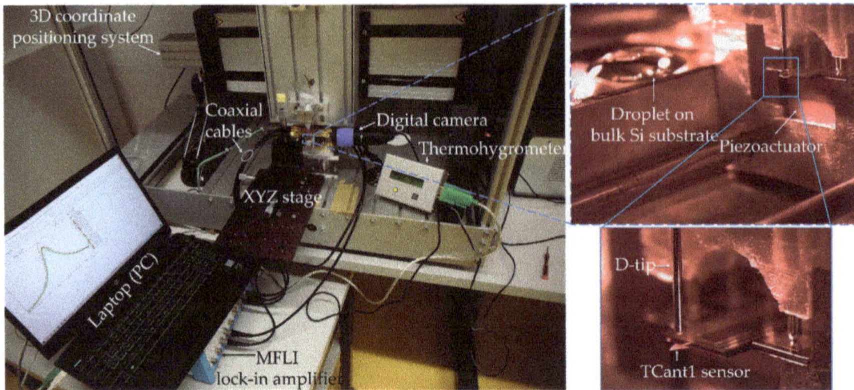

Figure 6. Optical image of the typical experimental setup utilized for tip-sensor contact process (particle deposition) and resonant frequency measurements (particles mass determination).

3.2. Particle Transport Process

The particle-coated tip was moved from the droplet to the target point (i.e., on the bulk silicon substrate or cantilever mass sensor surface) utilizing the micro-positioning system. Prior to this, the target surface e.g., the cantilever sensor (as depicted in Figure 6) was first mounted on a sampler (fixed on a workbench) followed by tip-alignment. The latter was performed by positioning the D-tip vertically above the cantilever at a defined point along the symmetry axis. The tip-on-sensor alignment process was optically aided by a camera (and was often repeated whenever the D-tip would be exchanged). The camera, in this case, helped to visualize the tip while moving and adjusting the tip apex as close as possible onto the sensor surface. This was necessary to mitigate and reduce the risk and possibility of breaking the fragile silicon cantilever sensor and therefore minimize the inaccuracies associated with tip misalignment which otherwise leads to an off-centered particle deposition/adsorption. Nonetheless, we estimated the minimum achievable tip-on-sensor position alignment accuracy to be about 10 μm. Taking due considerations, therefore, the coated tip was retracted from the droplet (to point z_1), then moved laterally (along the y-axis), and moved vertically downwards (to point z_0) to contact and imprint the particles on the sensing surface.

3.3. Particle Imprinting and Adhesion

The next step involved the transfer of the particles solution from the tip to the target (sensing) surface by mechanical contact, as depicted in Figure 5d. In this case, the particle-coated tip (apex) would contact the surface for a defined time duration, hereafter referred to as contact time t_c. It is worth noting that tip-droplet dipping time corresponded to the tip-sensor contact time. During tip-sensing surface contact, the liquid is detached (from the tip) and attached onto the target surface. In case of a cantilever sensor, slight deflection is observed during tip contact; hence, avoiding the risk of breaking the sensor.

After tip-sensor contact, the particles (along with the carrier fluid i.e., water) move radially outwards and inwards as depicted in Figure 7. During this process, particles arrange themselves on the surface with the solvent flow; and adsorption happens as a result of water evaporation and the inter-particle forces owing to surface tension.

The amount of tip-adherent particles solution transferable to the target surface was mainly influenced by changing the wettability of the substrate/sensing surfaces, size of the dispensing tip, and tip-sensor contact time t_c, using different types of particle solutions and particle concentration levels. Based on these dynamics, we present (in the next section) the typical outcomes of particle adsorption and arrangements on assorted bulk silicon substrates.

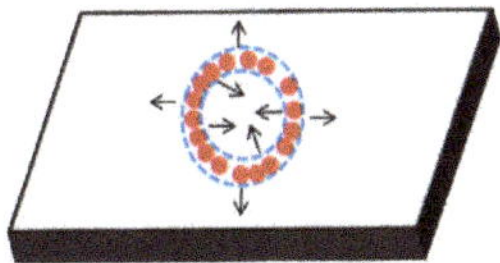

Figure 7. Schematic illustration of a particle-imprinted substrate sample. The inner radial liquid–air interface initiates inwards flow (in addition to outwards flow) to suppress the ring-clustering phenomenon.

3.4. Particles Adsorption on Bulk Silicon Surfaces

Here, we investigate the influence of wettability of substrate surfaces on particle adsorption. Hydrophilic bulk silicon surfaces were basically realized after general cleaning of the substrates (i.e., boiling in a 1:1 oxidant mixture solution of H_2SO_4 (96%) and H_2O_2 (30%), as discussed in Section 2.3). Nevertheless, for enhanced hydrophilicity, these samples were further treated with O_2 plasma (for ~30 s). For hydrophobic silicon substrates, the cleaned samples were dipped in a buffered 6 % to 7 % HF solution for about 10 s.

In Figure 8, we show different substrate samples and their wettability conditions (i.e., with hydrophobic and hydrophilic surface treatments) and the particles-imprinting outcomes. In this case, we used a stainless-steel tip with inner and outer diameters of about 0.117 mm and 0.236 mm, respectively, and a contact time of ~1 s. In both surface wettability conditions, all the samples were evidently adsorbed with the particles. Nevertheless, for hydrophobic silicon substrates, segments of adsorbed particles that consists of both mono- and multilayers were observed (Figure 8a); which otherwise seem to occupy a smaller surface area compared to the highly hydrophilic substrates. In contrast, the latter have a larger particles distribution area (Figure 8b). This primarily results from high surface energy of hydrophilic surfaces which consequently attracts the water (i.e.., particles carrier), thereby facilitating surface wetting. In Figure 8c, which is a magnified view of Figure 8b, we show a segment of monolayer particles arrangements arising from tip-substrate contact on a hydrophilic surface. It should be noted that the multi-layer cluster, observed on the hydrophobic surface (Figure 8a), is a consequence of the water-repulsive nature of these surfaces, which clearly limits their wettability. It is therefore evident that the particle-imprint approach can favorably work well on hydrophilic surfaces to realize monolayer particle assembly which facilitates and guarantees accurate particle counting/estimation.

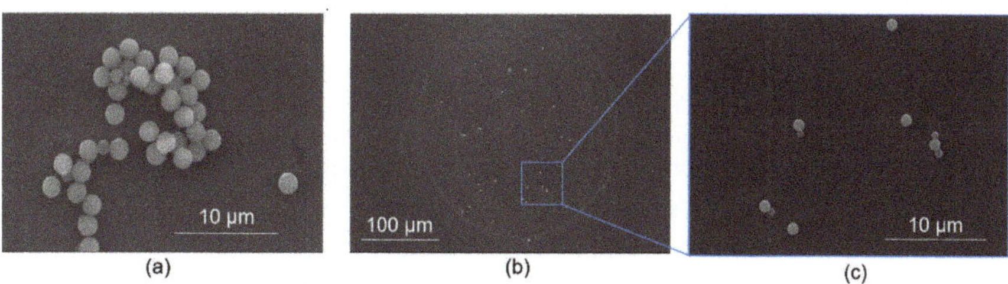

(a)   (b)   (c)

Figure 8. SEM images of particle-imprinted bulk silicon samples. (**a**) Partial segment of spherically shaped PMMA particles adsorbed on hydrophilic surfaces, (**b,c**) respectively depict a whole and magnified partial segment of PMMA particles adsorbed on a hydrophilic silicon surface. The particle-imprinting process was accomplished using a stainless-steel dispensing tip of inner/outer diameter $\approx 0.117/0.236$ mm with a contact time of about 1 s.

4. Cantilever-Based Particle Mass Detection

4.1. Cantilever Sensor Cleaning/Preparartion

As a pre-requisite, the cantilever sensors were cleaned to guarantee a contaminants-free surface for accurate determination of the initial (bare) cantilever mass m_0 and resonant frequency f_0. This consequently enhances the accuracy of determining the adsorbed particles mass, Δm. The actual cantilever mass m_0 and corresponding standard uncertainty for the fabricated sensors ranged from 4.26 µg ± 0.10 µg, 15.01 µg ± 0.10 µg to 23.01 µg ± 0.19 µg for TCant2, RCant1, and TCant1 cantilever types, respectively. Moreover, their typical mass sensitivities S_m were approximately 0.13 Hz/pg (TCant2), 0.03 Hz/pg (RCant1), and 0.02 Hz/pg (TCant1). The parameter $S_m \approx \frac{f_0}{2m_{eff}}$, in which m_{eff} denotes the effective mass of the cantilever; and m_{eff} is about $\frac{1}{4}$ of the static mass of a bare cantilever [35]. The surfaces of our cantilevers are not functionalized but contain a native oxide which makes them hydrophilic and suitable for particle adsorption.

Due to the strong van der Waals forces between magnetic polystyrene (MPS) microparticles and the (hydrophilic) silicon surface, the sticky magnetic polystyrene particles would not ordinarily be desorbed by soaking in acetone but through a controlled sonication process. It should however be noted that removing of MPS µPs from silicon substrate surfaces is nevertheless possible through exclusive wet cleaning processes. These processes primarily involve the use of alkaline and acidic solutions. For instance, the particle removal efficiency using alkaline solutions was demonstrated by Itano et al. to be superior to acid solutions [36]. However, these alkaline solutions (e.g., NH_4OH-H_2O_2-H_2O) etch the silicon surface (by ~0.25 nm/min or more) to lift off the particles, and then dislodge them from the silicon surface by electrical repulsion. On the other hand, acidic solutions such as H_2SO_4-H_2O_2 solution oxidize absorbed particles and decompose them by the strong oxidizing force [36]. Nonetheless, both alkaline and acidic solutions are etch-based particle removal techniques with a high risk of damaging the cantilever features e.g., the piezoresistive Wheatstone bridge and the electrical connection lines.

Contrarily, ultrasonic cleaning technology utilizes ultrasonic energy to agitate the particles (or contaminants) on the solid surface and a liquid (solvent) to rinse the loosened particles away. This method can most conveniently be used to remove contaminants in hard-to-reach areas. Moreover, the process takes considerably shorter cleaning time compared to the normal wet cleaning processes. Furthermore, it is a thorough process which yields high-quality cleaning. Nevertheless, the use of high intensity of vibrations (in which cantilever sensor is subjected to during cleaning) can potentially damage or destroy the sensor.

To mitigate this challenge, we devised and assembled a metallic adaptor (shown Figure 9a) onto which the cantilever sensor (adsorbed with MPS µPs) was mounted and clipped (Figure 9b). Such clipping was necessary to ensure that the sensor does not randomly vibrate and move with the liquid and accidently hit the wall of the glass beaker. The whole assembly was then immersed into a glass containing acetone for sonication (which lasted barely less than 3 min). After cleaning, optical inspection (Figure 9c) was done to assess the effectiveness of the particle removal process. Furthermore, the cantilever resonant frequency was again measured to ascertain, compare, and average with the pre-desorption resonant frequency value.

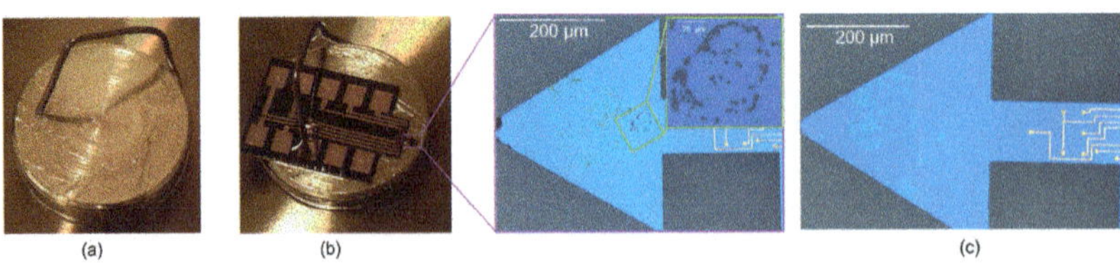

Figure 9. Optical images of (**a**) a specially designed metallic adaptor, (**b**) TCant1 cantilever (adsorbed with magnetic polystyrene particles) mounted and clipped on a metallic adaptor (prior to immersing it in acetone for sonication). The surface of the cantilever after sonication cleaning is shown in (**c**).

4.2. Resonant-Based Mass Measurement

4.2.1. Gravimetric Mass Sensing

In determining the mass of the adsorbate, a gravimetric measurement setup was deployed as schematically shown in Figure 10. The cantilever was excited in its fundamental in-plane mode using a piezo actuator (P-121.01, from PI Ceramics GmbH, Lederhose, Germany), while mechanical vibrations thereto were detected piezoresistively by means of a U-shaped Wheatstone bridge embedded in the sensor during the fabrication process. The direct current (DC) voltage to the Wheatstone bridge (1 V) and sinusoidal actuation signal (up-to 9.9 V_{pk}) to the piezo actuator were supplied by a lock-in-amplifier instrument (MFLI, Zurich Instruments Ltd, Zurich, Switzerland). All connections to and from the MFLI instrument to the piezo actuator and cantilever were accomplished via SMA connectors and coaxial cables.

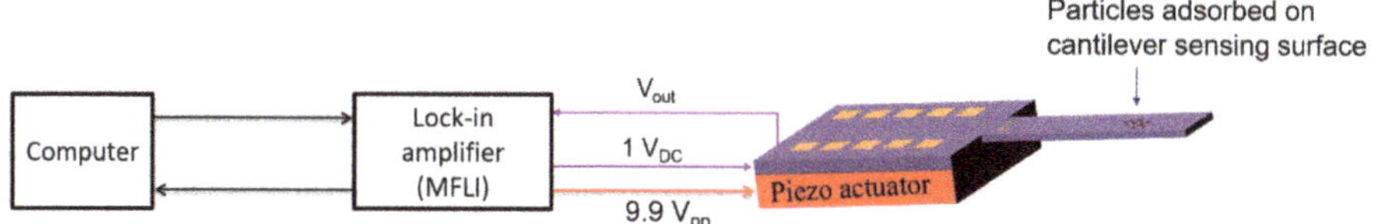

Figure 10. A schematic setup for the measurement of the cantilever resonant frequency f_0 responses to determine the mass Δm of the adsorbed particles. The lock-in-amplifier (MFLI) supplies voltage to the in-plane piezo actuator and Wheatstone bridge of the piezoresistive silicon cantilever sensor.

Initially, the resonant frequency f_0 of a pre-cleaned bare cantilever sensor was measured. Upon depositing, and after vaporizing the particle solution from the sensing surface, the particle-induced fundamental resonant frequency $f_0{'}$ was then measured under ambient conditions. In this case, a shift in resonant frequency between the bare and the mass-loaded cantilever was obtained:

$$\Delta f = f_0' - f_0. \tag{1}$$

This frequency shift Δf (depicted in Figure 11a) is linked to the adsorbate mass (Δm) in accordance with Equation (2). Given the knowledge of the adsorption position x (i.e., the distance between the loaded mass and the fixed end of the microcantilever with beam length L) and cantilever mass m_0, the value of Δm can most conveniently be calculated using [37]:

$$\Delta m = - \frac{2m_0}{U^2(x_{\Delta m})} \frac{\Delta f}{f_0}, \tag{2}$$

where,

$$U(x_{\Delta m}) = (\cos \lambda + \cos h \, \lambda)\left(\cos(\lambda \tfrac{x}{L}) - \cos h(\lambda \tfrac{x}{L})\right) + (\sin \lambda - \sin h \, \lambda)\left(\sin(\lambda \tfrac{x}{L}) - \sin h\left(\lambda \tfrac{x}{L}\right)\right), \tag{3}$$

is the mode shape function of the cantilever, in which λ denotes the vibration modal constant. The value of λ is simply the product of the modal wavenumber β_n and the length L of the cantilever beam, i.e., $\lambda = \beta_n L$; and it depends on the vibrational mode number (which is an integer number n $\geq$ 1). In our dynamic frequency response measurements, only the fundamental flexural mode (i.e., first vibrational mode, n = 1) was mainly involved; hence, $\lambda \approx 1.8751$ [35]. The calculated fundamental mode shape function in relation to the (normalized) particle adsorption point $x_{\Delta m}/L$ is shown in Figure 11.

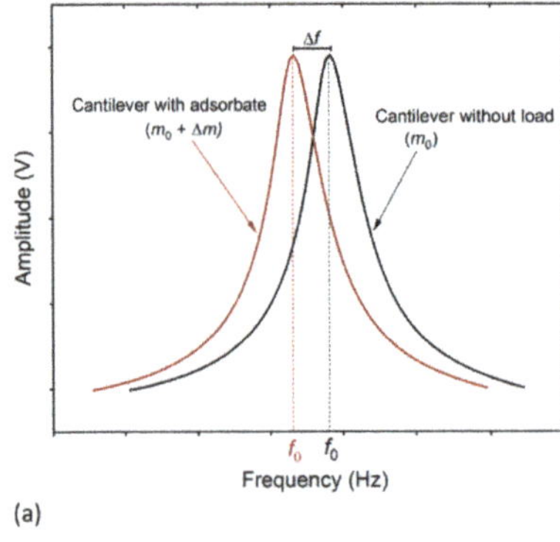
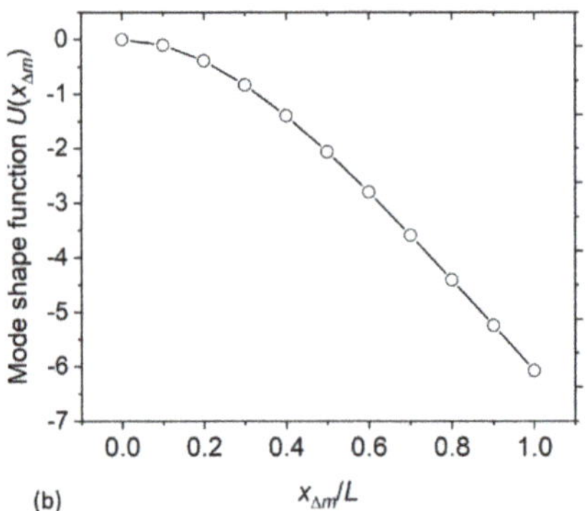

Figure 11. (**a**) Schematic representation of the shift Δf in resonant frequency of a cantilever with/without an adsorbate (load, Δm). (**b**) Mode shape function (calculated based on Equation (3)) at a point $x_{\Delta m}/L$ along the cantilever beam of length L vibrating in the first (fundamental) mode of which $\lambda \approx 1.8751$. In our experimental work, typical particle adsorption positions ranged from $x_{\Delta m}/L \approx 0.4$ to 0.8.

In determining the position $x_{\Delta m}$, SEM images were captured for each successfully prepared cantilever sample and $x_{\Delta m}$ was estimated using ImageJ [38]. The position of the adsorbed magnetic particles was determined from SEM images with an accuracy of approximately ± 0.01 µm. The SEM particles imaging was mainly done after measuring the resonant frequencies of each of the involved samples.

Nonetheless, if we assume a distributed mass condition of the adsorbed particles, then:

$$\Delta m = 2m_{\text{eff}}\frac{\Delta f}{f_0}, \tag{4}$$

where, $m_{\text{eff}} \approx m_0/4$ denotes the effective mass of the cantilever sensor.

4.2.2. Particle Mass Determination

The frequency shift Δf (determined in accordance with Equation (1)) was used to compute the particle mass Δm using Equations (2) and (3) or Equation (4). This was done by averaging at least five frequency sweeps (both before and upon particle adsorption on the sensor, and after particle removal i.e., cleaning of the sensor).

Typical resonance frequency f_0 responses of a cantilever with and without adsorbed magnetic polystyrene particles are, respectively, depicted by the brown/open and black/full lines/circles in Figure 12. The particles-induced resonant-frequency shift $\Delta f = -18.09$ Hz $\pm$ 0.86 Hz and -1.00 Hz $\pm$ 0.14 Hz (delineated in Figure 12a,b) corresponds to measurements from TCant1 sensors with $f_0 \sim$ 183.5 kHz and 179.8 kHz, respectively; while, $\Delta f = -38.91$ Hz $\pm$ 0.61 Hz and -11.77 Hz $\pm$ 0.76 Hz (shown in Figure 12c,d) resulted from TCant2 sensors with $f_0 \sim$ 271.9 kHz and 268.3 kHz, respectively. These Δf values together with their uncertainties were computed from repeated frequency response measurements. The initial resonant frequency of the fabricated cantilevers of the same category (e.g., TCant2 sensors with $f_0 = 271.9$ kHz and 268.3 kHz as depicted in Figure 12c,d) are slightly different, supposedly due to small variations in the sensor dimensions arising from the fabrication process. Primarily, for the in-plane excited cantilever sensors, $f_0 \propto w/L^2$. Typically, small changes in the beam width w and/or length L of the cantilever may emanate from the resolution of our photolithography (~1 µm). Correspondingly, therefore, this affects the expected resonant frequency f_0 by approximately ± 1.2 kHz, for triangular cantilever sensors. On the other hand, variations in cantilever thickness t may also result mainly from the membrane etching process. But this affects the effective cantilever mass (and its sensitivity S_m).

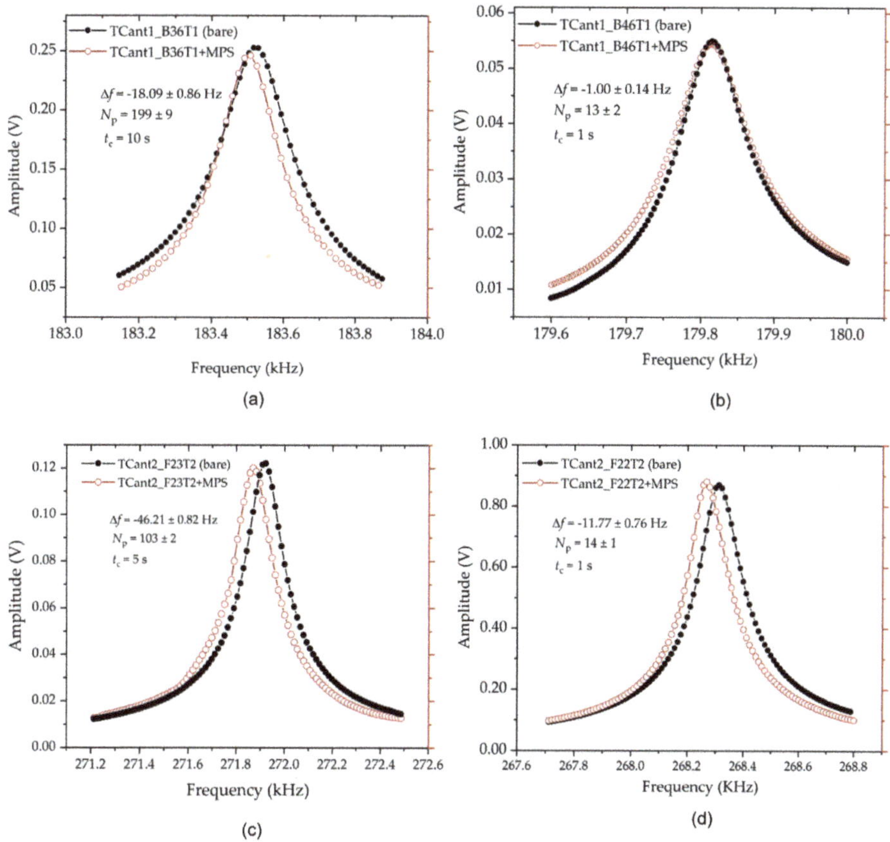

Figure 12. Plot of the resonant frequency responses before and after adsorbing magnetic polystyrene (MPS) microparticles on assorted fabricated triangular cantilevers. (**a,b**) denote measurements from TCant1 cantilever sensors, while (**c,d**) are measurements from TCant2 cantilevers. The MPS µPs (whose number N_p—determined based on point mass-condition i.e., from Equations (2) and (3)) were deposited on the sensing area through the particle-imprint approach by observing tip-sensor-contact time t_c of about 10 s, 5 s, and 1 s using stainless steel tips with nominal internal diameters of ~0.10 mm (**a,c,d**) and ~0.15 mm (**b**).

If we consider point-mass condition (Equations (2) and (3)) and thereby compute the particles mass Δm, the resonant frequency shifts (delineated in Figure 12a–d), translate to the particles concentration N_p of about 199 ± 9 and 13 ± 2, from TCant1 sensors; and 103 ± 2 and 14 ± 1, from TCant2 sensor. This clearly shows that the number of particles, for each sensor category, increased with resonant frequency shift. The values of N_p were determined from the mass ratios of Δm to a single magnetic polystyrene particle mass, i.e., 3.53 pg ± 0.25pg determined from its measured volume (diameter) and given density. Besides MPS particles, PMMA µPs were similarly imprinted on the cantilever sensor ($f_0 \approx 181.0$ kHz) and yielded $N_p \approx 35$, resulting from a resonance shift $\Delta f \sim -9.88$ Hz. To calculate the mass Δm using Equations (2) and (3), the position of the adsorbate (i.e., $x_{\Delta m} = 470$ µm to 750 µm) on the sensor was measured from SEM images (using ImageJ as earlier discussed in Section 4.2.1).

It is worth noting that the vibration amplitudes of the cantilever sensors with and without load (as depicted in Figure 12) were closely in agreement, with a small difference of less than 2%. This may connotate a small change in mechanical quality factor (i.e., damping); but it does not however affect the cantilever resonant frequency and the shift thereto nor the adsorbate mass (computed therefrom). Nevertheless, the measured Q values for TCant2 sensors ($\rightarrow$ Figure 12c,d) were relatively higher ($Q \sim 3000 \pm 150$, which depicts better stability) compared to TCant1 cantilevers ($\rightarrow$ Figure 12a,b) with $Q \sim 1800 \pm 200$. Furthermore, as expected—in accordance with Equation (2), different cantilever frequency bands were observed (in Figure 12a–d). This was typically due to the differences in the number of adsorbed particles (N_p or Δm) on each sensor, small variations in adsorbate position (i.e., $x_{\Delta m}$), and cantilever mass sensitivity (S_m).

After resonant frequency measurements, the particles-imprinted cantilever samples were microscopically analyzed (using SEM) to examine the nature of particle arrangement on the sensing surface. In all these cases, for instance, as depicted in Figure 13, the adsorbed particles resulted in monolayer particles assembly. Consequently, for the assorted SEM images delineated in Figure 13a–e, a particles-count of about 13, 11, 160, 25, and 18 was determined, respectively. In Figure 13a, for instance, an SEM-particles count of about 13 magnetic polystyrene particles was obtained (on TCant1 sensor) i.e., 12 equally sized ($\sim$1.83 µm $\pm$ 0.03 µm) and 1 oddly sized ($\sim$1.07 µm $\pm$ 0.03 µm) particles. Considering the density and volume of the particles, this SEM-particle count translates to a calculated mass of about 43.06 pg $\pm$ 0.16 pg, which agrees well with the particles mass $\Delta m \approx 44.39$ pg $\pm$ 6.07 pg (i.e., $N_p \approx 13 \pm 2$) determined from the corresponding resonant-frequency response (delineated in Figure 12b). A small difference between the two mass estimates was nevertheless observed. Besides, the frequency shift ($\Delta f \approx -1.00$ Hz) was notably small. Consequently, this necessitated an enhanced mass sensitivity of our sensors. Considerably, this limit was fairly extended based on our TCant2 cantilever design; from which, we realized a mass sensitivity $S_m \sim 0.13$ Hz/pg, and gravimetrically detected (Figure 12d) and determined (based on point mass-condition) about 14 MPS µPs, with a better frequency resolution. This number of MPS particles (i.e., $N_p \approx 14$) is similarly in good agreement with the corresponding observed particles-count from the SEM image (in Figure 13b), i.e., $N_p = 11$. Furthermore, in our recent works [39], miniaturized sensors for an enhanced airborne particles detectability has been demonstrated (with $m_0 = 2$ ng to 5 ng and a mass sensitivity $\sim$0.13 Hz/fg); and their use in liquid-based particles detection is further intended. Besides, special consideration is intended to apply our particle sampling and imprinting approach on commercial piezoresistive silicon cantilever sensors i.e., CAN30-1-2 sensor, from CiS Forschungsinstitut für Mikrosensorik GmbH, with $m_0 \approx 20$ µg; and, PRSA-L300-F80-TL sensor, from SCL-Sensor. Tech. Fabrication GmbH, with $m_0 \sim 0.5$ µg (a factor of 30 to 50 lower than our in-house fabricated cantilevers (Table 1)).

In Figure 14, we additionally show a correlation plot that compares the SEM particle-analysis results with the calculated number of particles from resonant-frequency responses (considering both point-mass and distributed-mass conditions). The latter condition assumes that the mass of the adsorbate is evenly distributed on the sensing surface; a factor that potentially leads to the poor correlation between the calculated number of particles (due to distributed mass condition) and the observed (SEM-) particle counts, as depicted in Figure 14. On the other hand, from the same plot, it is apparently clear that the resonance-based number of particles due to point-mass condition (cf. Equations (2) and (3)) is highly correlated with the (SEM)-particle counts. With a correlation coefficient of nearly 0.99 ($\rightarrow$ error-weighted linear fitting in Figure 14) and a minimum detection limit of about 0.05pg (exhibited from TCant2 sensors), it shows that the measurements are pretty much in agreement and our (TCant2) sensors offer reasonably high sensitivity, respectively. Notably though, some deviations from ideal correlation of the resonance-based and SEM-particle estimations were observed (Figure 14) and their possible causes will further be considered in the subsequent section, in which we will discuss an exemplar of assorted possible measurement influences.

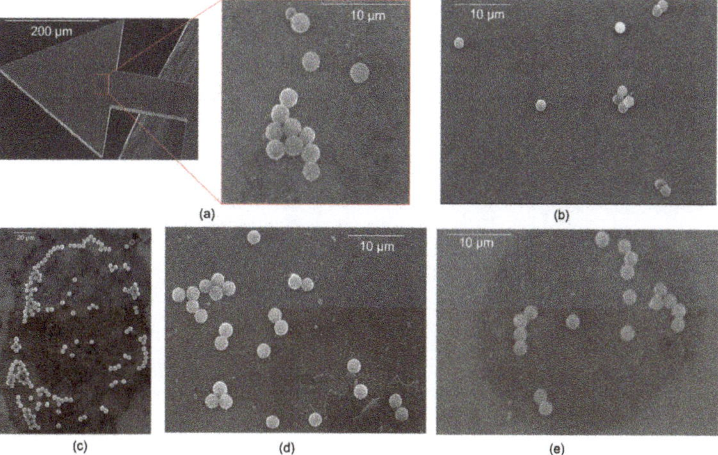

Figure 13. Typical SEM images of the adsorbates of magnetic polystyrene (MPS) microparticles on our various silicon-based piezoresistive TCant1 and TCant2 triangular cantilever sensors with (**a–e**) respectively denoting B46T1, F22T2, B24T1, A16T1, and A14T1 cantilevers. The SEM-counted number of particles in (**a–e**) are ~13, 11, 160, 25, and 18, respectively. From all the cantilever samples, a monolayer particle assembly was clearly observed. These MPS μPs were deposited on the cantilevers utilizing the particle-imprint method.

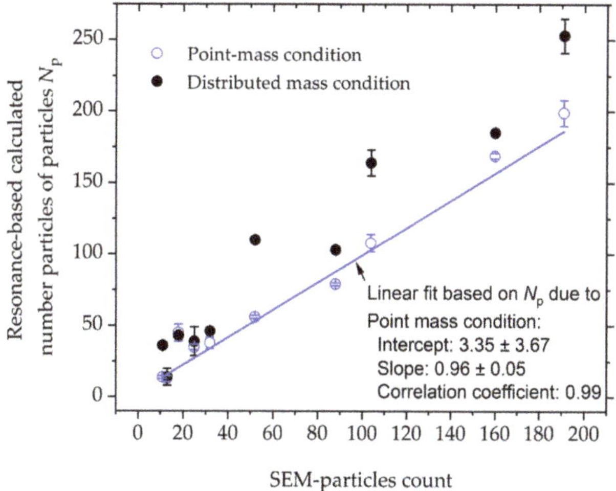

Figure 14. Comparison of the number of magnetic polystyrene microparticles on both TCant1 and TCant2 cantilever sensors. The number of particles N_p was determined from the measured resonant frequency responses (considering both point-mass and distributed mass conditions) and is compared with the SEM-particle counting. The blue line depicts an error-weighted linear fit for N_p at point-mass condition.

Nonetheless, considering the small amount of liquid realizable through the on-cantilever particle imprinting process coupled with the depicted high sensitivities, it is worth noting that the measurement approach presented herein can plausibly be applied in the testing or detection of liquid-borne viruses (e.g., coronavirus—whose primary mode of transmission from person to person is through virus-laden droplets). In testing for coronavirus, for instance, a test fluid e.g., sputum (from a suspected or infected person) is sampled. Apparently, most coronavirus rapid-testing methods detect the antibodies produced in response to viral infection [40]. Contrastingly, to detect the presence of coronavirus itself, a MEMS-cantilever-based sensor is desirable. This has recently been demonstrated by Digital Diagnostics AG by functionalizing the surface of the cantilever sensor with a capture layer of antibodies, which binds antibodies in a test sample fluid [41]. Given the flexibility and ease of adaptation of the on-cantilever particle imprinting approach, we believe it can potentially and cost-effectively be utilized to detect liquid-borne viruses such as coronavirus.

4.3. Assessment of Measurement Uncertainty

Resonant-based mass sensing, like conventional macroscale mass measurements [42], is influenced by the loading position and environmental conditions such as relative humidity (*RH*) and temperature (*T*). A temperature change (ΔT) correspondingly changes f_0 by multiplying it with $\frac{(\alpha + \alpha_E)}{2}\Delta T$ [35], where α = 2.6 ppm/K and α_E = −44 ppm/K [35,43] denote the linear coefficient of thermal expansion and temperature coefficient of Young's modulus E of silicon, respectively; while the change in relative humidity (ΔRH) is bound to decrease f_0 due to the moisture that cling to the sensor surface. In our measurement process, these variables were monitored for every measurement cycle and found to be small. For instance, the observed maximum temperature change ΔT for TCant1 sensor—delineated in Figure 15a—was ~0.1 °C within one hour, while the change in relative humidity ΔRH during the same period was ~1%. Similarly, for TCant2 sensor (Figure 15b), $\Delta T \approx 0.2$ °C, and $\Delta RH \approx 1\%$.

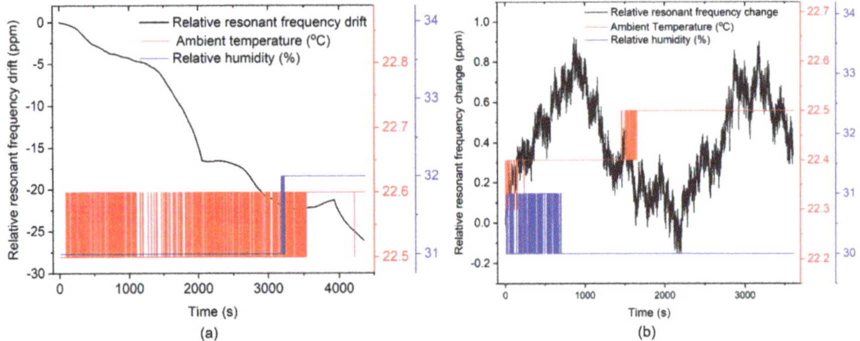

Figure 15. Relative resonant frequency drift (black line) of bare, silicon triangular cantilevers (**a**) TCant1 ($f_0 \approx 181.265$ kHz) and (**b**) TCant2 ($f_0 \approx 262.253$ kHz) under typical ambient temperature (red line) and relative humidity (blue line) conditions in the laboratory over measurement time (~1 h).

Furthermore, we define the relative resonant frequency drift $y(t)$ (shown in Figure 15) as a ratio of the measured resonance frequency $f_0(t)$, minus its initial value at $t = t_0$, and $f_0(t_0)$, i.e.,:

$$y(t) = \frac{\Delta f}{f_0} = \frac{f_0(t) - f_0(t_0)}{f_0(t_0)}, \tag{5}$$

which yields an overall relative frequency drift of about −26.1 ppm and 0.9 ppm for TCant1 and TCant2 sensors, respectively. It should however be noted that the typical duration for measuring the resonance frequency responses for each cantilever sensor (with/without magnetic particles) was about 1 to 5 min. Within this short measuring period (i.e., within 5 min), a maximum relatively relative frequency drift of 1.2 ppm (TCant1 sensor) and 0.4 ppm (TCant2 sensor) was observed, i.e., more than an order of magnitude lower than the overall (~1 h) observed drift values of ~−26.1ppm (Figure 15a) and ~0.9 ppm (Figure 15b), respectively. Furthermore, the effect of humidity can be assumed to be negligible (yielding less than 1ppm for $\Delta RH \approx 1\%$ [44]).

Further measurement influences (based on Equation (2)) included uncertainties associated with the determination of the mass of the cantilever sensor (primarily from the thickness of the sensor), particle(s) adsorption position $x_{\Delta m}$ along the cantilever beam (mainly due to tip alignment; of which, we consider the resolution of the positioning system) and the repeatability of resonant-frequency shift. Moreover, for the calculated number of particles (cf. Figure 14), particle-diameter estimation was a critical-influencing parameter. All particles were essentially assumed to be identical in shape (spherical) and size (~2 µm). This was typically observed in all the analyzed samples except one (Figure 13a) in which only one oddly sized particle (~1 µm) was observed. Besides, to minimize particle mass measurement uncertainties, further work involving calibration of the in-plane cantilever stiffness is intended.

5. Conclusions

In this study, we have designed and fabricated a triangular cantilever and utilized it as a sensing surface upon which a novel particle-imprinting approach has been implemented. With this approach, liquid-borne particles were sampled by dipping a dispensing tip into an arbitrary-sized particle-laden droplet (on a hydrophobic silicon substrate) followed by tip-sensor contact. We have also examined a few crucial parameters that regulated the number of particles adsorbed and assembled on the silicon sensors/substrates i.e., surface wettability, tip size, and contact time. With the particle-imprint approach, herein presented, the need for a dispensing-air pressure (essentially utilized in liquid dispensing) was eliminated. Furthermore, no complicated instrumentation is required to deposit numerable particles; hence, making it fairly cost-effective. Using on-cantilever imprinting method, a monolayer-particle assembly has been realized on hydrophilic silicon cantilever sensors with the lowest particles-count of about 11. Typically, from our fabricated triangular microcantilever mass sensors (i.e., TCant1 and TCant2), we can realize a minimum detectable frequency shift Δf_{min} of about 8 mHz and 7 mHz, respectively. We have also realized a higher resonant frequency of the TCant2 sensors, which was 271 kHz compared to a rectangular (regular) cantilever sensor (220 kHz) of equivalent total length (L = 1 mm) and mass ($m \approx 5.94$ µg). This effectively offered the TCant2 sensor superior advantages (over RCant1) with a higher mechanical quality factor ($Q \sim 3000$), an enhanced mass sensitivity ($S_m \approx$ 0.13 Hz/pg), and a lower minimum detectable mass ($\sim\Delta f_{min}/S_m \approx 0.05$ pg). In this regard, therefore, we have further envisaged the possibility of utilizing the on-cantilever imprint approach to measure/detect liquid-borne viruses (e.g., coronavirus) and single particles using our sensors. We have also briefly examined main sources of uncertainty which affected our particle sampling and resonant frequency response measurements.

Author Contributions: Conceptualization and methodology, W.O.N.; cantilever fabrication and characterization, W.O.N. and A.S. (Andi Setiono); validation, W.O.N., A.S. (Andi Setiono) and E.P; data analysis, W.O.N.; investigation, W.O.N.; resources, A.S. (Angelika Schmidt), H.B. and E.P.; writing—original draft preparation, W.O.N.; writing—review and editing, W.O.N., A.S. (Andi Setiono), A.S. (Angelika Schmidt), H.B. and E.P.; supervision, H.B. and E.P.; project administration, E.P.; and funding acquisition, E.P. All authors have read and agreed to the published version of the manuscript.

Funding: This project has received funding from the EMPIR program co-financed by the Participating States and from the European Union's Horizon 2020 research and innovation program under No. 17IND05 MicroProbes.

Acknowledgments: W.O.N. is grateful to the German Federal Ministry for Economic Cooperation and Development (BMZ) for a doctoral scholarship within the Braunschweig International Graduate School of Metrology (B-IGSM). A.S. acknowledges support from the Ministry of Research, Technology and Higher Education of the Republic of Indonesia under no. 343/RISET-Pro/FGS/VII/2016 (World Bank Loan No. 8245-ID). Authors are grateful to funding support from "Niedersächsisches Vorab", Germany, through the Quantum- and Nanometrology (Quanomet) initiative within the project NP 2-2. The authors gratefully thank Ulrich Neuschaefer-Rube (PTB) for his support and fruitful discussions on optical and CT measurements. The technical support given by Juliane Breitfelder, Aileen Michalski, Andreas Heidemann and Karl-Heinz Lachmund is also gratefully acknowledged.

Conflicts of Interest: The authors declare no conflict of interest.

References

1. Boisen, A.; Dohn, S.; Keller, S.S.; Schmid, S.; Tenje, M. Cantilever-like micromechanical sensors. *J. Micromech. Microeng.* **2011**, *74*, 36101. [CrossRef]
2. Burg, T.P.; Godin, M.; Knudsen, S.M.; Shen, W.; Carlson, G.; Foster, J.S.; Babcock, K.; Manalis, S.R. Weighing of biomolecules, single cells and single nanoparticles in fluid. *Nature* **2007**, *446*, 1066–1069. [CrossRef] [PubMed]
3. Stockslager, M.A.; Olcum, S.; Knudsen, S.M.; Kimmerling, R.J.; Cermak, N.; Payer, K.R.; Agache, V.; Manalis, S.R. Rapid and high-precision sizing of single particles using parallel suspended microchannel resonator arrays and deconvolution. *Rev. Sci. Instr.* **2019**, *90*, 85004. [CrossRef] [PubMed]
4. Raiteri, R.; Butt, H.-J.; Grattarola, M. Changes in surface stress at the liquid/solid interface measured with a microcantilever. *Electrochimica Acta* **2000**, *46*, 157–163. [CrossRef]
5. Park, K.; Kim, N.; Morisette, D.T.; Aluru, N.R.; Bashir, R. Resonant MEMS mass sensors for measurement of microdroplet evaporation. *J. Microelectromech. Syst.* **2012**, *21*, 702–711. [CrossRef]
6. Velanki, S.; Ji, H.-F. Detection of feline coronavirus using microcantilever sensors. *J. Micromech. Microeng.* **2006**, *17*, 2964–2968. [CrossRef]
7. Nyang'au, W.O.; Hamdana, G.; Setiono, A.; Bertke, M.; Xu, J.; Fahrbach, M.; Puranto, P.; Wasisto, H.S.; Peiner, E. Adsorption and detection of microparticles using silicon microcantilevers. *J. Phys. Conf. Ser.* **2019**, *1319*, 12010. [CrossRef]
8. Nyang'au, W.O.; Setiono, A.; Puranto, P.; Bertke, M.; Wasisto, H.S.; Viereck, T.; Bosse, H.; Peiner, E. Droplet-on-cantilever approach for determining the mass of magnetic particles. In Proceedings of the 20. GMA/ITG-Fachtagung Sensoren und Messsysteme 2019, Nürnberg, Germany, 25–26 June 2019; pp. 222–229.
9. Patil, N.D.; Bhardwaj, R.; Sharma, A. Self-sorting of bidispersed colloidal particles near contact line of an evaporating sessile droplet. *Langmuir* **2018**, *34*, 12058–12070. [CrossRef]
10. Hu, H.; Larson, R.G. Evaporation of a Sessile Droplet on a Substrate. *J. Phys. Chem. B* **2002**, *106*, 1334–1344. [CrossRef]
11. Cheng, W.; Hartman, M.R.; Smilgies, D.-M.; Long, R.; Campolongo, M.J.; Li, R.; Sekar, K.; Hui, C.-Y.; Luo, D. Probing in real time the soft crystallization of DNA-capped nanoparticles. *Angew. Chem. Int. Ed. Engl.* **2010**, *49*, 380–384. [CrossRef]
12. Wong, T.-S.; Chen, T.-H.; Shen, X.; Ho, C.-M. Nanochromatography driven by the coffee ring effect. *Anal. Chem.* **2011**, *83*, 1871–1873. [CrossRef] [PubMed]
13. Harris, D.J.; Conrad, J.C.; Lewis, J.A. Evaporative lithographic patterning of binary colloidal films. *Philos. Trans. A Math. Phys. Eng. Sci.* **2009**, *367*, 5157–5165. [CrossRef] [PubMed]
14. Deegan, R.D.; Bakajin, O.; Dupont, T.F.; Huber, G.; Nagel, S.R.; Witten, T.A. Capillary flow as the cause of ring stains from dried liquid drops. *Nature* **1997**, *389*, 827–829. [CrossRef]
15. Yunker, P.J.; Still, T.; Lohr, M.A.; Yodh, A.G. Suppression of the coffee-ring effect by shape-dependent capillary interactions. *Nature* **2011**, *476*, 308–311. [CrossRef]
16. Soltman, D.; Subramanian, V. Inkjet-printed line morphologies and temperature control of the coffee ring effect. *Langmuir* **2008**, *24*, 2224–2231. [CrossRef]
17. Sempels, W.; de Dier, R.; Mizuno, H.; Hofkens, J.; Vermant, J. Auto-production of biosurfactants reverses the coffee ring effect in a bacterial system. *Nat. Commun.* **2013**, *4*, 1757. [CrossRef] [PubMed]
18. Nyang'au, W.O.; Setiono, A.; Bertke, M.; Bosse, H.; Peiner, E. Cantilever-droplet-based sensing of magnetic particle concentrations in liquids. *Sensors* **2019**, *19*, 4758. [CrossRef] [PubMed]
19. Lindemann, T.; Zengerle, R. Droplet dispensing. In *Encyclopedia of Microfluidics and Nanofluidics*; Springer: New York, NY, USA, 2015; pp. 641–652.
20. Barbulovic-Nad, I.; Lucente, M.; Sun, Y.; Zhang, M.; Wheeler, A.R.; Bussmann, M. Bio-microarray fabrication techniques—A review. *Crit. Rev. Biotechnol.* **2006**, *26*, 237–259. [CrossRef]
21. Piner, R.D.; Zhu, J.; Xu, F.; Hong, S.; Mirkin, C.A. "Dip-Pen" nanolithography. *Science* **1999**, *283*, 661–663. [CrossRef]
22. Liu, G.; Hirtz, M.; Fuchs, H.; Zheng, Z. Development of Dip-Pen Nanolithography (DPN) and Its Derivatives. *Small* **2019**, *15*, e1900564. [CrossRef]

23. Quist, A.P.; Pavlovic, E.; Oscarsson, S. Recent advances in microcontact printing. *Anal. Bioanal. Chem.* **2005**, *381*, 591–600. [CrossRef] [PubMed]

24. Bettencourt, A.; Almeida, A.J. Poly(methyl methacrylate) particulate carriers in drug delivery. *J. Microencapsul.* **2012**, *29*, 353–367. [CrossRef] [PubMed]

25. Khan, F.A.; Akhtar, S.; Almohazey, D.; Alomari, M.; Almofty, S.A.; Badr, I.; Elaissari, A. Targeted delivery of poly (methyl methacrylate) particles in colon cancer cells selectively attenuates cancer cell proliferation. *Artif. Cells Nanomed. Biotechnol.* **2019**, *47*, 1533–1542. [CrossRef] [PubMed]

26. Hamdana, G.; Südkamp, T.; Descoins, M.; Mangelinck, D.; Caccamo, L.; Bertke, M.; Wasisto, H.S.; Bracht, H.; Peiner, E. Towards fabrication of 3D isotopically modulated vertical silicon nanowires in selective areas by nanosphere lithography. *Microelectron. Eng.* **2017**, *179*, 74–82. [CrossRef]

27. Ferreira, H.A.; Graham, D.L.; Freitas, P.P.; Cabral, J.M.S. Biodetection using magnetically labeled biomolecules and arrays of spin valve sensors (invited). *J. Appl. Phys.* **2003**, *93*, 7281–7286. [CrossRef]

28. Hejazian, M.; Li, W.; Nguyen, N.-T. Lab on a chip for continuous-flow magnetic cell separation. *Lab. Chip.* **2015**, *15*, 959–970. [CrossRef]

29. Viereck, T.; Kuhlmann, C.; Draack, S.; Schilling, M.; Ludwig, F. Dual-frequency magnetic particle imaging of the Brownian particle contribution. *J. Magn. Magn. Mater.* **2017**, *427*, 156–161. [CrossRef]

30. Engelmann, U.M.; Buhl, E.M.; Draack, S.; Viereck, T.; Ludwig, F.; Schmitz-Rode, T.; Slabu, I. Magnetic relaxation of agglomerated and immobilized iron oxide nanoparticles for hyperthermia and imaging applications. *IEEE Magn. Lett.* **2018**, *9*, 1–5. [CrossRef]

31. Kudr, J.; Haddad, Y.; Richtera, L.; Heger, Z.; Cernak, M.; Adam, V.; Zitka, O. Magnetic nanoparticles: From design and synthesis to real world applications. *Nanomaterials* **2017**, *7*. [CrossRef]

32. Yu, L.; Hao, G.; Gu, J.; Zhou, S.; Zhang, N.; Jiang, W. Fe_3O_4/PS magnetic nanoparticles: Synthesis, characterization and their application as sorbents of oil from waste water. *J. Magn. Magn. Mater.* **2015**, *394*, 14–21. [CrossRef]

33. Wasisto, H.S.; Merzsch, S.; Uhde, E.; Waag, A.; Peiner, E. Handheld personal airborne nanoparticle detector based on microelectromechanical silicon resonant cantilever. *Microelectron. Eng.* **2015**, *145*, 96–103. [CrossRef]

34. Setiono, A.; Bertke, M.; Nyang'au, W.O.; Xu, J.; Fahrbach, M.; Kirsch, I.; Uhde, E.; Deutschinger, A.; Fantner, E.J.; Schwalb, C.H.; et al. In-plane and out-of-plane MEMS piezoresistive cantilever sensors for nanoparticle mass detection. *Sensors* **2020**, *145*, 618. [CrossRef] [PubMed]

35. Schmid, S.; Villanueva, L.G.; Roukes, M.L. *Fundamentals of Nanomechanical Resonators*; Springer International Publishing: Cham, Switzerland, 2016.

36. Itano, M.; Kern, F.W.; Miyashita, M.; Ohmi, T. Particle removal from silicon wafer surface in wet cleaning process. *IEEE Trans. Semicond. Manufact.* **1993**, *6*, 258–267. [CrossRef]

37. Dohn, S.; Svendsen, W.; Boisen, A.; Hansen, O. Mass and position determination of attached particles on cantilever based mass sensors. *Rev. Sci. Instr.* **2007**, *78*, 103303. [CrossRef]

38. Rueden, C.T.; Schindelin, J.; Hiner, M.C.; DeZonia, B.E.; Walter, A.E.; Arena, E.T.; Eliceiri, K.W. ImageJ2: ImageJ for the next generation of scientific image data. *BMC Bioinform.* **2017**, *18*, 529. [CrossRef]

39. Bertke, M.; Xu, J.; Fahrbach, M.; Setiono, A.; Wasisto, H.S.; Peiner, E. Strategy toward miniaturized, self-out-readable resonant cantilever and integrated electrostatic microchannel separator for highly sensitive airborne nanoparticle detection. *Sensors* **2019**, *19*, 901. [CrossRef]

40. Pang, J.; Wang, M.X.; Ang, I.Y.H.; Tan, S.H.X.; Lewis, R.F.; Chen, J.I.-P.; Gutierrez, R.A.; Gwee, S.X.W.; Chua, P.E.Y.; Yang, Q.; et al. Potential rapid diagnostics, vaccine and therapeutics for 2019 novel coronavirus (2019-nCoV): A systematic review. *J. Clin. Med.* **2020**, *9*, 623. [CrossRef]

41. Kloppstech, K.; Devic, M. New 5 Minute Corona Virus Test Delivers Clear Results on Infection. Germany. 2020. Available online: https://digid.com/en/new-5-minute-corona-virus-test-delivers-clear-results-on-infection/ (accessed on 17 April 2020).

42. Shaw, G.A. Current state of the art in small mass and force metrology within the International System of Units. *J. Micromech. Microeng.* **2018**, *29*, 72001. [CrossRef]

43. Setiono, A.; Nyang'au, W.O.; Fahrbach, M.; Xu, J.; Bertke, M.; Wasisto, H.S.; Peiner, E. Improvement of frequency responses of an in-plane electro-thermal cantilever sensor for real-time measurement. *J. Micromech. Microeng.* **2019**, *29*, 124006. [CrossRef]
44. Yang, J.; Xu, J.; Wu, W.; Bertke, M.; Wasisto, H.S.; Peiner, E. Piezoresistive silicon cantilever covered by ZnO nanorods for humidity sensing. *Procedia Eng.* **2016**, *168*, 1114–1117. [CrossRef]

Review

Scanning and Actuation Techniques for Cantilever-Based Fiber Optic Endoscopic Scanners—A Review

Mandeep Kaur [1,2,3], Pierre M. Lane [2,3] and Carlo Menon [1,2,*]

1   MENRVA Research Group, Schools of Mechatronic Systems Engineering and Engineering Science, Simon Fraser University, Surrey, BC V3T 0A3, Canada; mka116@sfu.ca
2   School of Engineering Science, Simon Fraser University, Burnaby, BC V5A 1S6, Canada; plane@bccrc.ca
3   Imaging Unit, Integrative Oncology, BC Cancer Research Center, Vancouver, BC, V5Z 1L3, Canada
*   Correspondence: cmenon@sfu.ca

Abstract: Endoscopes are used routinely in modern medicine for in-vivo imaging of luminal organs. Technical advances in the micro-electro-mechanical system (MEMS) and optical fields have enabled the further miniaturization of endoscopes, resulting in the ability to image previously inaccessible small-caliber luminal organs, enabling the early detection of lesions and other abnormalities in these tissues. The development of scanning fiber endoscopes supports the fabrication of small cantilever-based imaging devices without compromising the image resolution. The size of an endoscope is highly dependent on the actuation and scanning method used to illuminate the target image area. Different actuation methods used in the design of small-sized cantilever-based endoscopes are reviewed in this paper along with their working principles, advantages and disadvantages, generated scanning patterns, and applications.

Keywords: endoscopes; medical imaging; MEMS actuators; piezoelectric; electrothermal; electrostatic; electromagnetic; shape memory alloys; scanning patterns

Citation: Kaur, M.; Lane, P.M.; Menon, C. Scanning and Actuation Techniques for Cantilever-Based Fiber Optic Endoscopic Scanners—A Review. *Sensors* **2021**, *21*, 251. http://doi.org/10.3390/s21010251

Received: 23 November 2020
Accepted: 30 December 2020
Published: 2 January 2021

Publisher's Note: MDPI stays neutral with regard to jurisdictional claims in published maps and institutional affiliations.

1. Introduction

An endoscope is an imaging device made up of a long and thin tube that can be inserted into the hollow openings of the body to image the inner sections in real time and in a less invasive manner. Advances in fiber optic systems led to the development of flexible endoscopes, enabling high-resolution images of narrow sections of the body and reducing the number of biopsies required for a specific diagnosis, with applications such as cancer detection, microvascular oxygen tension measurement, chronic mesenteric ischemia, subcellular molecular interactions, etc. Earlier developed standard white light endoscopes (WLEs) had limited ability to differentiate metaplasia from dysplasia. Such limitations were surpassed by enhancing the image contrast through the use of dyes in chromoendoscopy or applying digital filters in narrow band imaging (NBI) [1,2]. The increased use of endoscopic devices highly improved the diagnostic rate of cancers by permitting the visualization of early dysplasias which may lead to cancer development [1]. Cancer is the second leading cause of death in the world. Approximately one out of every six deaths are due to cancer, killing 9.6 million individuals worldwide in 2018 [3]. It has been observed that early detection has significantly improved life expectancy and reduced the mortality rate by 30–40% over the last two decades [4].

One of the recently advanced imaging devices was based on confocal laser endoscopy (CLE). Two variations of CLE that are used commercially in medical applications are the so-called e-CLE, which is an integrated confocal endoscope developed by Pentax Medical (Tokyo, Japan), and the probe-based confocal microscope (p-CLE) developed by Mauna Kea Technologies (Paris, France) [5]. Optical coherence tomography (OCT) is another frequently used technique to image tissue, where a change in the refractive index of the scattering coefficient alters the intensity of the backscattered light and is used to provide contrast in the image [6]. Photoacoustic (PA) imaging is another imaging technology that images a tissue surface using short pulsed light waves and detects the ultrasonic waves generated from the optical absorption. It integrates the benefits of high contrast in optical imaging and deep penetration of the ultrasonic imaging [7]. On the basis of imaging depth, PA imaging can be classified into PA microscopy (penetration depth < 10 mm), PA computed tomography (penetration depth between 10 and 100 mm), and minimally invasive PA imaging (penetration depth ≥ 100 mm) [8]. In addition to these techniques, there is visible light spectroscopy (VLS), where the tissue surface is exposed to visible light and the absorbance spectrum provides structural and functional information about the tissue. This technique is largely used to monitor the microvascular hemoglobin oxygen saturation, which can be further used to evaluate local ischemia situations by measuring the difference between oxygenated and deoxygenated hemoglobin [9,10].

The advancement in optics along with the development of micro-electro-mechanical systems (MEMS) and microfabrication techniques led to the fabrication of sub-millimeter-sized flexible endoscopes that can image the narrow cavities in the body, providing information about early-stage pre-cancerous tissues. The optimized final design of an endoscope is often the result of an optical design optimized using special software-based computer simulations [11]. The rapid evolvement of the software and fabrication technologies enabled endoscopes to capture the tissue images in three-dimensional space, providing in depth information about the target surface as well [12]. Most of the preliminary video endoscopes used coherent optical fiber bundles (CFBs) to transport light from a light source, such as a xenon lamp or a laser light, to the imaged surface and used charge-coupled devices (CCDs) to image the tissue surface [13]. Those CCD devices contained approximately 200,000 pixels, which provided limited resolution of the image [2]. The image resolution and optical magnification in newly developed endoscopes was enhanced using complementary metal oxide semiconductor (CMOS) chips, which provided images with over 1.3 MPixels of diffraction-limited resolution [14]. However, a minimum center-to-center distance between the optical fibers in a CFB and the honeycomb effect produced by the non-imaging area between fibers still limited the resolution in devices having a diameter smaller than 3 mm, independently of the imaging chip used [15].

It is possible to obtain high-spatial-resolution images with flexible endoscopes having sub-millimeter diameters by scanning the laser light at the proximal end and capturing the image on a temporal basis, i.e., acquiring one pixel at a time. Seibel et al. from the University of Washington developed such a cantilever-based scanning fiber endoscope (SFE), where a single-mode fiber was excited at resonance to scan the light beam on the target area, and an outer ring of optical fibers captured the backscattered light [16]. Since that development, a large number of cantilever-based imaging devices have been fabricated, which will be discussed later in the paper. In such devices, the tip displacement of an optical fiber acting as a cantilever beam dictates the field of view (FOV) and the resolution of the obtained image. Scanning and actuation techniques used to excite a cantilever beam play a critical role in the performance of such a scanning device. In addition, the small size and the distortion-free imaging requirements need to be considered during the design of an endoscopic device for medical purposes as the size of an imaging probe sets the targeted imageable area, and the motion of organs can lead to the generation of artifacts in the image.

A large number of state-of-the-art reviews on endoscopic imaging devices are available in the literature which are mainly focused on different imaging modalities. The review work in [17] provided a general description of confocal microscopy, OCT, and two-photon imaging modalities. Similarly, the review work in [18] investigated various imaging modalities with some information on their actuation mechanisms. The previous work [6] and the review article by Hwang et al. in [19] provided a general description of various endoscopic imaging technologies, modalities, packaging/scanning configurations, and actuation mechanisms. However, a detailed review of the different actuation mechanisms used in fiber optic endoscopic scanners has not been provided previously. For this reason, the scanning and actuation techniques used in the recently developed cantilever-based fiber optic scanners for medical purposes are described in this paper in detail. In the current review, the mathematical models and applications of MEMS actuators in fiber optic cantilever-based scanners are reported to provide information about the underlying working physics of these devices and can provide a foundation for the development of miniaturized and more efficient MEMS scanners.

The paper is organized as follows: Section 2 provides an overview of the optical components used in endoscopic devices. Scanning techniques used in such small-sized devices are described in Section 3. Section 4 describes the various actuation methods used to excite the miniaturized optical cantilever beams in detail. A discussion and conclusions about available cantilever-based scanning devices are reported in Sections 5 and 6, respectively.

2. Overview of Optical Components Used in Endoscopes

The key element used in cantilever-based endoscopic devices is an optical fiber acting as a waveguide through which light is transmitted using the principle of total internal reflection at the interface of two different dielectric media. The inner cylindrical portion is called the core, while the outer region is named the cladding. Both materials are characterized by a slight change in refractive index; it is from this that the name of step-index fibers comes. There are different classifications of optical fibers available based on normalized wave number V, defined as:

$$V = (2\pi a NA)/\lambda \tag{1}$$

where a is the core radius, λ is the wavelength of the light, and NA is the numerical aperture of the fiber [20]. NA is calculated from the refractive index of the core and cladding material of the fiber as [20]:

$$NA = \sqrt{n^2_{core} - n^2_{clad}} \tag{2}$$

The V-parameter determines the number of spatial modes of the electromagnetic wave that can propagate within the fiber. An optical fiber is defined as a single mode fiber (SMF) if $V < 2.405$.

For a large V-parameter, the total number of allowed modes N through the fiber can be approximated as [20]:

$$N \approx V^2/2 \tag{3}$$

Such fibers are called multimode fibers (MMF).

Other than simple step-index fibers, there are other fibers having different refractive index profiles with radius and they are increasingly finding use in communication fields to avoid the dispersion of optical waves during propagation. Among these fibers, the most commonly used ones are fibers having a double-step-shaped profile or a nearly parabolic variation of index designated double-clad fibers (DCF) and gradient-index (GRIN) fibers, respectively. In GRIN fibers, light propagates due to the profile of the refractive index instead of the total internal reflection and follows a sinusoidal path [20].

Different kinds of optical imaging modalities used in endoscopic imaging devices are described in detail in this section. These technologies are compared with a CCD/CMOS camera in Table 1. Following the definitions in [15], pixel density for the CFB, CCD, and CMOS technologies is the number of imaging elements (pixels or fibers) per square millimeter that can be achieved in practice, while image resolution is the number of these elements in a 1mm diameter. Pixel density and image resolution for SFE assume a single-mode fiber and are based on specific sampling considerations [15].

Table 1. Comparison between system performances of imaging technologies used in endoscopes.

	Pixel Density (Pixels/mm^2)	Image Resolution (Pixels)	Pixel Size	Advantages	Disadvantages
CFB	113 k	30.0 k/64.0 k	2 μm Ø	Small form factor, low cost	Cross-coupling and honeycomb effect degrade image resolution Aging effect results in non-working pixels due to fractured fibers within the bundle
CCD	238 k	95.0 k	0.5 μm × 0.5 μm	Small pixel size, low cost, no aging effect	Rectangular geometry limits the usable area, low dynamic range, poor light collection in low illumination area
CMOS	476 k	190 k	1.45 μm × 1.45 μm	Higher image resolution, low cost	Rectangular geometry limits the usable area, poor resolution in devices with diameter < 1 mm
SFE	345 k	282 k	Dependent on scanning pattern and sampling rate	Higher sampling rate and resolution in sub-millimeter-sized devices	Performance dependent on actuation method and sampling rate. Spatial point spread dependent on objective lens at the tip and illumination properties.

2.1. Fiber Bundles in Endoscopes

Fiber bundles comprise thousands of step-index (or GRIN) fibers contained within a very small area, ranging from a few micrometers to millimeters in range. A fiber bundle is used to carry the illumination light from the proximal end to the distal end, and vice versa, of a device. In a coherent fiber bundle, fibers at both ends are placed at the same relative position, so that the image is transmitted from one end to the other without any distortion. In miniaturized scanners, a coherent fiber bundle is predominantly used due to the perfect alignment of fibers, which facilitates the decoding of the signal [21]. As the light intensity information is transmitted from one end of the fiber bundle to the other, it is possible to place the scanning mechanism at the proximal end of the device, which is a major advantage of such devices.

Every core of the step-index fiber in the bundle acts as a pixel of the imaged data. Thus, the resolution of the image captured using a fiber bundle depends on the core size of the fiber and the core-to-core distance between the fibers [22]. For a coherent fiber bundle with the core separation distance among adjacent fibers represented by Δ_{core}, the cross-sectional core density or pixel density is given by [15]:

$$Core\ density = \frac{2}{\sqrt{3}} \left(\frac{1}{\Delta_{core}} \right)^2 \tag{4}$$

From the core density and the active area covered by the fibers A_{cfb}, neglecting the space occupied by outer protective jacket and sheath, the image resolution can be obtained as [15]:

$$Image\ resolution = core\ density * A_{cfb} \tag{5}$$

This equation indicates a major drawback of using fiber bundles for miniaturized devices as the resolution will be poor due to the honeycomb effect generated by the space between the consecutive fiber cores representing the non-imaged area [15].

Another disadvantage of using fiber bundles is the crosstalk phenomenon. The cladding of the step-index fibers constituting the bundle is rendered to a thin layer around the core during the fabrication of a flexible fiber bundle. A very thin cladding surface enables the leakage of the light as it travels through the core, reducing the contrast and the resolution of the image generated. It is possible to reduce the effect of crosstalk by increasing the thickness of the cladding section, enlarging contrast between the refractive indices of the core and cladding material, and/or by reducing the number of modes propagating through the fiber. However, these approaches make the device rigid, and the honeycomb effect will be more intensified, thus degrading the resolution [23,24].

It is possible to refine the resolution of the obtained images by post-processing the data using certain algorithms. Using specific transformation models based on the point spread function of the cores of each fiber from the bundle, it is possible to smooth the light gradient, reducing the pixilation effect given by the honeycomb pattern [25–27]. The limitation of these transformation methods to be able to improve the resolution due to under-sampling is alleviated through the use of pixel super-resolution techniques. In such methods, the source/image probe is shifted slightly a number of times for the acquisition of different images, and these images are further combined to enhance the resolution of the image [28].

In endoscopes that use fiber bundles, a pair of mirrors placed at the proximal end of the device permit the illumination of a single fiber from the bundle at a time. The fast-axis scanning (axis with high scanning frequency) is performed using the resonant scanner (mirror surface vibrated at resonance using one of the actuators described later in the paper), while a galvo scanner (optical mirror scanned using a galvanometer-based motor) is used for the slow-axis scanning. A schematic diagram of a confocal micro-endoscope developed using such a technique is shown in Figure 1.

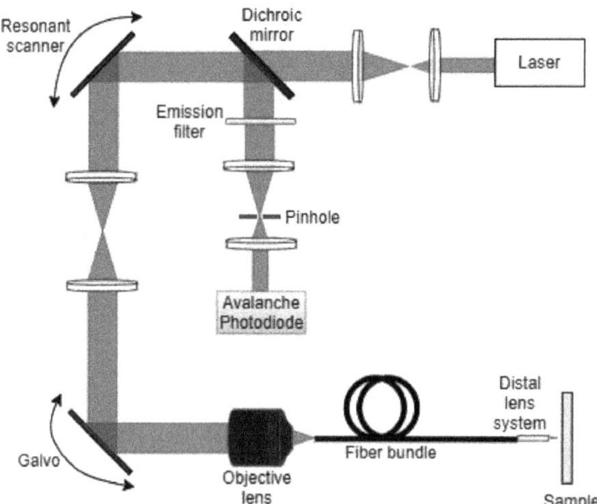

Figure 1. Block diagram of a confocal micro-endoscope.

Sung et al. developed an early fiber-optic-based confocal reflectance microscope to image epithelial cells and tissues in real time, showing the cell morphology and tissue architecture without the use of any fluorescent stains. The image guide contained 30,000 fibers and the device had an overall diameter of 7 mm and rigid length of 22 mm to image human tissue with a lateral and axial resolution (smallest resolvable feature) of 2 μm and 10 μm, respectively [29]. Knittel et al. developed a similar confocal endoscope where tissue images with a lateral and axial resolution of 3.1 μm and 16.6 μm, respectively, were obtained using a similar fiber bundle contained in a diameter of 1 mm [30]. Lane et al. developed a similar endoscopic probe for imaging bronchial epithelium having a diameter and length of 1.27 mm and 10 mm, respectively. This probe had a lateral resolution of 1.4 μm and an axial resolution of 16 µm–26 µm [31]. A commercially available endoscope based on this technique was developed by Mauna Kea Technologies [32,33].

2.2. Single Fiber Endoscopy

A single-mode fiber (SMF), characterized by having a step-shaped index of refraction change from the core of the fiber to the cladding surface, permits the propagation of only one mode of light through the fiber. A single spatial mode enables a diffraction-limited spot to be projected on the sample plane, resulting in a high-resolution image with the use of an SMF. Due to this property and high flexibility, such fibers find use in miniaturized optical scanners.

An SMF may find a use in a scanning fiber optic microscope, acting as a spatial filter [34] or as a pinhole detector [35]. The same fiber is used for laser light illumination and the collection of the reflected light [34]. An SMF serves as a pinhole in a confocal system and is used in spectrally encoded confocal microscopes [36,37]. By moving the fiber in a plane perpendicular to its axis using mechanical systems or a galvanometer, it is possible to obtain a 2D image.

The SFE described earlier uses an SMF vibrated in resonance to scan the light beam across the target tissue surface, and a peripheric ring of fibers detects the time-multiplexed backscattered light. In this case, the sample resolution depends on the scanning motion and sampling rate, which are not fixed a priori during fabrication. In an SFE, the smallest resolvable feature is determined by its point spread function. A wider tip displacement will provide a higher FOV and higher image resolution in terms of the number of pixels in the scanned area, as described in Table 1.

In contrast to an SMF, a multimode fiber can transmit a large number of spatial modes at the same time. These fibers have core sizes much larger than the single-mode fibers, usually in the range of 50 µm–2000 µm. Multimode fibers can be classified into step-index multimode fibers and graded-index (GRIN) multimode fibers on the basis of the change in refractive index from the core to the cladding, which can be sharp or gradual, respectively [38].

A multimode fiber can be considered as an alternative to a fiber bundle and supports the miniaturization of optical devices. As each fiber in the fiber bundle represents a pixel for the acquired image, each pixel can be represented by a propagating mode in the fiber. Thus, it is possible to increase the pixel density of a device by up to 1–2 orders of magnitude by replacing the fiber bundle with a multimode fiber [39]. A side-viewing endoscopic probe for PA and ultrasound (US) imaging is developed using an MMF to deliver laser pulses to the target tissue, and a coaxial US transducer detects the PA and US echo signals. The light and acoustic signal is deflected 45° by a scanning mirror placed at the distal end of the probe, which is rotated by magnets or a micromotor to provide a rotational scanning [40,41].

The main limitation of using a multimode fiber in an imaging device is modal dispersion, which causes multipath artifacts. Several methods have been explored to provide an image without image artifacts. For example, Papadopoulos et al. used a digital phase conjugation technique to generate a sharp focus point. In this technique, the phase of the distorted wavefront was calculated and an unmodulated beam of this phase was propagated in a backward direction to cancel out the distortions and to generate the original signal [42]. Some other groups proposed wave-front shaping methods to focus the light passing through a multimode fiber. Even though these methods successfully focused the light, they required continuous recalculation of the optimal wave due to the fiber motion [39,42]. These methods do not work in the case of reflection mode detection of objects. In reflection mode imaging, the transmission matrix describing the response between the modes at the input and output planes can be used to overcome the distortion [39].

The modal dispersion effect is avoided using GRIN fibers, where the refractive index change along the section of the fiber equalizes the travel time of different modes. Thus, different spatial modes propagate at similar velocities. Sato et al. used a GRIN fiber for the fabrication of a single-fiber endoscope used for reflectance imaging. However, this device had some problems related to nonuniform image quality, background distortion, etc. [43]. High-quality photoacoustic images using a GRIN fiber are reported in [44], where the light focusing property of the GRIN fibers permitted the propagation of spatially distributed Gaussian beams through the fiber, which enhanced the focusing of the spot at the output. This, in turn, permitted high-resolution imaging [44].

Double-clad fibers consisting of a central core and two outer cladding layers are another type of frequently used fibers in endoscopes. These fibers possess the unique feature of allowing the propagation of both single-mode and multimode light through the fiber. The single-mode light travels through the central core, while the multimode light is transmitted through the inner cladding material. Such a fiber is principally used in fluorescence imaging devices having single-mode illumination and multimode signal collection. Thus, the advantages of single-mode illumination and multimode collection are combined in these fibers [45–47].

It is possible to combine OCT and fluorescence imaging in a single endoscope using a DCF. In this case, OCT illumination and fluorescence excitation light is projected on the sample through the core of the DCF. The backscattered OCT signal is collected through the core, and the fluorescence emission from the sample is collected through the inner cladding of the fiber. The OCT source light and fluorescence excitation light are combined using a wavelength division multiplexer (WDM) before sending it through the core of DCF. The recollected light is separated using a DCF coupler, where the recollected OCT signal from the core of the DCF is submitted to an SMF, while the fluorescence signal from the inner cladding is forwarded to an MMF [48,49].

Buenconsejo et al. developed a device that combined narrowband reflectance, OCT, and autofluorescence imaging in a single-fiber endoscope using a DCF. This device worked analogously to other OCT devices, except for the difference that the red/green/blue (RGB) light was emitted from the central core, while the collection of the reflected light was performed through the inner cladding. The separation of the various light signals from three modalities was done using an additional WDM [50].

In a single-fiber-based micro-endoscope, the light beam can be steered at either the proximal or distal tip of the fiber. The possibility of using a light beam for a proximal scan allows the separation of large-sized beam scanning devices from the distal end of the endoscopic device used to monitor the target sample. Thus, it is possible to develop small-sized endoscopic devices that can image the deep tissue systems within the body. Proximal scanning is usually performed using side-viewing imaging probes, where a drive mechanism rotates the fiber to scan the light beam along the circumference of the target sample [51]. On the other hand, distal scanning is preferentially used in cantilever-based single-fiber endoscopes where the fiber tip is displaced mechanically using a variety of actuators. Usually, the fibers are excited at resonance to obtain high tip displacements using piezoelectric [52–60], electrostatic [61,62], electromagnetic [63,64], electrothermal [65–67], micromotor mirror [68–70], or shape memory alloy [71] actuators. The working principle of these actuators will be discussed in detail later in the paper. Pan et al. developed a fiber optic scanner where the beam was steered at the distal end using a pair of micromirrors [72]. The only commercially available single-fiber endoscope was developed by Pentax to image the upper and lower gastrointestinal (GI) tract. The optical scheme of this device is shown in Figure 2a [73]. In this case, the confocal images showing the subcellular and cellular structures of the upper and lower GI tract are imaged after the administration of the contrast agent. An in-vivo confocal image of rectal mucosa in human colon collected using the Pentax endoscope is shown in Figure 2b, where crypt lumens can be clearly identified [73].

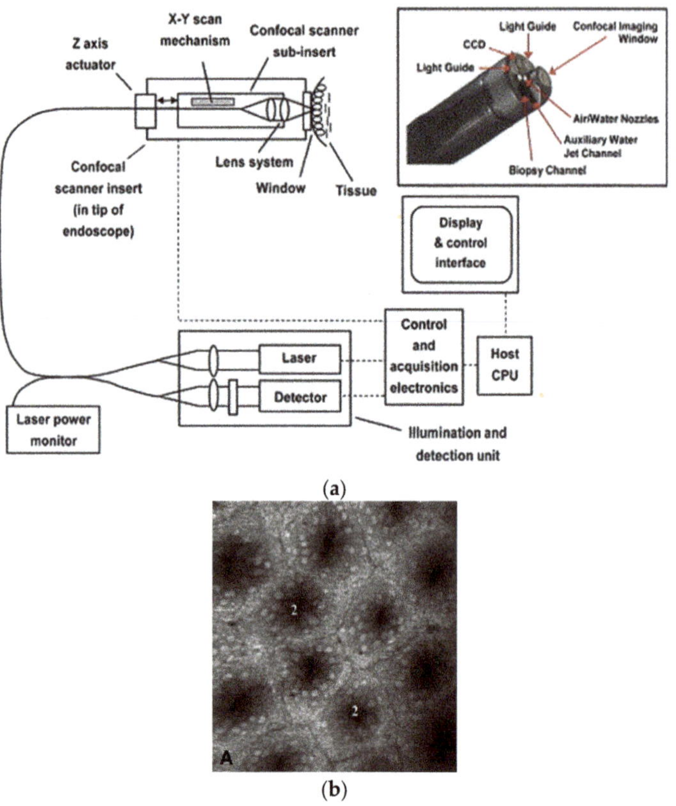

(a)

(b)

Figure 2. Fluorescence confocal imaging developed by Pentax: (**a**) schematic design of the micro-endoscope; (**b**) an en face image of rectal mucosa showing the crypt lumens (taken with the permission of [73]).

2.3. Graded-Index (GRIN) Lens Scanner

In lens scanners, light is deflected due to a non-planar interface between the air and the lens. The lateral motion of the lens perpendicular to the direction of the incident beam in these scanners causes variation in the refraction angle, allowing light scanning.

Wu et al. developed a paired-angle rotating scanning OCT (PARS-OCT) probe to image the gill structure of a Xenopus laevis tadpole, where the beam steering the distal end of the probe was obtained by the rotary motion of the two angle polished GRIN lenses. A schematic diagram of this system is shown in Figure 3 [74]. The OCT images of the gill structure of a tadpole obtained using this device are shown in Figure 4. The photograph of an OCT probe relative to the tadpole is shown in Figure 4a, while the OCT images in Figure 4b,c enable to clearly identify the gill pockets [74]. Sarunic et al. integrated a gear-based linear scan mechanism with the PARS-OCT device to control the rotational speed of inner and outer GRIN lenses. They were able to identify vitreous, retina, and choroid surfaces in the OCT images of an ex vivo porcine retina [75].

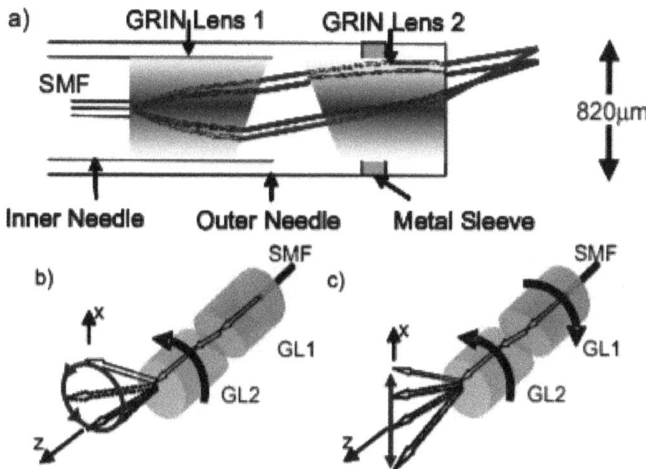

Figure 3. Schematic diagram of a PARS-OCT probe: (**a**) distal tip; (**b**) circular motion generation by rotating just one lens; (**c**) linear scan by rotation of both lenses (taken with permission of [74]) © The Optical Society.

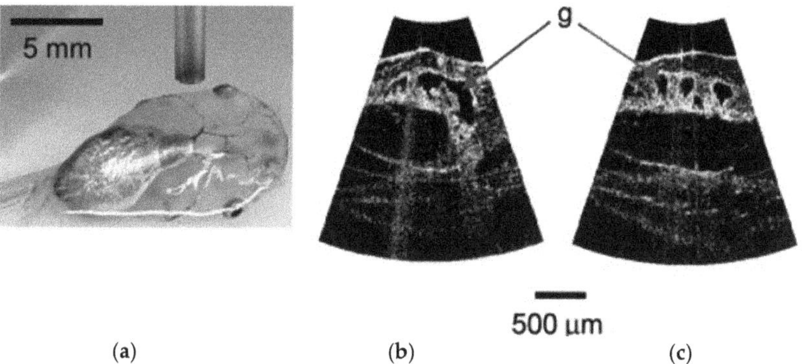

Figure 4. Endoscopic OCT image of the gill structure of a tadpole: (**a**) photograph of probe relative to the tadpole; (**b**,**c**) OCT images showing gill pockets indicated with g (taken with the permission of [74]). © The Optical Society.

3. Cantilever Beam Mechanics

In small cantilevered optical scanners, the image is obtained by scanning the light beam at the distal end of the device, as stated earlier. In most of the earlier endoscopes using CFBs to transport the light to the tissue surface, the beam is scanned using micromirrors placed at the proximal end of the device. In these so-called proximal scanners, the large-dimensioned scanning components can be separated from the distal end of the endoscopic device. Thus, it is possible to fabricate very compact-sized scanning devices. However, as the beam sweeps light across the CFB, a portion of the light enters through the cladding of the fibers as well, which results in poor contrast in the image. On the other hand, the scanning device is placed at the distal end of the endoscopic scanner to illuminate the light on the target sample in distal scanners. These single-fiber-based endoscopes require distal scanning to sweep the light across the target sample [24].

Among the distal scanners, it is possible to have two different configurations of devices based on the scanning direction [6]. In a side-view imaging device, the light from the fiber tip is deflected at a certain angle with the help of reflecting mirrors, or prisms. Such imaging devices provide circumferential images of the target surface, and 2-D cylindrical images can be obtained by moving the device along its axis [76]. However, in the forward-view imaging devices, the laser light is laterally scanned using special actuators and provides the image of the tissue surface at the front of the device [77].

The cantilever-based endoscopic scanners belong to the category of forward-view imaging devices. In such devices, an optical fiber is fixed at a distance of a few millimeters from its distal end. The free end of the fiber acts as a cantilever beam, which is vibrated, using certain actuators described later in the paper, to illuminate the target tissue area. The backscattered reflected light is used to reconstruct the image of the area using certain image processing algorithms. These cantilevered optical fibers can be vibrated in resonance, or at a frequency different from their resonant frequency.

Almost all the cantilevered-fiber optic endoscopes can be considered as cylindrical-shaped beams. The first resonant frequency (also called natural frequency) of a cylindrical-shaped cantilevered beam (where one side is rigidly blocked for any movement and the other end is free to move) is given by:

$$f_n = \frac{1.875^2}{4\pi} \sqrt{\frac{E}{\rho} \frac{R}{L^2}} \tag{6}$$

with E, ρ, R, and L being the Young's modulus, density, radius, and length of the cantilever beam, respectively [78]. From this equation, the driving frequency in resonant scanners depends on the inherent properties and dimensions of the optical fiber acting as the cantilever beam. In nearly all such scanners, the fiber is a standard 125 μm diameter fiber. Thus, the resonant and driving frequencies can be adjusted by changing the length of the cantilevered section.

The deflection of a cantilevered beam in the transverse direction can be obtained considering the Euler–Bernoulli beam. The Euler–Bernoulli beam theory describes the relationship between the beam deflection $w(x,t)$ and the applied load $f(x,t)$, assuming small deformations in the beam such that the planes perpendicular to the x-y axis do not bend after the deformation. The equation describing the deflection $w(x,t)$ of the beam in the y direction, in time (t) and along the length (x), can be derived considering the force and moment equilibrium of an infinitesimal element dx of the beam as in Figure 5. The equilibrium of forces in the y direction yields:

$$\left(V(x,t) + \frac{\partial V(x,t)}{\partial x} dx \right) - V(x,t) + f(x,t)dx = \rho A(x)dx \frac{\partial^2 w(x,t)}{\partial t^2} \tag{7}$$

where $V(x,t)$ is the shear force and $f(x,t)$ the total applied external force per unit length, while the term on the right-hand side describes the inertial force of the element, with $A(x)$ being the cross-section of the beam. Similarly, the equilibrium of moment acting on the element can be written as:

$$\left[M(x,t) + \frac{\partial M(x,t)}{\partial x} dx \right] - M(x,t) + \left[V(x,t) + \frac{\partial V(x,t)}{\partial x} dx \right] dx + [f(x,t)dx]\frac{dx}{2} = 0 \tag{8}$$

with $M(x,t)$ being the bending moment related to beam deflection $w(x,t)$ and flexural stiffness $EI(x)$ of the cantilever beam, where E is the Young's modulus, and $I(x)$ is the cross-sectional area moment of inertia [78]. $M(x,t)$ is given by:

$$M(x,t) = EI(x)\frac{\partial^2 w(x,t)}{\partial x^2} \tag{9}$$

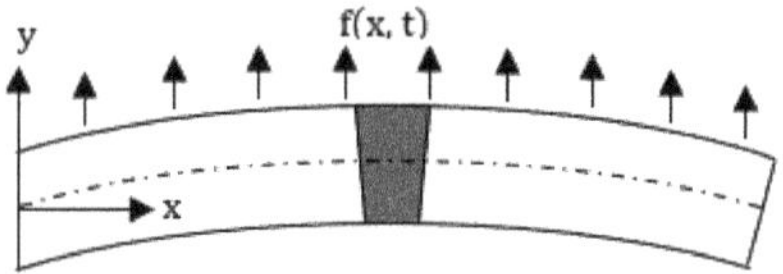

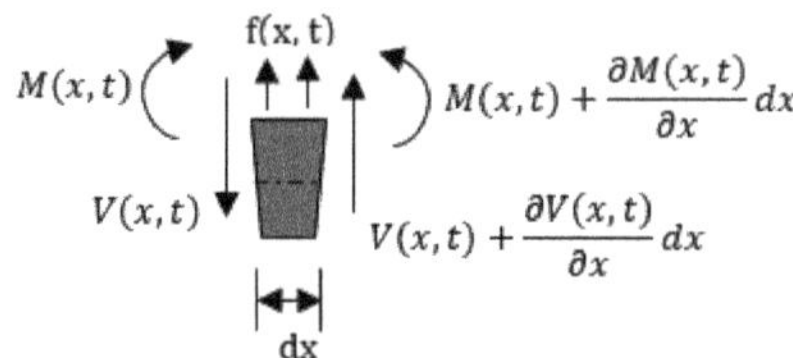

Figure 5. Euler–Bernoulli beam and a free body diagram of an element of the beam.

Simplifying and neglecting higher order terms in Equation (8), and combining it with (7) and (9), gives:

$$\rho A(x)\frac{\partial^2 w(x,t)}{\partial t^2} + \frac{\partial^2}{\partial x^2}\left[EI(x)\frac{\partial^2 w(x,t)}{\partial x^2}\right] = f(x,t) \tag{10}$$

The beam deformation under free vibration can be attained by considering $f(x,t) = 0$. For beams with a uniform cross-section, Equation (10) can be further simplified by having $A(x) = A$, and $I(x) = I$. Thus,

$$\frac{\partial^2 w(x,t)}{\partial t^2} + \frac{EI}{\rho A}\frac{\partial^4 w(x,t)}{\partial x^4} = 0 \tag{11}$$

The beam deflection can be solved using this equation with four boundary conditions and two initial conditions. The initial conditions are the specified initial deflection $w_0(x)$ and velocity $\dot{w}_0(x)$ profiles causing the motion:

$$w(x,0) = w_0(x) \ and \ w_t(x,0) = \dot{w}_0(x) \tag{12}$$

For a cantilever beam, the boundary conditions are the zero bending moment and the shear force at the free end, and no deflection and slope at the fixed end. In other words,

$$w(0,t) = 0 \tag{13}$$

$$\frac{\partial w(0,t)}{\partial x} = 0 \tag{14}$$

$$EI\frac{\partial^2 w(0,t)}{\partial x^2} = 0 \tag{15}$$

$$\frac{\partial}{\partial x}\left[EI\frac{\partial^2 w(0,t)}{\partial x^2}\right] = 0 \tag{16}$$

Equation (11) can be solved by separating variables as in $w(x,t) = X(x)T(t)$. This approach permits the separation of Equation (11) into two sub-equations, which can be solved separately to yield temporal and spatial results. The total solution can be obtained by combining the two results. As stated above, the temporal solution depends upon the initial conditions, which vary from case to case. Given the boundary conditions, the spatial part yields:

$$X(x) = \cosh(\beta_n x) + \cos(\beta_n x) + \sigma_n[\sinh(\beta_n x) + sin(\beta_n x)] \tag{17}$$

where β_n and σ_n are the coefficients depending on the mode considered. For the first resonant mode, $\beta_n l$ is 1.875, while σ_n is 0.7341 [78]. The first mode shape of a cantilevered beam actuated at resonance is shown in Figure 6 along with the beam in an initial state.

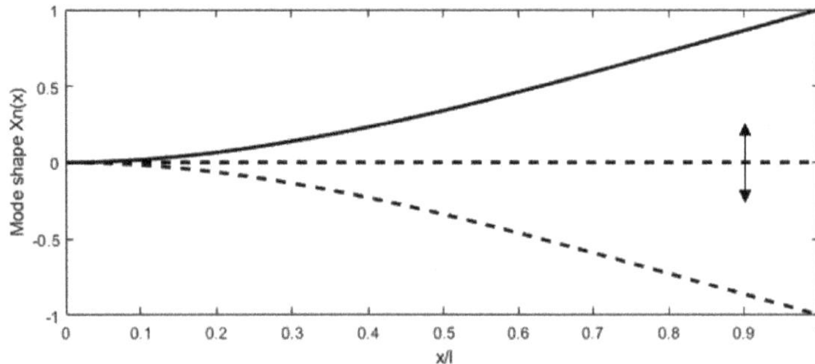

Figure 6. Deformation of a cantilever beam at resonance.

3.1. Resonantly Actuated Scanners

In scanning fiber endoscopes, cantilevered fibers are usually excited at resonance to scan the light beam. The main advantage of using the driving frequency equal to the resonant frequency consists in obtaining a higher tip displacement of the free end of the fiber, which results in high-resolution images (higher number of pixels in the FOV). It is possible to excite the cantilevered fiber using a variety of micro-actuators, such as piezoelectric [51,79–81], electromagnetic [82–84], electrothermal [85–87], shape memory alloys [88], or electroactive ionic polymer [89] actuators. The distal end of the fiber follows the mode shape shown in Figure 6. Each actuation method is better suited for the excitation frequency in certain ranges based on their working principle, which will be described in detail later in the paper. Various resonant scanners in the literature are compared in [90].

Depending upon the actuation technique used to excite the cantilever beam, it is possible to observe the development of 2D motion of the fiber tip by exciting the fiber along one direction. The so-called whirling motion, causing the fiber tip to follow an elliptical-shaped pattern instead of its linear motion, is caused by the cross-coupling of the motion between the planes perpendicular to the beam axes. It is possible to obtain a stable whirling motion within a small frequency range [91,92]. The cross-coupling motion can be avoided by exciting the cantilever beam along certain eigendirections [93]. On the other hand, Wu et al. developed a fiber optic scanner able to obtain 2D scanning using nonlinear cross-couplings [94].

3.2. Non-Resonantly Actuated Scanner

Some actuation methods such as electrothermal actuators are unable to generate motion at very high frequency. It is difficult to generate resonant scanners characterized with a low resonant frequency as they require long and slender beams, compromising their mechanical stability. In addition, the fiber tip displacement occurs symmetrically to the optical axis and is difficult to offset in resonant scanners. In such cases, the cantilevered optical fibers are excited at a frequency different from their resonant frequency. Even though there is less tip displacement for a given excitation power, it is possible to achieve beam scanning at low frequency and offset the center of the scan by adding a bias voltage [90].

In such scanners, the deflection of the distal tip of the fiber (δ_{tip}) is related to the displacement of the actuator exciting the vibration (δ_s) by:

$$\delta_{tip} = \delta_s \left[1 + \frac{3(1-a)}{2a} \right] + \frac{qgL^4}{8EI} \tag{18}$$

with L, a, q, E, I, g being the length of the cantilevered portion of the fiber, length ratio of the fixed end, mass per unit length of the fiber, Young's modulus, the moment of the inertia of the fiber, and acceleration of the gravity, respectively [95].

Zhang et al. proposed a similar scanner where a 45-mm-long fiber was electrothermally actuated using a micro-electro-mechanical (MEMS) actuator operating at no more than 6 V [95]. A similar scanner was developed by Park et al., where a 40-mm-long fiber, used as an endoscopic OCT probe, was actuated using a 3 V power source [96]. Some researchers were able to develop cantilevered scanners working at a frequency not too far from their resonance frequency. In such semi-resonant scanners, the fiber tip provided an intermediate displacement, and no nonlinear whirling effects were present. Moon et al. developed an OCT probe where the cantilever scanner was excited at 63 Hz using a piezo-tube actuator [97].

4. Actuators in Cantilever-Based Endoscopic Devices

The miniature size of MEMS devices, along with their light weight and stable performance characteristics, makes them attractive for micro and nano applications, among which are endoscopic optical devices. On the basis of the working principle, MEMS actuators can be subdivided into piezoelectric, electrostatic, electrothermal, electromagnetic, and shape memory alloy actuators. The piezoelectric actuators are widely used in endoscopic catheter design due to their compact size, low power consumption, and large output force. On the other hand, the actuation displacement is limited in such devices. Electrostatic actuators are the second most used actuation method in medical scanning devices due to their fast response and ease of fabrication. However, it is difficult to produce such devices at very small dimensions, which limits their use in systems requiring a distal actuation. Electrothermal actuators generate high actuation displacement and force, but the elevated working temperature and low working frequency limit their use in some cases. Electromagnetic and shape memory alloy actuators find limited applications in cantilevered fiber optic endoscopic scanners [98]. All these actuation methods are described in this section in great detail and compared in Table 2, where the number of ticks qualitatively indicates the intensity of a certain pattern [6].

4.1. Piezoelectric Actuators

The working principle of piezoelectric actuators is based on the so-called piezoelectric effect. Piezoelectric materials have the ability to change the material polarization in the presence of a mechanical stress and conversely generate strain or force in the presence of an external electric field. Among the various crystalline, ceramic, and polymeric materials, aluminum nitride (AlN) and lead zirconate titanate (PZT) are most frequently used in MEMS devices [99]. Piezoelectric actuators are characterized by providing fast response, low driving voltage, and low power consumption [100].

The relationship between the electric field applied to the material and the mechanical deformation exhibited by the material is nonlinear due to the presence of hysteresis and drift. For a small variation in electric field, the material behavior is almost linear and can be described by:

$$\varepsilon = Ed + c^{-1}\sigma \tag{19}$$

where ε is the strain tensor, E is the electric field vector, σ is the stress tensor, d is the piezoelectric tensor (vector of strain coefficients), and c is the elastic tensor. In the case of no external force, the second component on the right-hand side of Equation (19) becomes zero. The piezoelectric strain depends upon the direction of the mechanical and electrical fields [101].

Piezoelectric materials often show a nonlinear hysteresis behavior, which causes the relationship curve between the displacement exhibited by the material and the applied electric field to be different in ascending and descending directions. Various models have been proposed to describe this hysteresis phenomena. Bahadur and Mills proposed a hysteresis model to characterize the symmetric and asymmetric rate-dependent hysteresis.

The output charge (q) is related to the endpoint displacement (x) and the applied voltage (V_p) by:

$$q = C_0 V_p + T_{em} x \tag{20}$$

with C_0 and T_{em} being the capacitance of the piezoelectric element and the electromechanical coupling factor, respectively [102].

Table 2. Comparison between different actuation methods used in cantilevered endoscopes [6].

	Electrostatic	Electro-Thermal	Piezoelectric	Electromagnetic	Shape Memory Alloy
Force	✓	✓	✓✓✓	✓✓	✓✓✓
Displacement amplitude	✓✓	✓✓✓	✓	✓✓✓	✓✓
Compactness	✓✓✓	✓✓✓	✓✓	✓	✓✓
Power consumption	✓✓	✓	✓	✓✓✓	✓✓✓
Working principle	Electrostatic force	Thermal expansion	Piezoelectric effect	Magnetization effect	Material deformation
Motion range	1D/2D	1D/2D	2D	1D	1D
Scanning pattern	Spiral	Lissajous	Spiral	Linear	Linear
Advantages	Fast response, low voltage required, easy fabrication, and no hysteresis	Large displacement, low operating voltage, small dimensions	Large force generated, wide operating frequency range, low power consumption	Large displacement obtained, quick and linear response, easy to control	Flexibility, large frequency response
Disadvantages	Large device dimensions, pull-in problem, complicated circuit	High working temperature, not operable at very high frequencies	Limited displacement	Large device dimensions, difficult to manufacture	Low displacement

Normally, piezoelectric devices are restricted for 1D operation with force/displacement occurring along the axis defined by the electric field. In these so-called longitudinally translating piezo chips, the electric field is applied parallel to the polarization direction of the material, which causes the displacement in the same direction of the field normal to the surface of the electrodes. In shear piezo elements, the polarization is obtained in the direction perpendicular to the field direction. Thus, there is an orthogonal relationship between the direction of the displacement and that of the electric field [103].

Piezoelectric actuators are available in single disc/plate and tubular configurations. These configurations are described below in detail.

4.1.1. Disc Piezoelectric Actuator

Flat disc piezoelectric actuators can be constructed using a single piezo element (unimorph actuators) or using two different piezo elements (bimorph actuators). In either case, piezo elements are connected to a base material. There is an expansion or contraction of the piezo material in the presence of an electric current, which provides the bending motion of the actuator. The schematic diagram showing the working principle of a piezo bending actuator is shown in Figure 7. When a positive voltage is applied to the piezoelectric ceramic layer, it elongates in x direction, while the base material does not change its length, resulting in convex bending of actuator towards the conductive layer, as in Figure 7b. Similarly, bending in the opposite direction occurs by applying a negative voltage to the piezoelectric sheet. Li et al. fabricated a scanning fiber probe for an OCT endoscope, where an optical fiber was placed in the middle of the two piezoelectric plates with a common copper substrate element. The bending of the fiber tip in the vertical direction was obtained by exciting the two piezo elements with the same voltage, i.e., the structure will bend along the positive or negative directions by elongating or contracting both elements. On the other side, the motion of the tip along the horizontal axis was obtained by applying an opposite polarity voltage to the two elements. A two-dimensional Lissajous scanning pattern was obtained by the fiber tip by controlling the voltages on the two layers [57].

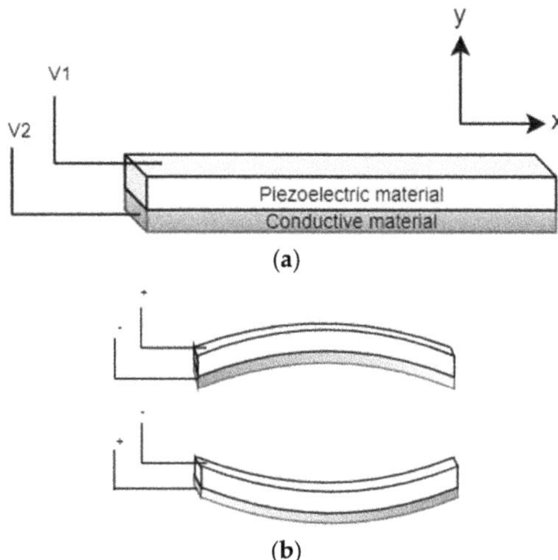

Figure 7. Piezoelectric bending actuator: (**a**) schematic diagram; (**b**) working principle.

Tekpinar et al. developed a piezoelectric fiber scanner where two planar piezo bimorph cantilever actuators were placed perpendicular to each other to generate the 2D motion of an optical fiber connected to them. In this case, the dimensions of the cantilevered fiber and the mode optimization allowed the fiber tip to interchangeably follow a raster, spiral, or Lissajous scanning pattern by simply changing the actuation parameters. A schematic diagram of such a model is shown in Figure 8A [104]. The different scanning patterns were projected on a United States Air Force (USAF) resolution target to evaluate the image uniformity. From Figure 8B, it can be seen that a raster scan provides a uniform illumination on the target ((a) within Figure 8B), while the spiral one is characterized by decreasing illumination along the radius ((b) within Figure 8B). The illumination in the case of a Lissajous scan is highly dependent on the fill factor ((c) within Figure 8B) [104].

Rivera et al. developed a compact multiphoton endoscope (with an outer diameter of 3 mm) where two bimorph piezoelectric actuators were used to excite a DCF. Two bimorph structures were placed in such a configuration to have perpendicular bending axes. A raster scanning pattern was generated by exciting the fiber in two directions. The fast-scanning motion was obtained by exciting one of the actuators at the resonant frequency of the extending cantilevered DCF, and slow axis scanning was performed by exciting the other actuator at a frequency much lower than the resonance frequency. The mechanical assembly of the described endoscope is illustrated in Figure 9a [105]. The developed prototype and the optical path diagram inside the endoscopic tip are shown in Figure 9b,c [105]. Using the developed probes, the authors were able to obtain the fluorescence images of an ex vivo mouse lung tissue at depths comparable to a commercial multiphoton microscope. The finer resolution (lateral and axial resolution of 0.8 and 9.4 µm, respectively) enabled the probe to clearly identify the alveolar walls and lumens in the unstained lung tissue, as in Figure 9d [105].

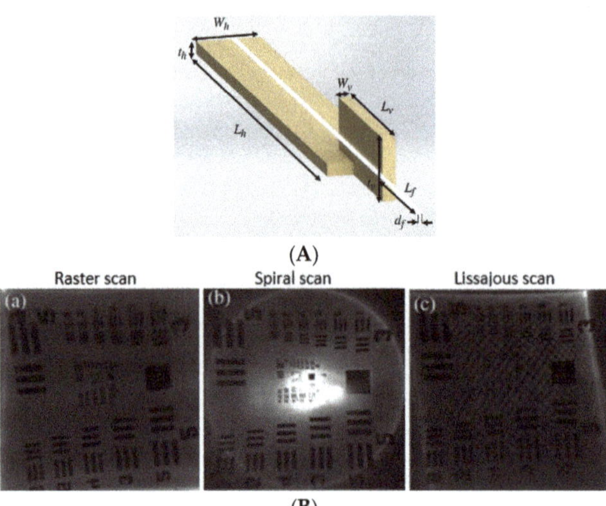

Figure 8. Multiple patterns generating fiber scanner: (**A**) schematic diagram of the scanner (**B**) achieved scanning patterns projected on a USAF target (taken with permission of [104]).

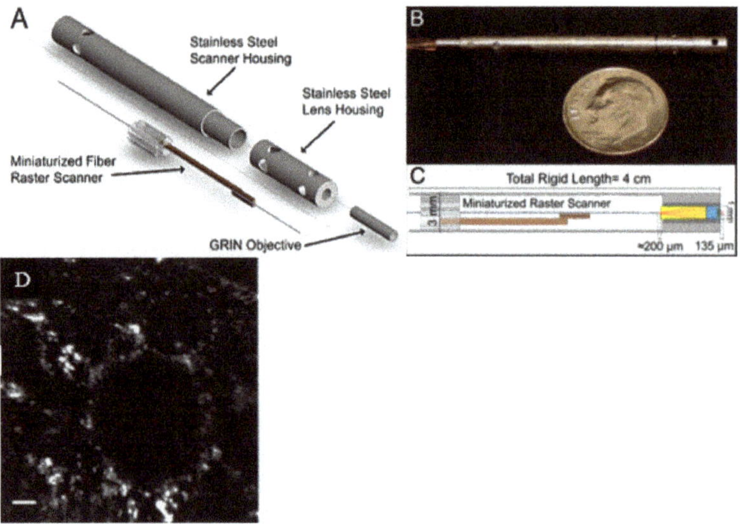

Figure 9. Raster scanning endoscope: (**A**) mechanical assembly; (**B**) developed prototype; (**C**) optical path diagram; (**D**) fluorescence image of an unstained ex vivo lung tissue (taken with the permission of [105]).

It is possible to enhance the cantilever fiber deflection with low power consumption by using very thin piezoelectric ceramic layers. Recent developments in the field showed the ability to form piezoelectric layers with thicknesses at submicron levels. A variety of methods were implemented to deposit the thin film of these materials. Such methods included arc discharged reactive ion-plating, epitaxial process, sol–gel spin-coating, and sputtering [100].

4.1.2. Tubular Piezoelectric Actuator

In the tubular structure, a thin ceramic sheet is bent into a cylindrical shape and can experience axial, radial, or bending motion. The tube shrinks radially and axially in the presence of a voltage difference between the inner and outer electrodes of the tube. In these actuators, the load can be mounted either on the curved surface for radial displacement or on the rims for axial displacement. These devices are very rapidly responsive, but the generated force is limited. Usually, the displacement produced by a piezoelectric device is very small. Thus, it is possible to stack up various piezoelectric disks or tubes to amplify the generated displacement [99].

As described earlier, piezoelectric actuators are most commonly used in cantilever-based endoscopic probes, especially the tubular piezoelectric actuators. In this case, the tube structure is divided into four electrodes and placed near the blocked end of the cantilever fiber. The base excitation of the fiber along a certain direction is obtained by applying voltage to two opposite electrodes. Seibel et al. used this configuration to obtain a 2D displacement of the fiber tip. In this small-sized endoscope, the drive voltage at two pairs of electrodes had an increasing amplitude and a phase shift of 90°, which resulted in a spiral pattern followed by the fiber tip [9,10]. The schematic diagram of the scanning fiber endoscope is shown in Figure 10 along with an enlarged view of the scanning portion showing the cantilevered fiber's connection with the tubular actuator [15]. Some in-vivo testing images of airways of a live pig taken with this endoscope (Figure 11b) were compared with the corresponding images from a conventional Pentax bronchoscope (Figure 11a). The two devices showed comparable images in terms of resolution and field of view [15].

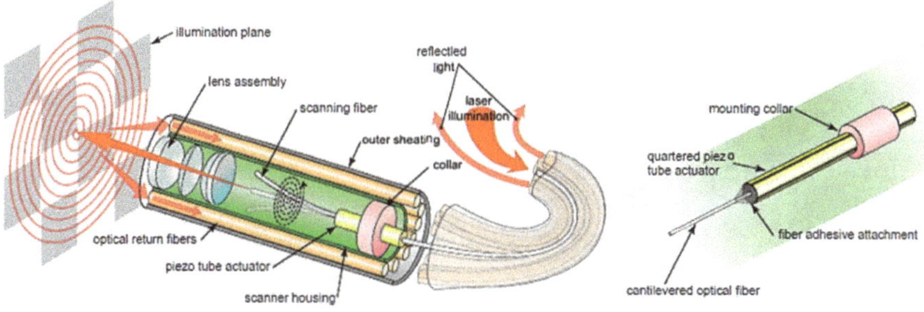

Figure 10. Schematic diagram of a scanning fiber endoscope (taken with permission of [15]).

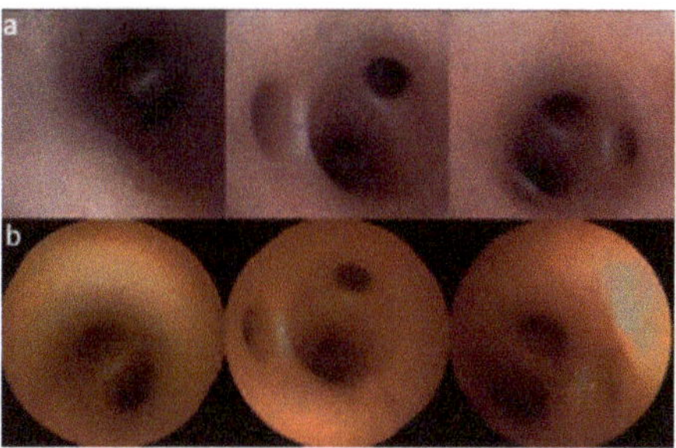

Figure 11. In vivo images of airways of a pig acquired with (**a**) conventional Pentax bronchoscope; (**b**) scanning fiber endoscope (taken with permission of [15]).

A similar fiber optic scanner using a tubular piezoelectric actuator, packaged within a 2 mm housing tube, was developed by Liang et al. as a two-photon and second harmonic endoscope. This novel endo-microscope enabled label-free histological imaging of tissue structures with subcellular resolution. The schematic diagram of the developed two-photon endoscope is illustrated in Figure 12a [106]. An overlaid two-photon and second harmonic generation image of a mouse liver acquired with such an endoscope is shown in Figure 12b, where the collagen fibers (in red) and vitamin A granules (in bright green) dispersed in the cytoplasm (dark green) can be clearly identified [106].

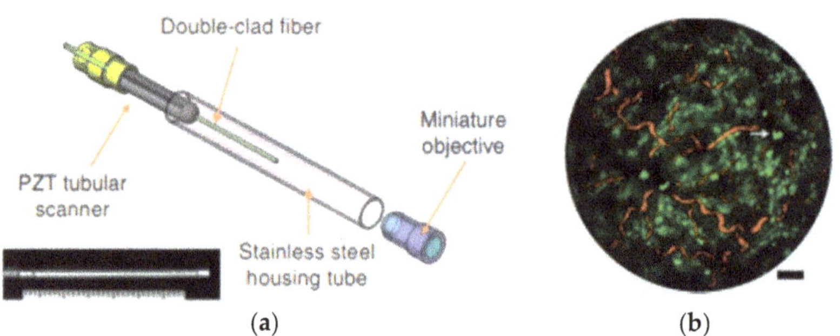

Figure 12. Nonlinear optical endoscope: (**a**) schematic diagram; (**b**) two-photon and second harmonic generation structural image of a resected mouse liver (taken with permission of [106]).

Vilches et al. developed a fiber scanner for OCT where a piezoelectric tubular actuator provided the base excitation motion to an optical fiber having a GRIN lens attached to its free end. In this configuration, schematized in Figure 13a, the beam scanning was attained by rotating the angle of the collimated beam, which provided high-resolution imaging while avoiding optical aberrations [107]. The cross-sectional tomogram of a human finger obtained with this scanner is shown in Figure 13b [107].

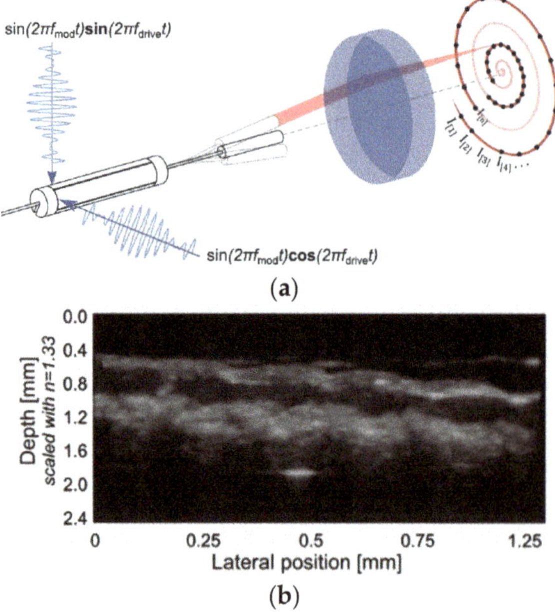

Figure 13. Fourier-plane fiber scanner: (**a**) schematic diagram; (**b**) cross-sectional tomogram of human finger (taken with permission of [107]).

4.2. Electrothermal Actuators

The working principle of an electrothermal actuator is based on the Joule effect. The electric resistivity causes an increase in temperature in the presence of the current flow through the actuator. The amount of heat generated in a material is directly proportional to the material's resistivity, current, and the length of the actuator, while it is inversely proportional to the cross-sectional area of the device. Generated heat causes thermal expansion and consequently the deformation of the material.

In electrothermal actuators, the cross-section is usually much smaller than the length of the actuator to make it more resistive and, consequently, cause higher temperature variations for a given input power. Thus, the temperature along the actuator can be calculated using a one-dimensional model. A correction factor can be included in the equation to consider this approximation. The nonlinear partial differential equation describing the temperature variation (T) in space and time can be obtained using the conservation of energy [108]. In the case of a rectangular section bar (with width w, and height h), the partial differential equation (PDE) describing the heat transfer along the length x becomes:

$$\rho c_p \frac{\partial T}{\partial t} dx = J^2 \rho_r dx - \left[\frac{d}{dx}\left(-k_p \frac{dT}{dx}\right) dx + (T - T_p)\left(\frac{S}{hR_T}\right) dx + \left(\frac{2h_{cs}}{w} + \frac{\lambda h_{cf}}{h}\right)(T - T_a) dx + \frac{\lambda \varepsilon_x \sigma}{h}\left(T^4 - T_a^4\right) dx \right] \quad (21)$$

with ρ, c_p, ρ_r, k_p being the density, specific heat, resistivity, and the thermal conductivity of the material characterizing the actuator, respectively. h_{cs} and h_{cf} are the convection coefficients for the side walls and the faces of the actuator element, respectively. λ is the coefficient describing the heat loss. ε_x and σ are the surface emissivity and the Stefan–Boltzmann constant for radiation heat transfer, respectively, while T_p and T_a are the substrate and ambient temperature, respectively. R_T is the thermal resistance between the actuator surface and the substrate material. J is the current density along the actuator given by the current passing through it per unit section of the actuator material. S is the shape factor and is a function of total heat flux defined as:

$$S = \frac{h}{w}\left(\frac{2t_v}{h} + 1\right) + 1 \quad (22)$$

where t_v is the air gap between the actuator material and the substrate [108,109].

The heat transfer through convection and radiation is evident at very high temperatures. In electrothermal actuators, the operable temperature is limited to avoid damage to the material. Thus, the corresponding terms in Equation (21) can be neglected [108,110]. Therefore, Equation (21) can be simplified to:

$$\rho c_p \frac{\partial T}{\partial t} = J^2 \rho_r + k_p \frac{d^2 T}{dx^2} - (T - T_p)\left(\frac{S}{hR_T}\right) \quad (23)$$

The temperature profile along the actuator can be determined knowing the initial temperature of the actuator and the two boundary conditions.

On the basis of configuration, electrothermal actuators can be divided into hot-and-cold arm, chevron, and bimorph actuators.

4.2.1. Hot-and-Cold Arm Actuators

These actuators are also called U-shaped actuators, folded beam actuators, heatuators, or pseudo bimorph actuators. As the name implies, the structure of the actuator is made up of at least one hot arm and one cold arm. The actuator is usually made up of a homogenous material with folded arms in a U-shaped pattern that are constrained by anchors. Usually, the anchor surfaces are characterized by having a large surface area, required to ensure heat dissipation. Two arms of the actuator characterized by different cross-sections are connected in series to an electric circuit. The current flows through the structure with different current densities within the two arms. Therefore, more heat is produced within the thin arm through the Joule heating principle compared to that of the wide arm. This differential thermal expansion of the material causes the thin arm to expand more and bend towards the wide section, generating the bending moment [98,111].

At the base of the cold arm, there is a thin section flexure arm which helps the bending deflection at the tip of the actuator in the shape of an arc in the actuator plane. The length of the flexure arm plays an important role in the value of the tip deflection. In the original model proposed by Guckel at al., the length of the flexure arm and the wide arm were equal to half of the length of the thin arm [112]. Huang and Lee developed the mathematical model describing the tip displacement of the actuator tip with respect to the function of the air gap between the two arms and the geometry of the arm structures. The smaller gap between the two arms led to a higher tip displacement, and the optimal length for the flexure arm was around 14–18% of the total length of the thin arm [113].

The temperature along the beam can be obtained by unfolding the beam and applying Equation (23). In the case of no external load acting on the tip, the lateral deflection of the tip at the free end of these actuators can be described by:

$$\delta y = \frac{1}{2} \frac{\left(a^4 - a^2 + 2a\right) A r \alpha \Delta T_{net} L^2}{5a^4 I + a^4 r^2 A - 2a^3 I + 5aI + r^2 aA + I + a^5 I - 2a^2 I} \tag{24}$$

where L, a, A, r, α, I are the actuator length, ratio of flexure arm length to hot arm length, cross-section area of the flexure and hot arm, center gap between the hot arm and the flexure component, coefficient of thermal expansion, and the moment of inertia of the hot arm (flexure arm), respectively. ΔT_{net} is the net temperature difference defined as the temperature which would cause the expansion of the hot arm alone and is the same as the net expansion in the real actuator case, where a small expansion of the cold arm corresponds to a decrease in the flexure component and results in a decrease in the net expansion between the two arms [114].

Hot-and-cold arm actuators are used widely in MEMS devices. There are large variations in the geometry of the actuators to achieve asymmetric thermal expansion. It is possible to obtain the in- plane deflection by changing the length of the arms instead of the cross-section [115], or using a combination of both the difference in the length and cross-section [116], or connecting the two arms of the actuator in parallel instead of series enabling higher current density in the thick arm, causing the tip deflection towards the thin arm [117], or changing the resistivity of one arm by selectively doping it.

Lara-Castro et al. designed an array of four electrothermal actuators based on the hot-and-cold arm configuration to obtain out-of-plane displacement. Four actuators are used to control the rotation of a MEMS mirror for endoscopic OCT purposes [118]. Some other changes in the geometry of the actuator include using two hot arms [66,119]. Seo et al used this kind of double hot arm electrothermal actuator to obtain the lateral in-plane displacement of the optical fiber for endoscopic purposes [66,120]. An optical fiber was firmly connected to the linking bridge connecting the two hot arms and the cold arm. The differential thermal expansion between hot and cold arms allowed the cantilevered fiber to move in a lateral direction (in-plane motion), while that between the actuator surface and the fiber gave the vertical motion, causing the fiber tip to follow a Lissajous scanning pattern [66]. The schematic diagram of a 1.65 mm diameter confocal endo-microscope catheter developed using this type of actuator is shown in Figure 14 [65].

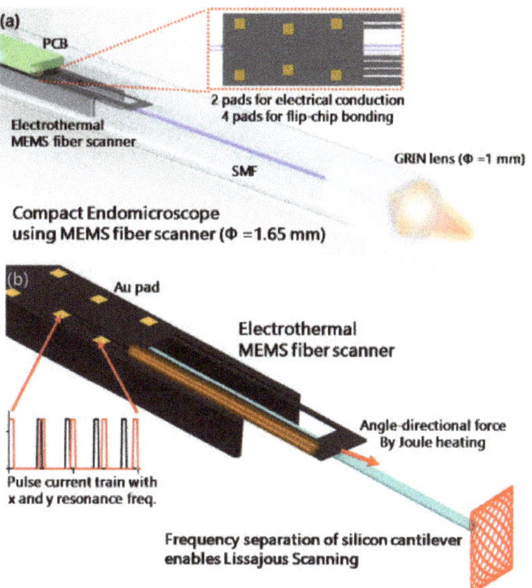

Figure 14. Electrothermally actuated confocal endo-microscope: (**a**) schematic diagram; (**b**) working principle (taken with permission of [65]).

4.2.2. Chevron Actuators

Chevron actuators are also called bent-beam actuators or V-shaped actuators and are the other type of in-plane electrothermal actuator, with a slightly different working principle. In this case, the in-plane displacement of the tip of the actuator is obtained from the total thermal expansion of the components instead of a differential expansion [98].

In a V-shaped electrothermal actuator, two symmetrical slanted beams are connected at a certain angle to a central shuttle beam at the apex to the base with anchors. The current passing through the actuator causes the thermal expansion of both slanted beams due to the Joule heating principle. As the movement of the beam is constrained by the anchors and the central shuttle, the thermal expansion causes a compression force and a bending moment, which gives rise to the lateral displacement of the shuttle beam [121].

Similar to the U-shaped beams, the temperature distribution along the arms of the actuator can be obtained using Equation (23). Enikov et al. described the analytical model for V-shaped thermal actuators. The analysis of the beam deformation was considered by taking into account the buckling effect in the beam due to the axial thermal load and transversely applied force, if any. The numerical solution of the thermoelastic buckling model of the beam led to the tip deflection of the beam or the central shuttle beam [121]. However, Sinclair presented a simplified model describing the tip displacement to be:

$$\delta = \left[l^2 + 2(l)l' - l\cos(\theta)^2 \right]^{1/2} - l\sin(\theta) \tag{25}$$

where θ is the initial tilt angle of the arm beam, l is the length of the single actuator arm, and l' is the elongation due to thermal expansion [122].

These actuators provide certain advantages over the bent-beam actuators described earlier, such as rectilinear displacement, larger exhibited force at the tip, and lower power consumption [98]. The displacement of the central tip of the actuator can be increased by using longer arm components or reducing the bending angle θ. The opposite changes increase the exhibited force. Moreover, it is possible to amplify the motion of the shuttle beam by connecting the two bent-beam actuators in cascade. In the cascaded configuration, two V-shaped electrothermal actuators are anchored to the substrate and connected together with secondary V-shaped beams. The current can be passed either through the primary units only or through all the structural components [123]. Similarly, it is possible to increase the output force without changing the displacement from the device by placing multiple V-shaped actuators in parallel [122]. It is even possible to combine the parallel and cascade configurations to obtain the desired displacement and force outputs [124].

Another variation in chevron actuators consists of changing the geometry of the actuator to obtain a wide range of output properties. Among these, the most frequently used are the electrothermal actuators with Z-shaped patterned arms. In this configuration, the thermal expansion of the beams is blocked due to symmetry constraints, leading to the bending of the beams and thus the in-plane displacement of the central shuttle element. Z-shaped actuators permit smaller feature sizes and larger displacement compared to the V-shaped electrothermal actuators [98,125]. Another alternative in chevron actuators is a combination of straight and bent beams, or the so-called kink actuator. This kind of actuator consists primarily of straight arms which undergo thermal expansion by the Joule effect, while the small kink in the middle serves to guide the motion of the actuator. Kink actuators provide higher displacement at lower power levels as compared to V-shaped actuators [126].

Chevron actuators find use in some cantilever-based optical scanners. Kaur et al. developed a sub-millimeter-sized cantilevered fiber optical scanner that can find use as a forward-viewing endoscopic probe. In this design, shown in Figure 15a, an electrothermal chevron actuator made with two parallel legs excites an SMF at resonance. In this case, the total thermal expansion of the actuating material provides a base excitation motion to the cantilevered fiber [127]. A resolution target image captured with this scanner is provided in Figure 15b.

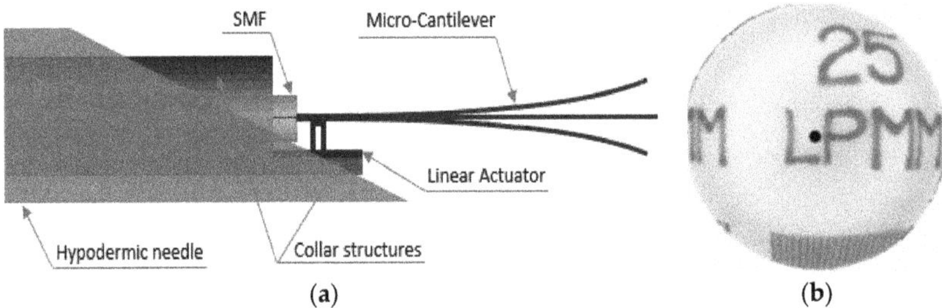

Figure 15. Cantilevered fiber scanner using chevron actuator: (**a**) schematic diagram; (**b**) reconstructed image of a resolution target.

4.2.3. Bimorph Actuators

A third class of electrothermal MEMS actuators is bimorph or bi-material type actuators. In this kind of actuator, two or more materials with different thermal expansion coefficients are stacked on top of each other. The different thermal expansion causes the actuator to bend or curl due to the induced strain generated by the Joule heating during actuation, which results in an out-of-plane motion [111].

The basic design of a bimorph electrothermal actuator consists of a cantilever-shaped micro-actuator fabricated using two layers of different materials connected to each other. The bending direction of the actuator tip during actuation will be dictated by the material with the higher thermal expansion coefficient compared to the one with the lower thermal expansion coefficient. The mathematical model for the tip deflection of such a micro-actuator is described by Chu et al. [128]. Assuming a constant curvature, the deflection at the free end of a cantilevered bi-material actuator is given by:

$$\delta = kL^2/2 \tag{26}$$

with L being the length of the cantilevered bi-metallic beam, and k being the curvature, which is:

$$k = \frac{1}{r} = \frac{6b_1b_2E_1E_2t_1t_2(t_1+t_2)(\alpha_2-\alpha_1)\Delta T}{(b_1E_1t_1{}^2)^2 + (b_2E_2t_2{}^2)^2 + 2b_1b_2E_1E_2t_1t_2(2t_1{}^2+3t_1t_2+2t_2{}^2)} \tag{27}$$

where r is the radius of curvature, b, t, E, α are the width, thickness, Young's modulus of elasticity, and the thermal expansion coefficient, respectively, of the two materials characterizing the actuator. ΔT is the change in temperature due to Joule heating [128].

As with other electrothermal actuators, it is possible to place the different bimorph actuators in a cascaded configuration to amplify the obtainable tip displacement. In such structures, the various bimorphs are placed together in a serpentine direction, which causes the tip deflection from each bimorph to be added in series, yielding the higher overall tip displacement. It is possible to use different geometries for the bimorph structures to adapt according to the required spacing limitations in the microdevices. Large numbers of bimorph structures can be connected in a vertical cascaded form to generate a large out-of-plane displacement. Many bimorph structures can also be placed in parallel to lift the high load mirror surface [129].

Bimorph actuators are the most frequently used electrothermal actuators and find use in a large number of scanning mirror MEMS devices. Zhang et al. developed a cantilevered fiber scanner excited non-resonantly using a MEMS stage. This platform, placed at a certain distance from the fixed end of the fiber, was connected to the fixed surface using three-segment bimorph actuators at four edges. Three segments of the Al-SiO$_2$ bimorph actuators were placed in a configuration, shown in the schematic of Figure 16, to cancel out the lateral motion generating large vertical motion at the fiber tip [95]. Such a device along with imaging optics was packaged in a 5.5 mm probe to use as an endoscopic OCT probe [96].

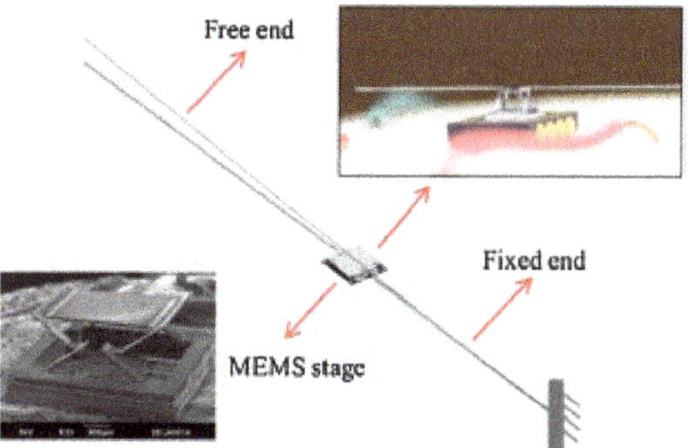

Figure 16. Schematic diagram of a non-resonant fiber scanner using bimorph electrothermal actuation technique (taken with the permission of [95]).

4.3. Electromagnetic Actuators

The working principle that governs the motion in these actuators is the so-called electromagnetic principle, where the conversion from electric/magnetic energy to mechanical energy takes place by means of a magnetic field. Similar to an electrostatic actuator, there are stationary and moving parts, named the stator and the rotor, respectively. Depending on whether the magnetic field is generated by the static or rotor component, there are two different configurations available for these actuators. Both configurations are described below in detail.

4.3.1. Moving Magnet Configuration

In this configuration of MEMS electromagnetic actuators, a bulk magnet is placed inside an electric coil. When the current flows inside the wire coil, it generates a magnetic field. The intensity of the generated magnetic field depends upon the current passing through the coil, the radius of the coil surface, and the distance from the coil. For a circular coil, the generated magnetic field is given by the Biot–Savart equation:

$$H(z) = \frac{\mu_0 N I \, r^2}{2(r^2 + z^2)^{3/2}} \tag{28}$$

where μ_0 is the permeability constant, N is the number of turns, I is the current, r is the mean radius, and z is the distance along central axis [130].

In the moving magnet configuration, a permanent magnet with net magnetization vector M is placed inside an external magnetic field H_{ext} created by one or more electric coils at angle α. The external field applies a torque on the moveable magnet given by:

$$T_H = \left| V_{mag} M \times H_{ext} \right| = V_{mag} M H_{ext} sin(\alpha) \tag{29}$$

with V_{mag} being the magnetic volume. Using a soft magnetic material, the generated torque T_H rotates M moving it away from the equilibrium position (easy axis) by an angle θ. An anisotropy magnetic torque T_a will be generated inside the magnet, tending to realign it to its initial position.

$$T_a = -K_a sin(2\theta) \tag{30}$$

K_a is the magnetic-anisotropy constant. An opposite torque T_a is exerted on the easy axis and thus on the magnet itself [64]. If the direction of H_{ext} remains constant at an angle γ ($\alpha = \gamma$ at the beginning) from the easy axis, the torque becomes:

$$T_H = V_{mag} M H_{ext} sin(\gamma - \theta) \tag{31}$$

In a permanent magnet, $M = M_s$ (saturated magnetization), and $\theta = 0$. The magnetization of soft magnets changes with the applied external magnetic field and M_s.

The external magnetization vector acting along the direction of M is:

$$H_a = H_{ext} cos(\gamma - \theta) \tag{32}$$

Thus, the change in M induces the poles at the end of magnet, which generates a demagnetized field H_d in the opposite direction of H_a:

$$H_d = -N_M M / \mu_0 \tag{33}$$

where N_M is the shape anisotropy coefficient. The net field (H_i) inside the sample changes to the sum of the applied and demagnetized field. The sample moves to reduce H_i and cause the magnetization vector to be:

$$M = \min\left[\frac{\mu_0 H_{ext} cos(\gamma - \theta)}{N_M}, M_s \right] \tag{34}$$

In equilibrium, the field torque T_H rotates M from easy axis and is balanced by the anisotropy torque T_a and tends to align M and vibrate the magnetic component [131].

Joos et al. developed an OCT probe for imaging based on this technique. In this probe an electromagnetic coil was placed at the outer surface in the center, in which a magnet was placed carrying a thin-walled 28-gauge tube. An SMF fiber was contained in a "S"-shaped 34-gauge stainless-steel tube placed within the 28-gauge tube, as in the schematic shown in Figure 17a. In the presence of an electric current at the coil, the electromagnetic force generated the sliding motion of the 28-gauge tube along the curved part of the inner S-shaped tube, allowing the fiber to move in the lateral direction [132]. The performance of the device was tested by imaging the ocular tissue structures. A real-time OCT image of ocular conjunctiva is shown in Figure 17b, where different tenons and sclera layers can be clearly identified [132].

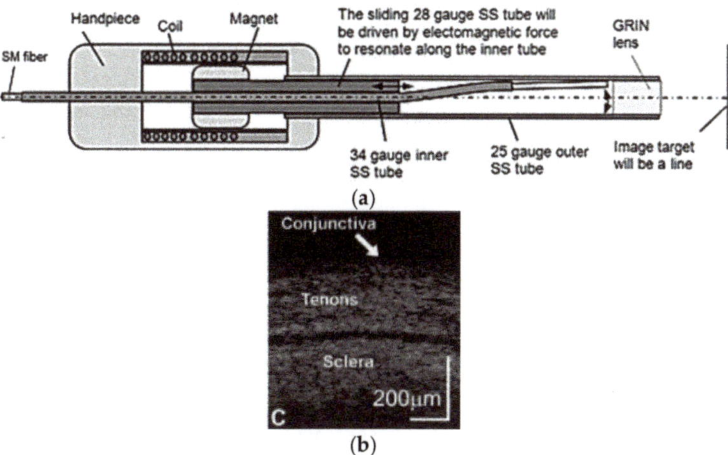

Figure 17. Forward-viewing OCT probe based on electromagnetic actuation: (**a**) schematic diagram; (**b**) real-time image of conjunctiva (taken with the permission of [132]) © The Optical Society.

Sun et al. developed a cantilevered fiber scanner for medical endoscopic applications, where an SMF with a collimating lens was excited at resonance using an electromagnetic actuator working on this principle. The researchers fixed a soft cylindrical magnet to an optical fiber using a 1 mm diameter polyimide pipe, and a tilted coil was fabricated using a microfabrication lithography technique [133]. The schematic of the design is illustrated in Figure 18. In the presence of an AC current applied to the coil, a magnetic field was generated within the coil and vibrated the magnet fixed to the fiber, resulting in excitation of the fiber [133]. Two tilted coils can be used to drive the fiber in two directions to obtain a 2D image [134].

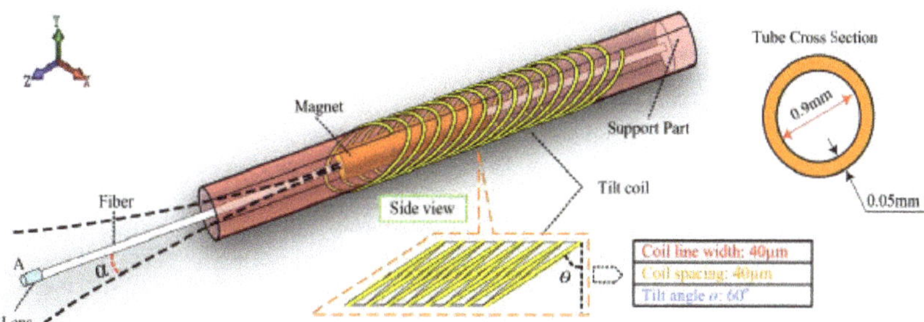

Figure 18. Schematic of an electromagnetically driven fiber scanner (taken with the permission of [133]).

A similar probe, schematized in Figure 19, was recently developed by Yao et al., where a cantilevered fiber containing a mass element and a lens at its tip was excited at a second resonance mode using a pair of flexible driving coils. The geometry of the cantilevered portion generated a 2D elliptical motion, with a larger scan angle at the fiber tip in the presence of a magnetic force generated by the soft magnet in the presence of a magnetic field [64].

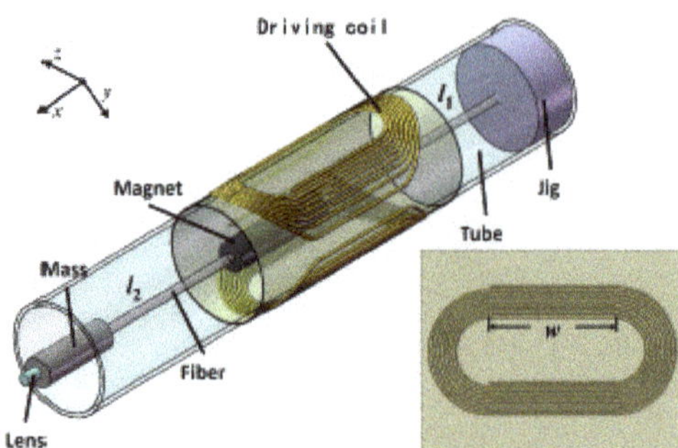

Figure 19. Schematic design of a fiber scanner excited at second resonance mode using an electromagnetic actuator (taken with the permission of [64]).

4.3.2. Moving Coil Configuration

In the moving coil configuration, an electric coil is fabricated on the scanner and is placed inside a static magnetic field created by external magnets. When the current flows through the coil in the presence of an external magnetic field, a force is exerted on the coil, designated as the Lorentz force. The force generated on the coil is given by:

$$F = |I\mathbf{L} \times \mathbf{B}| = BILsin(\theta) \tag{35}$$

where $\mathbf{B}$ is the external magnetic field, I is the current, L is the length of the conductor, and θ is the angle between the direction of the current and magnetic field [135]. The force produced can be written in terms of Equation (28) as well. Usually, the conductor is placed perpendicularly to the magnetic field to obtain the maximum exerted force. Equation (35) will be simplified to:

$$F = BIL \tag{36}$$

In the case of a coil with N turns, the generated magnetic torque on the coil is:

$$T_{mag} = 2\sum_{n=1}^{N} BILr_n \tag{37}$$

with r_n being the distance of the nth coil turn from the center [90].

As in the previous case, the actuator (coil surface) deforms due to the generated torque, and a restorative torque will arise in the coil to bring it to its initial state, causing the vibration of the moving coil. This technique is frequently used to actuate micromirror surfaces [136,137] and finds limited use in cantilevered fiber scanners.

4.3.3. Magnetostrictive Actuation

Magnetic materials are characterized by a special property which allows them to change their dimensions in the presence of a magnetic field. This effect is called magnetostriction. The material can undergo a change in dimension until it reaches the value of saturation magnetostriction, which depends on the magnetization and, therefore, on the applied magnetic field [138].

Bourouina et al. developed a 2D optical scanner based on the magnetostrictive effect. In this case, a silicon cantilever was coated with a magnetostrictive film. Due to the uniaxial nature of the magnetostrictive material, bending and torsion vibrations were generated simultaneously in the presence of an AC magnetic field generated by the electric coils placed in its surroundings. Later, a piezoresistive detector was incorporated in the device to measure the bending and torsional vibrations [139,140].

A slightly different fiber optic scanner was developed by a group of researchers from the University of Texas. In this design, an optical fiber was coated with a ferromagnetic gel which experienced a bending motion in the presence of an external magnetic field generated by a magnet placed at the outer surface. The schematic configuration of this device is shown in Figure 20 [141].

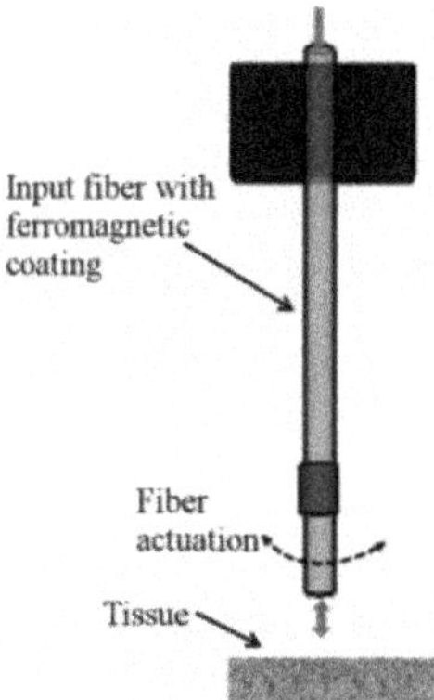

Figure 20. Schematic of a magnetically actuated fiber-based imaging system (reprinted with permission from [141]).

4.4. Electrostatic Actuators

An electrostatic actuator includes at least two pairs of electrodes attached to two plates separated by a gap. One of these plates is fixed by anchors and is named the stator, while the other plate is able to move and is designated the shuttle. In the presence of a voltage difference between the two plates, an attractive electrostatic force generates among them, causing the movement of the shuttle plate towards the stator. The amount of the electrostatic force generated between the two components depends on the gap and the dielectric constant of the media separating the two plates. The generated electrostatic force is given by:

$$F_{es} = \frac{A\varepsilon V^4}{2g^2} \tag{38}$$

where A is the electrode area, ε is the dielectric constant of the air, V is the total voltage difference applied to the plates, and g is the air-gap distance [142]. The maximum voltage that can be applied to a pair of electrostatic electrodes is delimited by the pull-in voltage. The electrostatic force increases with the applied voltage until the point when the force causes the two plates to collapse together. The maximum applicable voltage without causing this phenomenon is called pull-in point voltage. The electrostatic actuators can be classified into parallel plate and comb drive, which are described below.

4.4.1. Parallel Plate Actuator

In a parallel plate configuration, two electrodes are placed parallel to each other in an interdigitated finger configuration. In optical scanners, the moving electrode is mostly represented by a polysilicon mirror used to deflect the light.

Another variation of a parallel plate actuator is a system where the moving electrode has a rotational degree of freedom. The application of the voltage to the electrodes in this case causes the rotation of the moving electrode with a tilt angle obtained from the equilibrium between the electrostatic torque generated by the electrostatic force and the restoring torque. The large deflection angle in this case requires a large air gap between the electrodes. The maximum deflection of the rotating electrode should be less than one third of the air gap to avoid the pull-in phenomenon [143].

Most of the torsional electrostatic actuators are divided into small-sized scanner arrays causing large tilting angles with small air gaps. There are some systems developed with tapered electrodes to allow large tilting angles.

One of the main drawbacks of electrostatic actuators is that a large driving voltage is required to obtain moderate deflection angles. It is possible to partially overcome this by using tapered electrodes instead of the parallel-shaped ones [144].

4.4.2. Comb Drive Actuator

In electrostatic comb actuators, multiple plates are connected to make interdigitated static and mobile rows. Such a configuration enables an increase in the interaction area between the two electrodes, and, consequently, high electrostatic forces are generated. As in the parallel plate configuration, the out-of-plane motion of the mobile mirror structure can be obtained by making a vertical offset between the torsional support and the driving arm [143].

In vertical comb drives, the moving comb motion is out-of-plane with the motion of the fixed comb, which avoids the pull-in phenomenon. Moreover, the deflecting mirror can be decoupled from the actuating part, permitting a large possible deflection of the mirror itself. The higher electrostatic torque generated by the comb structure leads to the possibility of higher driving frequencies and thus a higher scan speed [145].

Vertical comb drives are used frequently to actuate micromirrors [146–148]. It is possible to place the moving comb structures at a certain angle with respect to the fixed ones to obtain an angular vertical comb drive. The initial angle between the comb structures determines the obtainable maximum angle rotation of the mirror connected to it [61].

A group of researchers at Fraunhofer University studied the design optimization for comb drive micro-actuators. It was more convenient to place the electrodes in a star-shaped pattern to obtain a higher deflection of the mirror surface [149].

Both types of actuators are mainly used to actuate micromirrors [100]. Munce et al. developed an electrostatically driven fiber optic scanner (shown in Figure 21a) where a single-mode fiber was placed in a platinum coil. The packaged probe had a diameter of 2.2 mm. There were two insulated wires placed around the optical fiber which acted as electrodes. An electrostatic force was generated around the fiber in the presence of a potential difference between the two electrodes, which allowed the fiber tip to vibrate [150]. An in-vivo Doppler OCT image of the heart of a Xenopus laevis tadpole, taken with this device (Figure 21b), enabled clear visualization of the left and right aortic arches [150].

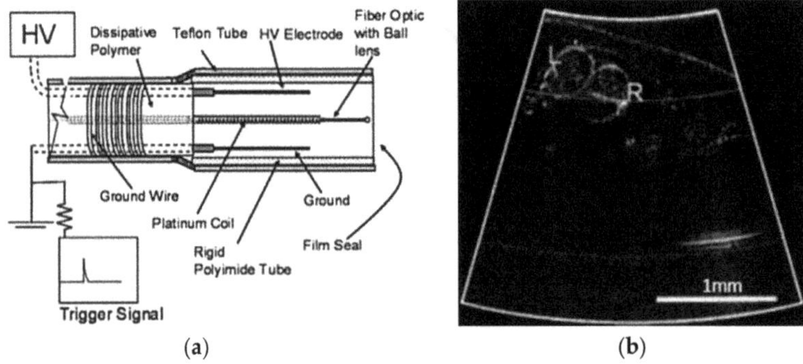

Figure 21. Electrostatically driven fiber scanner: (**a**) schematic diagram; (**b**) Doppler OCT image of a tadpole heart (taken with permission of [150]) © The Optical Society.

4.5. Shape Memory Alloy Actuators

Shape memory alloys (SMAs) are unique metallic alloys having the ability to return to their original shape after being deformed plastically. The deformation recovery is usually obtained by increasing the temperature of the material, which releases the state of stress.

SMA material is available in either a wire or a sheet shape. Frequently used SMA actuators take the form of a coil structure as they can provide a larger stroke as compared to a straight wire per unit length. The shear modulus and the spring constant of a SMA coil/spring depend on the composition, temperature, strain, and shape memory treatment applied to the material. In the design of an SMA actuator, the material is deformed at a low temperature and thermally treated to remember its shape. The shape recovery of the coil is obtained by increasing the temperature via the Joule effect by passing a current through the wire [151].

SMA coil actuators show a one-way shape memory effect. Thus, a bias spring or a second SMA coil spring is combined with an SMA coil actuator to obtain a two-way actuation. When a current is passed through the SMA coil, it tends to return to its original shape, which exerts a force on the second spring/coil, permitting motion in one direction. When the current is stopped, the SMA coil cools down and it is re-deformed by the force applied by a bias spring or activating the second coil [152].

SMA coils are largely used in endoscopes to bend the distal tip for a long time. Maeda et al. fabricated a 2 mm diameter endoscope head where an actuation ring connected to two SMA springs guided the bending motion of the endoscope tip which contained the optical guide fibers by pulling or releasing the pull wire connected to it. The schematic diagram showing the structure of the described design is shown in Figure 22 [153]. When an SMA coil spring (1) is heated, it tends to recover its shape and rotates the actuation ring to the right, which in turn pulls the wire, causing the bending motion of the tip. By stopping the current in that coil and heating the other coil, the tip returns to its original position [153].

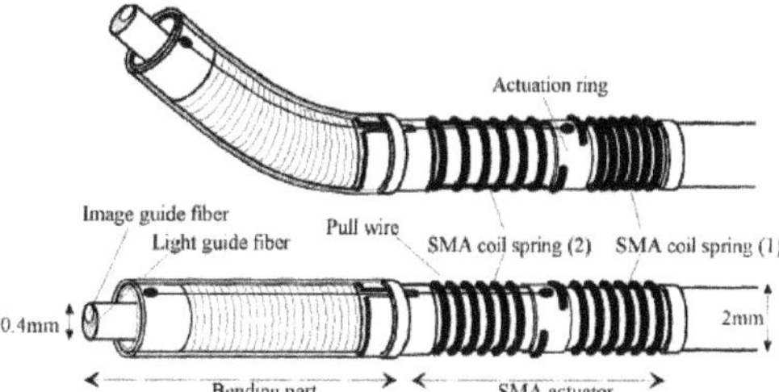

Figure 22. Schematic design of endoscopic tip guided using SMA coils (taken with the permission of [153]).

Haga et al. used three SMA coil actuators at equilateral triangle locations between two links to form a joint. The bending motion generated by these actuators allowed the snake-like movement of the central working channel of the endoscope, i.e., guided the endoscope [154].

A similar endoscope was designed by Makishi et al., where the active bending motion of the endoscopic tip, containing a CCD imager, was obtained using three SMA coil actuators [155]. The endoscope design showing the structure of the device is illustrated in Figure 23 [155]. Another similar endoscope was developed by Kobayashi et al., where the bending motion of the endoscopic central channel containing a CMOS imager and three LEDs was obtained using three SMA wires and a stopper coil. In this case, a large bending angle was obtained at low cost by allowing the SMA wires to follow an arc-shaped deformation [71].

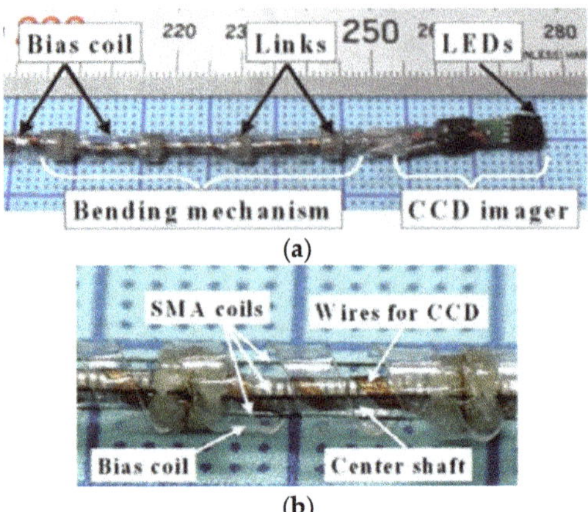

Figure 23. Active bending endoscope using SMA coil actuators: (**a**) endoscope design; (**b**) enlarged view of the actuation mechanism (taken with the permission of [155]).

5. Discussion

Optical endoscopic imaging enabled the visualization of cellular and subcellular structures in real time, enabling early interventions and improved diagnostic yield from fewer biopsies. Moreover, the less invasive nature of the imaging device results in reduced tissue trauma, low risk of complications and operational costs, and fast recovery times. In cantilever fiber endoscopes, an optical fiber is rigidly fixed to an actuator's substrate surface, leaving a few millimeters of free end at its distal end. Most often, this free end is vibrated at resonance to obtain a large tip displacement, which in these imaging devices is directly related to the resolution of the obtained image. Various imaging applications, such as confocal endo-microscopy, OCT, photoacoustic imaging, etc., can be found using such endoscopes.

Frequently used MEMS actuators in fiber optic endoscopic devices comprise piezo-electric, electrothermal, electromagnetic, electrostatic, and shape memory alloy actuators. Piezoelectric actuators are available in plate or tubular structure, among which the latter is largely used as it provides the base excitation to the fiber held in the center along two directions. By linearly increasing the amplitude of sinusoidal voltages at two pairs of electrodes with a 180° phase shift, one can obtain a spiral scanning pattern.

Electrothermal actuators are available in three different types including hot-and-cold arm, chevron, and bimorph actuators. In a hot-and-cold arm actuator, the different geometries of the two legs produce asymmetric heat generation and, consequently, an asymmetric thermal expansion, generating a bending motion at the actuator tip. Similarly, in bimorph actuators, the different material properties cause the bending of the actuator tip. These techniques can be combined with other actuation methods to obtain a 2D Lissajous scanning pattern. Chevron actuators have symmetrical legs and provide linear 1D motion at the tip.

Electrostatic actuators are available in parallel plate and comb drive configurations. These actuators find limited use in fiberoptic scanners due to their large dimensions but are frequently used in scanners having proximal scanning devices such as mirrors.

The shape recovery property of shape memory alloy materials makes them an excellent alternative for providing a large actuation force in compact dimensions.

Various imaging endoscopic devices using these actuators are described in this paper. The performance of some recently developed fiber optic cantilever-based scanners is summarized and compared with the clinically available endoscopes in Table 3.

Table 3. Fiber-optic cantilevered scanners developed for endomicroscopy.

Actuation Principle	Imaging Modality	Scanning Direction	Resolution	FOV	Frequency	Driving Voltage	Frame Rate	Scanner Dimensions	Scanning Pattern	Cantilever Fiber	Refs.
Piezoelectric bimorph	Multiphoton	Forward	0.8 μm (lateral), 10 μm (axial)	110 μm × 110 μm	1.05 kHz (fast axis), 4.1 Hz	50 Vpp (resonant bimorph) 200 Vpp (non-resonant)	4.1 fps (512 × 512 pixels)	3 mm (diameter) 40 mm (rigid length)	Raster scan	DCF	[105]
Piezoelectric tube	Two-photon	Forward	0.61 μm ÷ 1.10 μm (from center of FOV to peripheric zone)	160 μm	3.1 kHz	48 Vpp	6 fps	~2 mm (diameter)	Spiral	DCF with GRIN lens	[156]
Piezoelectric tube	Two-Photon	Forward	10 μm	70°	5 kHz	25 V	15 fps	1.6 mm (diameter)	Spiral	SMF	[16]
Piezoelectric thin film	E-OCT	Side-view	5 μm (axial)	152°	394 Hz	1.3 Vpp	-	3.4 mm × 2.5 mm	radial	-	[157]
Electrothermal	Confocal en-domicroscope	Forward	~1.7 μm	378 μm × 439 μm	239 Hz (x-axis) 207 Hz (y-axis)	16 Vpp	1 fps	1.65 mm (diameter) 28 mm (rigid length)	Lissajous	SMF	[65]
Electrothermal	OCT (A-scan)	Forward	~17 μm (lateral) ~9 μm (axial)	~3 mm (beam scanning)	~100 Hz	3 Vac_pp, 1.5 V DC offset	200 fps	5.5 mm (diameter) 55 mm (rigid length)	raster	-	[96]
Electromagnetic	OCT (B-scan)	Forward	4–6 μm (axial) 25–35 μm (lateral)	2 mm	5 Hz	±10 V (triangle wave)	-	0.51 mm (diameter)	Linear	SMF	[132]
Micromotor	PA and US endoscopy	Side-view	~58 μm (PA radial) ~30 μm (US radial) ~100 μm (PA transvers) ~120 μm (US transverse)	~310°	~4 Hz	~3.2V DC	4 fps	2.5 mm (diameter) ~35 mm (Rigid length)	Radial	MMF	[40]
Electromagnetic	Confocal	Forward	0.8 μm (lateral)	390 μm × 390 μm	700 Hz (fast scan) 1–2 Hz (slow scan)	-	1 fps	8 mm (diameter)	Raster	SMF	[158]
Cellvizio	Confocal	Forward	5–15 μm (axial) 2–5 μm (lateral)	600 μm × 500 μm	-	-	12 fps	2.5 mm	-	-	[33]

As stated earlier, scanning fiber endoscopes were introduced recently in the field. Most of the devices mentioned in this paper find limited use in clinical application but are more in the transition from research to clinical phase. The only commercially available cantilever-based endoscopes for clinical use are those developed by Pentax [73] and Mauna Kea [32] Technologies.

The choice of an endoscopic device depends on the target imaging region, ease of use and cost of the device. The cost of an endoscopic device comprises fabrication and operating cost. The operating costs mainly consist in decontamination and sterilization, which can be performed at high temperature such as autoclaving or hot air oven or at low temperature using chemical agents. The fabrication cost of an endoscope highly depends on the cost of the laser source, actuation, and detection mechanisms. Among the various actuators studied in this paper, electrothermal actuators are economic ones. However, the high working temperature can limit their usage at a high frequency. Otherwise, piezoelectric tubular actuators are cost-effective in terms of possible bidirectional actuation at high actuation frequency. The development of MEMS actuators enabled batch production of miniaturized actuators, reducing the fabrication costs to great extent and permitting the fabrication of disposable endoscopic scanners.

6. Conclusions

The actuation and scanning devices for cantilever-based endoscopic probes are described in this review paper along with their design and working principles. The size of an actuator is an important factor in a cantilevered endoscopic device. The developments in the MEMS field permit the mass production of small-sized actuators, enabling the fabrication of small-sized imaging probes. The endoscopic technology is aiming towards the design of low-cost disposable imaging devices as the reprocessing and sanitation of an endoscopic device is a complex and expensive process. The fabrication of small-sized actuation mechanisms helps to reduce the price of device. Moreover, small-sized probes can be advanced further to image the small body cavities, enabling early detection of pre-cancerous surfaces and helping in the diagnostic procedure to control the lesion at a preliminary stage.

Author Contributions: Conceptualization, C.M. and M.K.; formal analysis, M.K.; investigation, M.K.; resources, M.K.; data curation, M.K.; writing—original draft preparation, M.K.; writing—review and editing, M.K., P.M.L., and C.M.; visualization, M.K.; supervision, C.M. and P.M.L.; project administration, C.M. and P.M.L.; funding acquisition, C.M. and P.M.L. All authors have read and agreed to the published version of the manuscript.

Funding: This work was supported by the Natural Sciences and Engineering Research Council (NSERC) of Canada, Canadian Institutes of Health Research (CIHR), the Canada Research Chair (CRC) program, and the Mitacs internship program.

Institutional Review Board Statement: The ethical review and approval are not applicable for this review paper.

Informed Consent Statement: The informed consent statement is not applicable.

Data Availability Statement: No new data were created or analyzed in this study. Data sharing is not applicable to this article.

Conflicts of Interest: The authors declare no conflict of interest.

References

1. Coda, S.; Thillainayagam, A.V. State of the art in advanced endoscopic imaging for the detection and evaluation of dysplasia and early cancer of the gastrointestinal tract. *Clin. Exp. Gastroenterol.* **2014**, *7*, 133–150. [CrossRef]
2. Mannath, J.; Banks, M.R. Emerging technologies in endoscopic imaging. *F1000 Med. Rep.* **2012**, *4*, 3. [CrossRef] [PubMed]
3. WHO, Cancer, World Health Organization, 12 09 2018. Available online: https://www.who.int/news-room/fact-sheets/detail/cancer (accessed on 6 October 2020).
4. Stewart, B.W.; Wild, C.P. *World Cancer Report 2014*; IARC: Lyon, France, 2014.
5. Conigliaro, R.; Pigò, F. New Techniques in Endoscopy: Confocal Laser Endomicroscopy. In *New Techniques in Gastrointestinal Endoscopy*; IntechOpen: London, UK, 2011; pp. 213–230.

6. Kaur, M.; Lane, P.M.; Menon, C. Endoscopic Optical Imaging Technologies and Devices for Medical Purposes: State of the Art. *Appl. Sci.* **2020**, *10*, 6865. [CrossRef]

7. Leng, X.; Chapman, W.; Rao, B.; Nandy, S.; Chen, R.; Rais, R.; Gonzalez, I.; Zhou, Q.; Chatterjee, D.; Mutch, M.; et al. Feasibility of co-registered ultrasound and acoustic-resolution photoacoustic imaging of human colorectal cancer. *Biomed. Opt. Express* **2018**, *9*, 5159–5172. [CrossRef] [PubMed]

8. Zhou, J.; Jokerst, J.V. Photoacoustic imaging with fiber optic technology: A review. *Photoacoustics* **2020**, *20*, 100211. [CrossRef] [PubMed]

9. Ubbink, R.; Van Dijk, L.J.; Van Noord, D.; Johannes, T.; Specht, P.A.C.; Bruno, M.; Mik, E.G. Evaluation of endoscopic visible light spectroscopy: Comparison with microvascular oxygen tension measurements in a porcine model. *J. Transl. Med.* **2019**, *17*, 65. [CrossRef] [PubMed]

10. Benaron, A.D.; Parachikov, H.I.; Cheong, W.-F.; Friedland, S.; Rubinsky, B.E.; Otten, D.M.; Liu, F.W.H.; Levinson, C.J.; Murphy, A.L.; Price, J.W.; et al. Design of a visible-light spectroscopy clinical tissue oximeter. *J. Biomed. Opt.* **2005**, *10*, 44005–44009. [CrossRef]

11. Gorevoy, A.; Machikhin, A.S.; Batshev, V.; Kolyuchkin, V.Y. Optimization of stereoscopic imager performance by computer simulation of geometrical calibration using optical design software. *Opt. Express* **2019**, *27*, 17819–17839. [CrossRef]

12. Geng, J.; Xie, J. Review of 3-D Endoscopic Surface Imaging Techniques. *IEEE Sens. J.* **2013**, *14*, 945–960. [CrossRef]

13. Udovich, J.A.; Kirkpatrick, N.D.; Kano, A.; Tanbakuchi, A.; Utzinger, U.; Gmitro, A.F. Spectral background and transmission characteristics of fiber optic imaging bundles. *Appl. Opt.* **2008**, *47*, 4560–4568. [CrossRef]

14. Scheffer, D. Endoscopes Use CMOS Image Sensors, Vision Systems Design, 30 07 2007. Available online: https://www.vision-systems.com/home/article/16750278/endoscopes-use-cmos-image-sensors (accessed on 28 July 2020).

15. Lee, C.M.; Engelbrecht, C.J.; Soper, T.D.; Helmchen, F.; Seibel, E.J. Scanning fiber endoscopy with highly flexible, 1 mm catheterscopes for wide-field, full-color imaging. *J. Biophotonics* **2010**, *3*, 385–407. [CrossRef] [PubMed]

16. Seibel, E.J.; Johnston, R.S.; Melville, C.D. A full-color scanning fiber endoscope. *Opt. Fibers Sens. Med. Diagn. Treat. Appl. VI* **2006**, *6083*, 608303. [CrossRef]

17. Piyawattanametha, W. A review of MEMS scanner based endoscopic optical imaging probe. In Proceedings of the 2013 International Conference on Optical MEMS and Nanophotonics (OMN), Kanazawa, Japan, 18–22 August 2013; pp. 53–54.

18. Qiu, Z.; Piyawattanamatha, W. New Endoscopic Imaging Technology Based on MEMS Sensors and Actuators. *Micromachines* **2017**, *8*, 210. [CrossRef] [PubMed]

19. Hwang, K.; Seo, Y.-H.; Jeong, K.-H. Microscanners for optical endomicroscopic applications. *Micro Nano Syst. Lett.* **2017**, *5*, 1–11. [CrossRef]

20. Powers, J.P. *An Introduction to Fiber Optic Systems*; Irwin: Chicago, IL, USA, 1997.

21. Ghatak, A.; Thyagarajan, K. Optical Waveguides and Fibers. In *Fundamentals of Photonics*; SPIE: Bellingham, WA, USA, 2009; pp. 249–292.

22. Göbel, W.; Kerr, J.N.D.; Nimmerjahn, A.; Helmchen, F. Miniaturized two-photon microscope based on a flexible coherent fiber bundle and a gradient-index lens objective. *Opt. Lett.* **2004**, *29*, 2521–2523. [CrossRef] [PubMed]

23. Huang, L.; Oesterberg, U.L. Measurement of cross talk in order-packed image fiber bundles. In *Proceedings of the SPIE's 1995 International Symposium on Optical Science, Engineering, and Instrumentation*; SPIE: Bellingham, WA, USA, 1995; pp. 480–488.

24. Flusberg, A.; Cocker, B.; Piyawattanametha, E.D.; Jung, W.; Cheung, J.C.; Schnitzer, E.L.M. Fiber-optic fluorescence imaging. *Nat. Methods* **2005**, *2*, 941–950. [CrossRef]

25. Shao, J.; Liao, W.-C.; Liang, R.; Barnard, K. Resolution enhancement for fiber bundle imaging using maximum a posteriori estimation. *Opt. Lett.* **2018**, *43*, 1906–1909. [CrossRef]

26. Dumripatanachod, M.; Piyawattananmetha, W. A Fast Depixation Method of Fiber Bundle Image for an Embedded System. In Proceedings of the 8th Biomedical Engineering International Conference, Pattaya, Thailand, 25–27 November 2015.

27. Shinde, A.; Perinchery, S.M.; Matham, M.V. Fiber pixelated image database. *Opt. Eng.* **2016**, *55*, 83105. [CrossRef]

28. Vyas, K.; Hughes, M.; Gil Rosa, B.; Yang, G.-Z. Fiber bundle shifting endomicroscopy for high-resolution imaging. *Biomed. Opt. Express* **2018**, *9*, 4649–4664. [CrossRef]

29. Sung, K.-B.; Liang, C.; Descour, M.; Collier, T.; Follen, M.; Richards-Kortum, R. Fiber-optic confocal reflectance microscope with miniature objective for in vivo imaging of human tissues. *IEEE Trans. Biomed. Eng.* **2002**, *49*, 1168–1172. [CrossRef]

30. Knittel, J.; Schnieder, L.; Buess, G.; Messerschmidt, B.; Possner, T. Endoscope-compatible confocal microscope using a gradient index-lens system. *Opt. Commun.* **2001**, *188*, 267–273. [CrossRef]

31. Lane, P.M.; Lam, S.; McWilliams, A.; Le Riche, J.C.; Anderson, M.W.; Macaulay, C.E. Confocal fluorescence microendoscopy of bronchial epithelium. *J. Biomed. Opt.* **2009**, *14*, 024008. [CrossRef] [PubMed]

32. Cellvizio, Cellvizio Targeted Biopsies, Mauna Kea Technolgies. Available online: https://www.maunakeatech.com/en/cellvizio/10-cellvizio-targeted-biopsies (accessed on 9 October 2020).

33. Osdoit, A.; Lacombe, F.; Cave, C.; Loiseau, S.; Peltier, E. To see the unseeable: Confocal miniprobes for routine microscopic imaging during endoscopy. *Biomed. Opt.* **2007**, *6432*, 64320. [CrossRef]

34. Giniunas, L.; Juskaitis, R.; Shatalin, S. Scanning fiber-optic microscope. *Electron. Lett.* **1991**, *27*, 724–726. [CrossRef]

35. Kimura, S.; Wilson, T. Confocal scanning optical microscope using single-mode fiber for signal detection. *Appl. Opt.* **1991**, *30*, 2143–2150. [CrossRef]

36. Pitris, C.; Bouma, B.E.; Shiskov, M.; Tearney, G.J. A GRISM-based probe for spectrally encoded confocal microscopy. *Opt. Express* **2003**, *11*, 120–124. [CrossRef]

37. Yelin, D.; Rizvi, I.; White, W.M.; Motz, J.T.; Hasan, T.; Bouma, B.E.; Tearney, G.J. Three-dimensional miniature endoscopy. *Nat. Cell Biol.* **2006**, *443*, 765. [CrossRef]

38. Zhao, J.-H.; Zeng, H. Advanced Spectroscopy Technique for Biomedicine. In *Biophysics of Skin and Its Treatments*; Springer Science and Business Media LLC: Geneva, Swizterland, 2012; pp. 1–54.

39. Choi, Y.; Yoon, C.; Kim, M.; Yang, T.D.; Fang-Yen, C.; Dasari, R.R.; Lee, K.J.; Choi, W. Scanner-Free and Wide-Field Endoscopic Imaging by Using a Single Multimode Optical Fiber. *Phys. Rev. Lett.* **2012**, *109*, 203901. [CrossRef]

40. Yang, J.-M.; Chen, R.; Favazza, C.; Yao, J.; Li, C.; Hu, Z.; Zhou, Q.; Shung, K.K.; Wang, L.V. A 25-mm diameter probe for photoacoustic and ultrasonic endoscopy. *Opt. Express* **2012**, *20*, 23944–23953. [CrossRef]

41. Wei, W.; Li, X.; Zhou, Q.; Shung, K.K.; Chen, Z. Integrated ultrasound and photoacoustic probe for co-registered intravascular imaging. *J. Biomed. Opt.* **2011**, *16*, 106001. [CrossRef]

42. Papadopoulos, I.N.; Farahi, S.; Moser, C.; Psaltis, D. Focusing and scanning light through a multimode optical fiber using digital phase conjugation. *Opt. Express* **2012**, *20*, 10583–10590. [CrossRef] [PubMed]

43. Sato, M.; Kanno, T.; Ishihara, S.; Suto, H.; Takahashi, T.; Nishidate, I. Reflectance Imaging by Graded-Index Short Multimode Fiber. *Appl. Phys. Express* **2013**, *6*, 052503. [CrossRef]

44. Zhang, Y.; Cao, Y.; Cheng, J.-X. High-resolution photoacoustic endoscope through beam self-cleaning in a graded index fiber. *Opt. Lett.* **2019**, *44*, 3841–3844. [CrossRef] [PubMed]

45. Fu, L.; Gan, X.; Gu, M. Nonlinear optical microscopy based on double-clad photonic crystal fibers. *Opt. Express* **2005**, *13*, 5528–5534. [CrossRef]

46. Yelin, D.; Bouma, B.E.; Yun, S.H.; Tearney, G.J. Double-clad fiber for endoscopy. *Opt. Lett.* **2004**, *29*, 2408–2410. [CrossRef]

47. Wang, L.; Choi, H.Y.; Jung, Y.; Lee, B.-H.; Kim, K.-T. Optical probe based on double-clad optical fiber for fluorescence spectroscopy. *Opt. Express* **2007**, *15*, 17681–17689. [CrossRef]

48. Mavadia, J.; Xi, J.; Chen, Y.; Li, X. An all-fiber-optic endoscopy platform for simultaneous OCT and fluorescence imaging. *Biomed. Opt. Express* **2012**, *3*, 2851–2859. [CrossRef]

49. Xi, J.; Chen, Y.; Zhang, Y.; Murari, K.; Li, M.-J.; Li, X. Integrated multimodal endomicroscopy platform for simultaneous en face optical coherence and two-photon fluorescence imaging. *Opt. Lett.* **2012**, *37*, 362–364. [CrossRef]

50. Buenconsejo, A.L.; Hohert, G.; Manning, M.; Abouei, E.; Tingley, R.; Janzen, I.; McAlpine, J.N.; Miller, D.; Lee, A.; Lane, P.M.; et al. Submillimeter diameter rotary-pullback fiber-optic endoscope for narrowband red-green-blue reflectance, optical coherence tomography, and autofluorescence in vivo imaging. *J. Biomed. Opt.* **2019**, *25*, 1–7. [CrossRef]

51. Tearney, G.J.; Brezinski, M.E.; Fujimoto, J.G.; Weissman, N.J.; Boppart, S.A.; Bouma, B.E.; Southern, J.F. Scanning single-mode fiber optic catheter–endoscope for optical coherence tomography. *Opt. Lett.* **1996**, *21*, 543–545. [CrossRef]

52. Liu, X.; Cobb, M.J.; Chen, Y.; Kimmey, M.B.; Li, X. Rapid-scanning forward-imaging miniature endoscope for real-time optical coherence tomography. *Opt. Lett.* **2004**, *29*, 1763–1765. [CrossRef] [PubMed]

53. Park, H.-C.; Seo, Y.-H.; Hwang, K.; Lim, J.-K.; Yoon, S.Z.; Jeong, K.-H. Micromachined tethered silicon oscillator for an endomicroscopic Lissajous fiber scanner. *Opt. Lett.* **2014**, *39*, 6675–6678. [CrossRef] [PubMed]

54. Huang, G.; Ding, Z. Rapid two-dimensional transversal scanning fiber probe for optical coherence tomography. In *Coherence Domain Optical Methods and Optical Coherence Tomography in Biomedicine XI*; SPIE: Bellingham, WA, USA, 2007; p. 64292W.

55. Liu, Z.; Fu, L.; Gao, F.; Zhang, X. Design and implementation of a 2-D endoscopic optical fiber scanner. In Proceedings of the Seventh International Conference on Photonics and Imaging in Biology and Medicine, Wuhan, China, 24–27 November 2008; Volume 7280, p. 72801D.

56. Wu, T.; Ding, Z.; Wang, K.; Chen, M.; Wang, C. Two-dimensional scanning realized by an asymmetry fiber cantilever driven by single piezo bender actuator for optical coherence tomography. *Opt. Express* **2009**, *17*, 13819–13829. [CrossRef] [PubMed]

57. Li, G.; Zhou, A.; Gao, H.; Liu, Z. Endoscope two dimensional scanning fiber probe and the driving method. In Proceedings of the International Conference on Optical Instruments and Technology (OIT2011), Beijing, China, 6–9 November 2011; Volume 8199, p. 819913.

58. Kretschmer, S.; Jäger, J.; Vilches, S.; Ataman, Ç.; Zappe, H. A bimodal endoscopic imager in a glass package. *J. Micromech. Microeng.* **2018**, *28*, 105009. [CrossRef]

59. Rajiv, A.; Zhou, Y.; Ridge, J.; Reinhall, P.G.; Seibel, E.J. Electromechanical Model-Based Design and Testing of Fiber Scanners for Endoscopy. *J. Med. Devices* **2018**, *12*, 041003. [CrossRef]

60. Akhoundi, F.; Qin, Y.; Peyghambarian, N.; Barton, J.K.; Kieu, K. Compact fiber-based multi-photon endoscope working at 1700 nm. *Biomed. Opt. Express* **2018**, *9*, 2326–2335. [CrossRef]

61. Aguirre, A.D.; Hertz, P.R.; Chen, Y.; Fujimoto, J.G.; Piyawattanametha, W.; Fan, L.; Wu, M.C. Two-axis MEMS Scanning Catheter for Ultrahigh Resolution Three-dimensional and En Face Imaging. *Opt. Express* **2007**, *15*, 2445–2453. [CrossRef]

62. Zara, J.M.; Yazdanfar, S.; Rao, K.D.; Izatt, J.A.; Smith, S.W. Electrostatic micromachine scanning mirror for optical coherence tomography. *Opt. Lett.* **2003**, *28*, 628–630. [CrossRef]

63. Kim, K.H.; Park, B.H.; Maguluri, G.N.; Lee, T.W.; Rogomentich, F.J.; Bancu, M.G.; Bouma, B.E.; De Boer, J.F.; Bernstein, J.J. Two-axis magnetically-driven MEMS scanning catheter for endoscopic high-speed optical coherence tomography. *Opt. Express* **2007**, *15*, 18130–18140. [CrossRef]

54. Yao, J.; Yang, Z.; Peng, T.; Sun, B.; Zhang, H.; Zhao, M.; Dai, B.; Liu, H.; Ding, G.; Sawada, R. A Single-Fiber Endoscope Scanner Probe Utilizing Two-Degrees-of-Freedom (2DOF) High-Order Resonance to Realize Larger Scanning Angle. *IEEE Trans. Compon. Packag. Manuf. Technol.* **2019**, *9*, 2332–2340. [CrossRef]

55. Seo, H.Y.; Hwang, K.; Jeong, K.-H. 1.65 mm diameter forward-viewing confocal endomicroscopic catheter using a glip-chip bonded electrothermal MEMS fiber scanner. *Opt. Express* **2018**, *26*, 4780–4785. [CrossRef] [PubMed]

56. Seo, Y.-H.; Hwang, K.; Park, H.-C.; Jeong, K.-H. Electrothermal MEMS fiber scanner for optical endomicroscopy. *Opt. Express* **2016**, *24*, 3903–3909. [CrossRef] [PubMed]

57. Tanguy, Q.A.; Gaiffe, O.; Passilly, N.; Cote, J.-M.; Cabodevila, G.; Bargiel, S.; Lutz, P.; Xie, H.; Gorecki, C. Real-time Lissajous imaging with a low-voltage 2-axis MEMS scanner based on electro-thermal actuation. *Opt. Express* **2020**, *28*, 8512–8527. [CrossRef] [PubMed]

58. Sun, J.; Guo, S.; Wu, L.; Liu, L.; Choe, S.-W.; Sorg, B.S.; Xie, H. 3D In Vivo optical coherence tomography based on a low-voltage, large-scan-range 2D MEMS mirror. *Opt. Express* **2010**, *18*, 12065–12075. [CrossRef] [PubMed]

59. Duan, C.; Tanguy, Q.; Pozzi, A.; Xie, H. Optical coherence tomography endoscopic probe based on a tilted MEMS mirror. *Biomed. Opt. Express* **2016**, *7*, 3345–3354. [CrossRef]

70. Li, H.; Oldham, K.; Wang, T.D. 3 degree-of-freedom resonant scanner with full-circumferential range and large out-of-plane displacement. *Opt. Express* **2019**, *27*, 16296–16307. [CrossRef]

71. Kobayashi, T.; Matsunaga, T.; Haga, Y. Active Bending Electric Endoscope Using Shape Memory Alloy Wires. In *Computational and Experimental Simulations in Engineering*; Springer Science and Business Media LLC: Geneva, Switzerland, 2015; pp. 131–139.

72. Pan, Y.; Xie, H.; Fedder, G.K. Endoscopic optical coherence tomography based on a microelectromechanical mirror. *Opt. Lett.* **2001**, *26*, 1966–1968. [CrossRef]

73. Polglase, A.L.; McLaren, W.J.; Skinner, S.A.; Kiesslich, R.; Neurath, M.F.; Delaney, P.M. A fluorescence confocal endoscope for in vivo microscopy of the upper- and the lower-GI tract. *Gastrointest. Endosc.* **2005**, *62*, 686–695. [CrossRef]

74. Wu, J.; Conry, M.; Gu, C.; Wang, F.; Yaqoob, Z.; Yang, C. Paired-angle-rotation scanning optical coherence tomography forward-imaging probe. *Opt. Lett.* **2006**, *31*, 1265–1267. [CrossRef]

75. Sarunic, V.M.; Han, S.; Wu, J.; Yaqoob, Z.; Humayun, M.; Yang, C. *Endoscopic Optical Coherence Tomography of the Retina at 1310 nm Using Paried-Angle-Rotating Scanning*; SPIE: San Jose, CA, USA, 2007.

76. Kim, K.J.; Choi, J.W.; Yun, S.H. 350-um side-view optical probe for imaging the murine brain in vivo from the cortex to the hypothalamus. *J. Biomed. Opt.* **2013**, *18*, 502–511. [CrossRef]

77. Lee, C.M.; Chandler, J.E.; Seibel, E.J. Wide field fluorescence imaging in narrow passageways using scanning fiber endoscope technology. *Endosc. Microsc. V* **2010**, *7558*, 755806. [CrossRef]

78. Inman, D.J. *Engineering Vibrations*; CRC Press: Pearson, NJ, USA, 2014.

79. Qiu, Z.; Piyawattanametha, W. MEMS Actuators for Optical Microendoscopy. *Micromachines* **2019**, *10*, 85. [CrossRef] [PubMed]

80. Benjamin, K.J.; David, I.J.; Delu, S.D.; Ebenezer, D.R.; Jennifer, L.S.; Aaishah, R.L.; Wei, P.L.; Gui-Shuang, Y.L.; Tomas, A.L.; Dunaief, J.L.; et al. Optical coherence tomography identifies outer retina thinning in frontotemporal degeneration. *Neurology* **2017**, *89*, 1604–1611.

81. Lumbroso, B.; Huang, D.; Chen, C.J.; Jia, Y.; Rispoli, M.; Romano, A.; Waheed, N.K. *Clinical OCT Angiography Atlas*; Jaypee Brothers Medical Publishers: New Delhi, India, 2015.

82. Boppart, S.A.; Bouma, B.E.; Pitris, C.; Tearney, G.J.; Fujimoto, J.G.; Brezinski, M.E. Forward-imaging instruments for optical coherence tomography. *Opt. Lett.* **1997**, *22*, 1618–1620. [CrossRef] [PubMed]

83. Paddock, S.W.; Fellers, T.J.; Davidson, M.W. Introductory Confocal Concepts, MicroscopyU. Available online: https://www.microscopyu.com/techniques/confocal/introductory-confocal-concepts (accessed on 23 June 2020).

84. Liu, L.; Wang, E.; Zhang, X.; Liang, W.; Li, X.; Xie, H. MEMS-based 3D confocal scanning microendoscope using MEMS scanners for both lateral and axial scan. *Sens. Actuators A* **2014**, *215*, 89–95. [CrossRef] [PubMed]

85. Dickensheets, D.L.; Kino, G.S. Micromachined scanning confocal optical microscope. *Opt. Lett.* **1996**, *21*, 764–766. [CrossRef]

86. Maitland, K.C.; Shin, H.J.; Ra, H.; Lee, D.; Solgaard, O.; Richards-Kortum, R. Single fiber confocal microscope with a two-axis gimbaled MEMS scanner for cellular imaging. *Opt. Express* **2006**, *14*, 8604–8612. [CrossRef]

87. Jung, I.W.; Lopez, D.; Qiu, Z.; Piyawattanametha, W. 2-D MEMS Scanner for Handheld Multispectral Dual-Axis Confocal Microscopes. *J. Microelectromech. Syst.* **2018**, *27*, 605–612. [CrossRef]

88. Li, R.; Wang, X.; Zhou, Y.; Zong, H.; Chen, M.; Sun, M. Advances in nonlinear optical microscopy for biophotonics. *J. Nanophotonics* **2018**, *12*, 033007. [CrossRef]

89. Piston, D.W.; Fellers, T.J.; Davidson, M.W. Multiphoton Microscopy, MicroscopyU. Available online: https://www.microscopyu.com/techniques/multi-photon/multiphoton-microscopy (accessed on 23 June 2020).

90. Holmstrom, S.T.S.; Baran, U.; Urey, H. MEMS Laser Scanners: A Review. *J. Microelectromech. Syst.* **2014**, *23*, 259–275. [CrossRef]

91. Haight, E.C.; King, W.W. Stability of Nonlinear Oscillations of an Elastic Rod. *J. Acoust. Soc. Am.* **1972**, *52*, 899–911. [CrossRef]

92. Hyer, M.W. Whirling of a base-excited cantilever beam. *J. Acoust. Soc. Am.* **1979**, *65*, 931–939. [CrossRef]

93. Kundrat, M.J.; Reinhall, P.G.; Lee, C.M.; Seibel, E.J. High performance open loop control of scanning with a small cylindrical cantilever beam. *J. Sound Vib.* **2011**, *330*, 1762–1771. [CrossRef] [PubMed]

94. Wu, L.; Ding, Z.; Huang, G. *Realization of 2D Scanning Pattern of a Fiber Cantilever by Nonlinear Coupling*; SPIE: Wuhan, China, 2007; p. 65340I.

95. Zhang, X.; Duan, C.; Liu, L.; Li, X.; Xie, H. A non-resonant fiber scanner based on an electrothermally-actuated MEMS stage. *Sens. Actuators A Phys.* **2015**, *233*, 239–245. [CrossRef]
96. Park, H.-C.; Zhang, X.; Yuan, W.; Zhou, L.; Xie, H.; Li, X. Ultralow-voltage electrothermal MEMS based fiber-optic scanning probe for forward-viewing endoscopic OCT. *Opt. Lett.* **2019**, *44*, 2232–2235. [CrossRef]
97. Moon, S.; Lee, S.-W.; Rubinstein, M.; Wong, B.J.F.; Chen, Z. Semi-resonant operation of a fiber-cantilever piezotube scanner for stable optical coherence tomography endoscope imaging. *Opt. Express* **2010**, *18*, 21183–21197. [CrossRef]
98. Yang, S.; Xu, Q. A review on actuation and sensing techniques for MEMS-based microgrippers. *J. Micro-Bio Robot.* **2017**, *13*, 1–14. [CrossRef]
99. Zhang, P. Sensors and actuators. In *Advanced Industrial Control Technology*; Elsevier BV: Heidelberg, Germany, 2010; pp. 73–116.
100. Pengwang, E.; Rabenorosoa, K.; Rakotondrabe, M.; Andreff, N. Scanning Micromirror Platform Based on MEMS Technology for Medical Application. *Micromachines* **2016**, *7*, 24. [CrossRef]
101. Emery, J. Piezoelectricity, 04 03 1997. Available online: http://stem2.org/je/piezoelc.pdf (accessed on 12 June 2020).
102. Bahadur, I.M.; Mills, J.K. A new model of hysteresis in piezoelectric actuators. In Proceedings of the 2011 IEEE International Conference on Mechatronics and Automation, Beijing, China, 7–10 August 2011; pp. 789–794.
103. ThorLabs. Piezoelectric Tutorial. Available online: https://www.thorlabs.com/newgrouppage9.cfm?objectgroup_id=5030 (accessed on 15 June 2020).
104. Tekpınar, M.; Khayatzadeh, R.; Ferhanoğlu, O. Multiple-pattern generating piezoelectric fiber scanner toward endoscopic applications. *Opt. Eng.* **2019**, *58*, 023101. [CrossRef]
105. Rivera, D.R.; Brown, C.M.; Ouzounov, D.G.; Pavlova, I.; Kobat, D.; Webb, W.W.; Xu, C. Compact and flexible raster scanning multiphoton endoscope capable of imaging unstained tissue. *Proc. Natl. Acad. Sci. USA* **2011**, *108*, 17598–17603. [CrossRef] [PubMed]
106. Liang, W.; Hall, G.; Messerschmidt, B.; Li, M.-J.; Li, X. Nonlinear optical endomicroscopy for label-free functional histology in vivo. *Light Sci. Appl.* **2017**, *6*, e17082. [CrossRef] [PubMed]
107. Vilches, S.; Kretschmer, S.; Ataman, Ç.; Zappe, H. Miniaturized Fourier-plane fiber scanner for OCT endoscopy. *J. Micromech. Microeng.* **2017**, *27*, 105015. [CrossRef]
108. Mayyas, M. Comprehensive Thermal Modeling of ElectroThermoElastic Microsturctures. In *Actuators*; Molecular Diversity Preservation International: New York, NY, USA, 2012; Volume 1, pp. 21–35.
109. Lin, L.; Chiao, M. Electrothermal responses of lineshape microstructures. *Sens. Actuators A* **1996**, *55*, 35–41. [CrossRef]
110. So, H.; Pisano, A.P. Electrothermal modeling, fabrication and analysis of low-power consumption thermal actuator with buckling arm. *Microsyst. Technol.* **2013**, *21*, 195–202. [CrossRef]
111. Buser, R.; De Rooij, N.; Tischhauser, H.; Dommann, A.; Staufert, G. Biaxial scanning mirror activated by bimorph structures for medical applications. *Sens. Actuators A Phys.* **1992**, *31*, 29–34. [CrossRef]
112. Guckel, H.; Klein, J.; Christenson, T.; Skrobis, K.; Laudon, M.; Lovell, E. Thermo-magnetic metal flexure actuators. In *Technical Digest IEEE Solid-State Sensor and Actuator Workshop*; Institute of Electrical and Electronics Engineers: New York, NY, USA, 2003.
113. Huang, Q.-A.; Lee, N.K.S. Analysis and design of polysilicon thermal flexure actuator. *J. Micromech. Microeng.* **1999**, *9*, 64–70. [CrossRef]
114. Hickey, R.; Kujath, M.; Hubbard, T. Heat transfer analysis and optimization of two-beam microelectromechanical thermal actuators. *J. Vac. Sci. Technol. A* **2002**, *20*, 971–974. [CrossRef]
115. Huang, Q.-A.; Lee, N.K.S. Analytical modeling and optimization for a laterally-driven polysilicon thermal actuator. *Microsyst. Technol.* **1999**, *5*, 133–137. [CrossRef]
116. Lee, C.-C.; Hsu, W. Optimization of an electro-thermally and laterally driven microactuator. *Microsyst. Technol.* **2003**, *9*, 331–334. [CrossRef]
117. Moulton, T.; Ananthasuresh, G. Micromechanical devices with embedded electro-thermal-compliant actuation. *Sens. Actuators A* **2001**, *90*, 38–48. [CrossRef]
118. Lara-Castro, M.; Herrera-Amaya, A.; Escarola-Rosas, M.A.; Vázquez-Toledo, M.; López-Huerta, F.; Aguilera-Cortés, L.A.; Herrera-May, A.L. Design and Modeling of Polysilicon Electrothermal Actuators for a MEMS Mirror with Low Power Consumption. *Micromachines* **2017**, *8*, 203. [CrossRef] [PubMed]
119. Venditti, R.; Lee, J.S.H.; Sun, Y.; Li, D. An in-plane, bi-directional electrothermal MEMS actuator. *J. Micromech. Microeng.* **2006**, *16*, 2067–2070. [CrossRef]
120. Seo, Y.-H.; Park, H.-C.; Jeong, K.-H. Electrothermal MEMS fiber scanner with lissajous patterns for endomicroscopic applications. In Proceedings of the 2016 IEEE 29th International Conference on Micro Electro Mechanical Systems (MEMS), Shanghai, China, 24–28 January 2016; Volume 24, pp. 367–370. [CrossRef]
121. Enikov, E.; Kedar, S.; Lazarov, K. Analytical model for analysis and design of V-shaped thermal microactuators. *J. Microelectromech. Syst.* **2005**, *14*, 788–798. [CrossRef]
122. Sinclair, M. A high force low area MEMS thermal actuator. In Proceedings of the ITHERM 2000—The Seventh Intersociety Conference on Thermal and Thermomechanical Phenomena in Electronic Systems (Cat. No.00CH37069), Las Vegas, NV, USA, 23–26 May 2002.
123. Que, L.; Park, J.-S.; Gianchandani, Y. Bent-beam electrothermal actuators-Part I: Single beam and cascaded devices. *J. Microelectromechan. Syst.* **2001**, *10*, 247–254. [CrossRef]

124. Iqbal, S.; Malik, A.A.; Shakoor, R.I. Design and analysis of novel micro displacement amplification mechanism actuated by chevron shaped thermal actuators. *Microsyst. Technol.* **2018**, *25*, 861–875. [CrossRef]
125. Guan, C.; Zhu, Y. An electrothermal microactuator with Z-shaped beams. *J. Micromechan. Microeng.* **2010**, *20*, 1–9. [CrossRef]
126. Rawashdeh, E.; Karam, A.; Foulds, I.G. Characterization of Kink Actuators as Compared to Traditional Chevron Shaped Bent-Beam Electrothermal Actuators. *Micromachines* **2012**, *3*, 542–549. [CrossRef]
127. Kaur, M.; Brown, M.; Lane, P.M.; Menon, C. An Electro-Thermally Actuated Micro-Cantilever-Based Fiber Optic Scanner. *IEEE Sens. J.* **2020**, *20*, 9877–9885. [CrossRef]
128. Chu, W.-H.; Mehregany, M.; Mullen, R.L. Analysis of tip deflection and force of a bimetallic cantilever microactuator. *J. Micromechan. Microeng.* **1993**, *3*, 4–7. [CrossRef]
129. Tanguy, Q.A.A.; Bargiel, S.; Xie, H.; Passilly, N.; Barthès, M.; Gaiffe, O.; Rutkowski, J.; Lutz, P.; Gorecki, C. Design and Fabrication of a 2-Axis Electrothermal MEMS Micro-Scanner for Optical Coherence Tomography. *Micromachines* **2017**, *8*, 146. [CrossRef]
130. Pawinanto, R.E.; Yunas, J.; Majlis, B.; Hamzah, A. Design and Fabrication of Compact MEMS Electromagnetic Micro-Actuator with Planar Micro-Coil Based on PCB. *TELKOMNIKA Telecommun. Comput. Electron. Control.* **2016**, *14*, 856–866. [CrossRef]
131. Judy, J.W.; Muller, R.S. Magnetically actuated, addressable microstructures. *J. Microelectromechan. Syst.* **1997**, *6*, 249–256. [CrossRef]
132. Joos, K.M.; Shen, J.-H. Miniature real-time intraoperative forward-imaging optical coherence tomography probe. *Biomed. Opt. Express* **2013**, *4*, 1342–1350. [CrossRef]
133. Sun, B.; Sawada, R.; Yang, Z.; Zhang, Y.; Itoh, T.; Maeda, R. Design and fabrication of driving microcoil with large tilt-angle for medical scanner application. In Proceedings of the 2014 Symposium on Design, Test, Integration and Packaging of MEMS/MOEMS, Cannes, France, 1–4 April 2014; pp. 1–6. [CrossRef]
134. Sun, B.; Nogami, H.; Pen, Y.; Sawada, R. Microelectromagnetic actuator based on a 3D printing process for fiber scanner application. *J. Micromechan. Microeng.* **2015**, *25*, 075014. [CrossRef]
135. Lv, X.; Wei, W.; Mao, X.; Chen, Y.; Yang, J.; Yang, F. A novel MEMS electromagnetic actuator with large displacement. *Sens. Actuators A* **2015**, *221*, 22–28. [CrossRef]
136. Barbaroto, P.R.; Ferreira, L.O.S.; Doi, I. Micromachined scanner actuated by electromagnetic induction. In *Optomechatronic Systems III*; International Society for Optics and Photonics: New York, NY, USA, 2002; pp. 691–698.
137. Tang, T.-L.; Hsu, C.-P.; Chen, W.-C.; Fang, W. Design and implementation of a torque-enhancement 2-axis magnetostatic SOI optical scanner. *J. Micromechan. Microeng.* **2010**, *20*, 025020. [CrossRef]
138. Cullity, B.D.; Graham, C.D. Magnetostriction and the Effects of Stress. In *Introduction to Magnetic Materials*; Wiley: Hoboken, NJ, USA, 2009; pp. 241–273.
139. Bourouina, T.; Lebrasseur, E.; Reyne, G.; Debray, A.; Fujita, H.; Ludwig, A.; Quandt, E.; Muro, H.; Oki, T.; Asaoka, A. Integration of two degree-of-freedom magnetostrictive actuation and piezoresistive detection: Application to a two-dimensional optical scanner. *J. Microelectromechan. Syst.* **2002**, *11*, 355–361. [CrossRef]
140. Garnier, A.; Bourouina, T.; Fujita, H.; Orsier, E.; Masuzawa, T.; Hiramoto, T.; Peuzin, J.-C. A fast, robust and simple 2-D micro-opticalscanner based on contactless magnetostrictive actuation. In Proceedings of the IEEE Thirteenth Annual International Conference on Micro Electro Mechanical Systems, Miyazaki, Japan, 23–27 January 2000.
141. Pandojirao-Sunkojirao, P.; Rao, S.; Phuyal, P.C.; Dhaubanjar, N.; Chiao, J.-C. A Magnetic Actuator for Fiber-Optic Applications. *Int. J. Optomechatron.* **2009**, *3*, 215–232. [CrossRef]
142. Collard, D.; Fujita, H.; Toshiyoshi, B.; Legrand, B.; Buchaillot, L. Electrostatic Micro-actuators. In *Microsystems Technology: Fabrication, Test & Reliability*; Kogan Page Science: London, UK, 2003; pp. 75–115.
143. Bourouina, T.; Fujita, H.; Reyne, G.; Motamedi, M.E. Optical Scanning. In *MOEMS: Micro-Opto-Electro-Mechanical Systems*; SPIE: Bellingham, WA, USA, 2005; pp. 323–367.
144. Henri, C.; Franck, L. Fabrication, Simulation and experiment of a rotating electrostatic silicon mirror with large angular deflection. In Proceedings of the IEEE Thirteen Annual International Conference on Micro Electro Mechanical Systems, Miyazaki, Japan, 23–27 January 2000.
145. Patterson, P.R.; Hah, D.; Fujino, M.; Piyawattanametha, W.; Wu, M.C. Scanning micromirrors: An overview. In *Optomechatronic Micro/Nano Components, Devices, and Systems*; SPIE: Bellingham, WA, USA, 2004; Volume 5604, pp. 195–208.
146. Piyawattanametha, W.; Barretto, R.P.J.; Ko, T.H.; Flusberg, B.A.; Cocker, E.D.; Ra, H.; Lee, D.; Solgaard, O.; Schnitzer, M.J. Fast-scanning two-photon fluorescence imaging based on a microelectromechanical systems two- dimensional scanning mirror. *Opt. Lett.* **2006**, *31*, 2018–2020. [CrossRef]
147. Hofmann, U.; Janes, J.; Quenzer, H.-J. High-Q MEMS Resonators for Laser Beam Scanning Displays. *Micromachines* **2012**, *3*, 509–528. [CrossRef]
148. Izawa, T.; Sasaki, T.; Hane, K. Scanning Micro-Mirror with an Electrostatic Spring for Compensation of Hard-Spring Nonlinearity. *Micromachines* **2017**, *8*, 240. [CrossRef] [PubMed]
149. Schenk, H.; Dürr, P.; Kunze, D.; Lakner, H.; Kück, H. A resonantly excited 2D-micro-scanning-mirror with large deflection. *Sens. Actuators A* **2001**, *89*, 104–111. [CrossRef]
150. Munce, N.R.; Mariampillai, A.; Standish, B.A.; Pop, M.; Anderson, K.J.; Liu, G.Y.; Luk, T.; Courtney, B.K.; Wright, G.A.; Vitkin, I.A.; et al. Electrostatic forward-viewing scanning probe for Doppler optical coherence tomography using a dissipative polymer catheter. *Opt. Lett.* **2008**, *33*, 657–659. [CrossRef] [PubMed]

151. Huebsch, W.; Hamburg, S.; Guiler, R. Aircraft morphing technologies. In *Innovation in Aeronautics*; Elsevier BV: Amsterdam, The Netherlands, 2012; pp. 37–55.
152. Ishii, T. Design of shape memory alloy (SMA) coil springs for actuator applications. In *Shape Memory and Superelastic Alloys*; Elsevier BV: Heidelberg, Germany, 2011; pp. 63–76.
153. Maeda, S.; Abe, K.; Yamamoto, K.; Tohyama, O.; Ito, H. Active endoscope with SMA (Shape Memory Alloy) coil springs. In Proceedings of the Ninth International Workshop on Micro Electromechanical Systems, San Diego, CA, USA, 24 January 2002.
154. Haga, Y.; Esashi, M. Small Diameter Active Catheter Using Shape Memory Alloy Coils. *IEEJ Trans. Sens. Micromach.* **2000**, *120*, 509–514. [CrossRef]
155. Makishi, W.; Matsunaga, T.; Esashi, M.; Haga, Y. Active Bending Electric Endoscope Using Shape Memory Alloy Coil Actuators. *IEEJ Trans. Sens. Micromach.* **2007**, *127*, 75–81. [CrossRef]
156. Liang, W.; Park, H.-C.; Li, K.; Li, A.; Chen, D.; Guan, H.; Yue, Y.; Gau, Y.-T.A.; Bergles, D.E.; Li, M.-J.; et al. Throughput-Speed Product Augmentation for Scanning Fiber-Optic Two-Photon Endomicroscopy. *IEEE Trans. Med. Imaging* **2020**, *39*, 3779–3787. [CrossRef]
157. Naono, T.; Fujii, T.; Esashi, M.; Tanaka, S. Non-resonant 2-D piezoelectric MEMS optical scanner actuated by Nb doped PZT thin film. *Sens. Actuators A* **2015**, *233*, 147–157. [CrossRef]
158. Thong, P.S.-P.; Olivo, M.; Kho, K.-W.; Zheng, W.; Mancer, K.; Harris, M.R.; Soo, K.-C. Laser confocal endomicroscopy as a novel technique for fluorescence diagnostic imaging of the oral cavity. *J. Biomed. Opt.* **2007**, *12*, 014007. [CrossRef]

Communication

Investigating the Trackability of Silicon Microprobes in High-Speed Surface Measurements

Min Xu [1,*], Zhi Li [1], Michael Fahrbach [2,3], Erwin Peiner [2,3] and Uwe Brand [1]

1   Physikalisch-Technische Bundesanstalt (PTB), Bundesallee 100, 38116 Braunschweig, Germany; Zhi.li@ptb.de (Z.L.); uwe.brand@ptb.de (U.B.)
2   Institute of Semiconductor Technology (IHT), Technische Universität Braunschweig, Hans-Sommer-Straße 66, 38106 Braunschweig, Germany; m.fahrbach@tu-braunschweig.de (M.F.); e.peiner@tu-bs.de (E.P.)
3   Laboratory for Emerging Nanometrology (LENA), Langer Kamp 6 a/b, 38106 Braunschweig, Germany
*   Correspondence: min.xu@ptb.de

Abstract: High-speed tactile roughness measurements set high demand on the trackability of the stylus probe. Because of the features of low mass, low probing force, and high signal linearity, the piezoresistive silicon microprobe is a hopeful candidate for high-speed roughness measurements. This paper investigates the trackability of these microprobes through building a theoretical dynamic model, measuring their resonant response, and performing tip-flight experiments on surfaces with sharp variations. Two microprobes are investigated and compared: one with an integrated silicon tip and one with a diamond tip glued to the end of the cantilever. The result indicates that the microprobe with the silicon tip has high trackability for measurements up to traverse speeds of 10 mm/s, while the resonant response of the microprobe with diamond tip needs to be improved for the application in high-speed topography measurements.

Keywords: roughness measurement; piezoresistive microprobe; high-speed surface measurement

Citation: Xu, M.; Li, Z.; Fahrbach, M.; Peiner, E.; Brand, U. Investigating the Trackability of Silicon Microprobes in High-Speed Surface Measurements. *Sensors* **2021**, *21*, 1557. https://doi.org/10.3390/s21051557

Academic Editor: Bruno Tiribilli

Received: 4 January 2021
Accepted: 18 February 2021
Published: 24 February 2021

Publisher's Note: MDPI stays neutral with regard to jurisdictional claims in published maps and institutional affiliations.

1. Introduction

Surface roughness plays a great role in diverse fields, such as semiconductor technology, automotive manufacturing, and medicine engineering. It is an important predictor of the performance of a mechanical component [1,2]. The international standard ISO 4287/4288 [3,4] specifies various parameters, R_a, R_q, R_z, R_{sk}, etc., for the evaluation of the surface roughness. One of the most widely used parameters, the arithmetical mean deviation of the surface height R_a is expressed mathematically by

$$R_a = \frac{1}{N} \sum_{i=1}^{N} |Z_i| \tag{1}$$

where N is the number of measured points in a sampling length, and Z_i is the ordinate values of the roughness profile.

The measurement of surface roughness demands a technology with high accuracy, high throughput, relatively large measurement range, and low probing force. The measurement methods of surface texture and roughness can be divided into two categories: contact stylus instruments and optical instruments such as vertical scanning interferometry (VSI). Although the optical methods have the advantages of high throughput and no probing force, their utilization is limited by the optical properties and surface structures of the artifacts. The optical instruments have difficulties in measuring the surfaces with slopes. Undesirable light reflection and diffraction effects decrease signal quality. A study performed by Jaturunruangsri [5] proves that the stylus method is more accurate than the VSI instrument in the roughness measurements for hard materials.

However, one drawback of the contact stylus instruments lies in the low throughput. During the measurement, the stylus scans line by line. It is especially time-consuming for large range measurements. The measurement throughput of the contact stylus instrument depends on the traverse speed of the motion stages during the measurement. The maximum traverse speeds of the state-of-the-art stylus instruments are in the range of 1–3 mm/s [6,7]. The work of Arvithe Davinci et al. [7] indicates that the traverse speed of the stylus profilometer affects the roughness measurement results significantly. When the speed is below 500 μm/s, the roughness measurement results are stable. If the speed is further increased, a sharp variation in the results happens. This points out that an improvement of the trackability of the stylus at high speed is necessary.

As to the research on more rapid stylus instruments, Morrison developed a prototype stylus profilometer that can measure with speeds up to 5 mm/s in 1995 [8,9]. Since then there was very little progress on developing stylus instruments with higher traverse speeds. One obstacle to further improving the traverse speed of the stylus lies in the fact that the stylus probe will lose contact to the surface and thus profile fidelity as it moves over steep features. No loss of tracking and signal fidelity at high traverse speeds above 10 mm/s is a hard requirement for stylus probes.

The Physikalisch-Technische Bundesanstalt (PTB) developed a slender piezoresistive silicon cantilever type microprobe together with the Institute for Semiconductor Technology of Technical University of Braunschweig and the Forschungsinstitut für Mikrosensorik GmbH (CiS) Erfurt [10–16]. Cantilevers of 5 mm length, 200 μm width, 50 μm thickness and a mass of about 0.1 mg, as shown in Figure 1 were used, which are commercially available as CAN50-2-5 from the CiS GmbH (https://www.cismst.de/en/loesungen/mikrotastspitzen/ (accessed on 4 January 2021)). At the free end of the microprobe is an integrated probing tip with a height of more than 100 μm, a cone angle of 40°, and a radius of about 0.1 μm. The long and sharp tip enables the microprobe for the roughness measurement on the surface with R_a value under 25 μm [17].

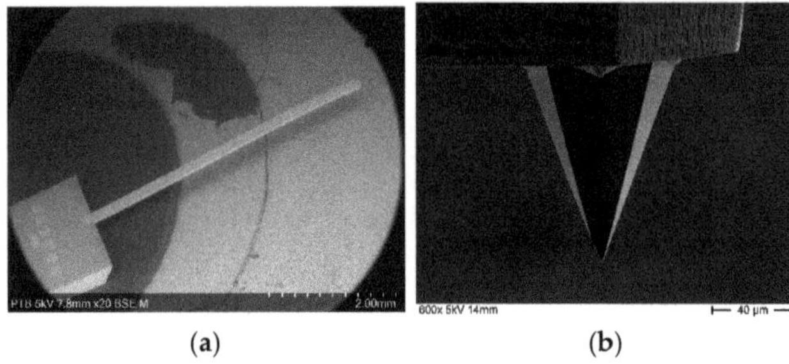

(a)   (b)

Figure 1. The scanning electron microscope images of the microprobe. (**a**) Sample of 5-mm long piezoresistive microprobe; (**b**) integrated silicon tip.

When a force acts vertically on the probing tip, the cantilever is deflected and a full bridge piezoresistive strain gauge on the cantilever close to its clamping measures the bending of the cantilever. The microprobe converts the deflection into a voltage output. The nonlinearity of the conversion influences the measurement accuracy directly. The nonlinearity between the output voltage variation and the deflection is about 0.3%. The deflection range of the microprobe is up to 200 μm. It means that the error caused by the conversion nonlinearity is less than 0.6 μm in measuring a height of 200 μm.

The signal fidelity of the contact stylus probe is influenced by many factors, such as:

1. The geometrical structure of the probe tip, such as the opening angle and the tip radius. Inappropriate tip geometry leads to contact positions other than the tip end and results in measurement deviations from the artifact surface. This is called tip-sample convolution effect [18].
2. The non-linearity between the probe output signal and the surface variation.

3. The dynamic behavior of the probe. The probe loses track as it traverses over steep features if the dynamics of the probe are not high enough.

4. The measurement bandwidth of the probe [19], usually defined by the first free resonant frequency of the probe.

Among the above factors, the demand on the last two factors, the dynamics and the measurement bandwidth of the probe, increases with the traverse speed. Hence these two factors become especially important in high-speed measurements. Compared to the existing conventional styli (with masses of several mg, the first free resonant frequency in the order of hundreds of Hz, and a cone angle of either 60° or 90°), the microprobe has a much lower mass, higher resonant frequency, and a sharper tip. It is suggested that the microprobe has a high potential for high-speed measurements.

This paper demonstrates the microprobe as a promising stylus probe candidate for high-speed roughness measurement at 10 mm/s. It is intended to give an uncomplicated method to evaluate the trackability of the microprobe. The analysis results can be proved with simple and feasible experiments, and the theoretical analysis and experimental results will indicate the improvement direction of the microprobe for better performance.

In the following, a theoretical model is proposed to investigate the dynamics of the microprobe. The steepest feature that the microprobe can measure without loss of tracking is examined, the resonant frequencies of the microprobes are analyzed and measured, and proof-of-principle experiments were performed and detailed.

2. Modelling the Dynamic Behavior of the Microprobe

Since the limited tracking fidelity is noticed first at steep features, we investigate the behavior of the microprobe on artifact surfaces with sharp variations.

The steepest feature that a probe can measure is restricted by the trackability, the tip form, and the mounting angle of the probe. In this section, we focus on the influence of the dynamic behavior of the microprobe.

It should be noted that the tip works as a mechanical low pass Gaussian filter and smoothens the sharp features in contact measurements [20]. If the cantilever tracks all the frequency components passing through the low pass filter formed by the tip, it can be considered that the cantilever can track the surface with fidelity. In other words, the dynamics of the cantilever are high enough for the measurement.

The microprobe is supposed to traverse across a falling edge and tip trajectory is drawn. The steepest feature on the surface should vary slower than the tip trajectory, otherwise, the tracking will be lost.

2.1. Theoretical Analysis

It is assumed that the artifact is an ideal step artifact with 90° inclined sidewalls and a height *H*, as presented in Figure 2. The falling edge of the top surface is set to be position 0, the origin of the x-z coordinate system.

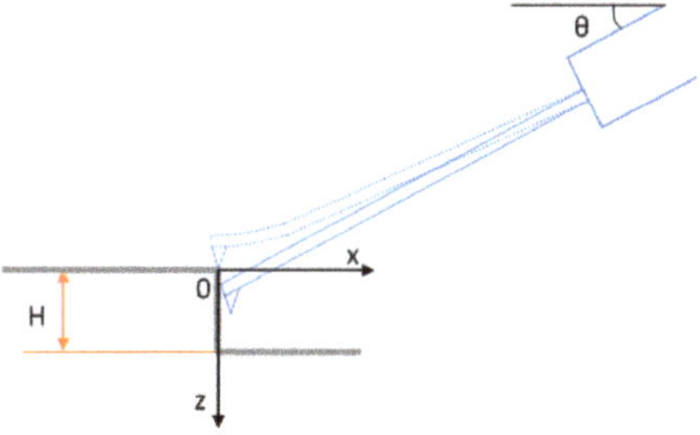

Figure 2. The deflection of the microprobe changes after passing the starting point of the falling edge (position 0) of the step artefact.

The microprobe is mounted with tilt angle θ.

The tip traverses across the artifact surface and moves over the edge of the top surface (position 0). It loses contact with the surface and the deflection of the cantilever decreases. In the case the cantilever doesn't touch the step artifact ground, i.e., under free-flight condition there is

$$m\ddot{z} + d\dot{z} + kz = F_z(0) = F(0)\cos\theta \tag{2}$$

where z is the vertical deflection variation of the cantilever, $F(0)$ is the deflection force on the cantilever at position 0, $F_z(0)$ is the vertical component of $F(0)$, k is the spring constant of the cantilever, and d is its damping factor. In this calculation it is assumed that the whole mass of the cantilever is concentrated in the tip and m is the effective mass.

The reduced effective mass of the beam $m_b = 0.24M$, M is the total mass of the beam [21]. The effective mass of the whole cantilever m is the sum of the tip mass m_t and the reduced effective mass of the beam m_b:

$$m = m_t + m_b = m_t + 0.24M \tag{3}$$

The vertical speed $\dot{z}$ and the deflection variation z can be calculated by single and double integration, respectively, of the acceleration:

$$\dot{z} = \dot{z}(0) + \int_0^t \ddot{z}\, dt \tag{4}$$

$$z = \int_0^t \dot{z}\, dt \tag{5}$$

where t is the time it takes to move the tip from position 0 to position z, $\dot{z}(0)$ is the vertical speed of the microprobe cantilever tip at position 0.

As the top surface of the step feature is flat, the initial condition exists:

$$\dot{z}(0) = 0 \tag{6}$$

Based on Equations (2)–(6), the trajectory of the cantilever tip under free-flight condition drawing the deflection variation z at the time t is only determined by the initial probing force $F(0)$ for a microprobe with the given tilt angle θ.

The angle β, the inclination of the steepest slope that the microprobe can track, is calculated through analysing the gradient of the tip trajectory $\nabla z(x)$ under free-flight condition:

$$\beta = \tan^{-1}(\nabla z(x)) \tag{7}$$

where x is the displacement along the x axis.

In the case with constant traverse speed v_x, there is

$$\nabla z(x) = \frac{\dot{z}}{v_x} \tag{8}$$

The steepest slope feature is determined by the ratio of the vertical speed to traverse speed. With the increment of the traverse speed v_x, the feature that the tip can track becomes less inclined.

The above analysis calculates the steepest slope feature that the microprobe can track. If the slope is known with angle γ and the trackability of the microprobe on the slope is evaluated, the effect of friction should be considered. The friction F_f is calculated by:

$$F_f = \mu\big(F(0)\cos\theta - kz - d\dot{z}\big)\cos(\theta + \gamma)/\cos\theta \tag{9}$$

where μ is the coefficient of friction between the microprobe tip and the artifact surface.

2.2. The Effect of Tip-Sample Convolution

The vertical speed of the tip at the origin is 0, the microprobe tip at this position cannot track any slope according to Equation (8). However, it does not necessarily mean that the microprobe will lose track at this position since tip-sample convolution effect should be taken into consideration.

When the tip traverses across an arbitrary surface, the imaged profile is the convolution result of the surface with the tip shape, and sharp features are smoothed. The artifact surface after tip-sample convolution S' is calculated by [22]

$$S' = S \oplus TIP \tag{10}$$

where S is the original artefact surface, and TIP is the tip surface.

It is defined that $S = s(x)$, $TIP = tip(u)$, $u_0 \leq u \leq u_1$, and $S' = s'(x)$. u_0 and u_1 are the lower and upper limits of abscissa values of the tip outline separately. There is

$$s'(x) = s(x) - \min_{u_0 \leq u \leq u_1} (tip(u) - s(x+u)) \tag{11}$$

When the artifact surface other than the sampling position touches the tip, or in other words, when the tip surface other than the tip end touches the artifact surface, the tip-sample convolution occurs.

If the inclination of the tip trajectory under free-flight condition is larger than that of an artefact surface after tip-sample convolution, we expect that the tip can track the surface with fidelity.

2.3. The Dynamic Behavior of the Microprobe with a Silicon Tip

For the microprobe with parameters listed in Table 1, the vertical cantilever tip movement versus time calculated using Equation (5) are drawn in Figure 3a with four different initial probing forces 32 µN, 64 µN, 96 µN, and 130 µN. It indicates that the larger the initial probing force is, the faster the tip drops and the larger the measurement range is. It costs 0.1 milliseconds for the tip dropping from $z = 0$ µm to $z = -2$ µm if the initial probing force is 32 µN. The time is reduced to 0.04 milliseconds if the initial probing force is increased to 130 µN.

In the following, the trackability of the microprobe is investigated with respect to step features with 30° and 70° inclined sidewalls.

First, the tip-sample convolution effect is simulated as shown in Figure 3b. The surfaces are convoluted according to Equation (11) by the tip with the radius of 0.1 µm, the nominal radius of the microprobe silicon tip.

The tip trajectories under free-flight condition at the traverse speed of 20 µm/s, 10 mm/s, and 50 mm/s, calculated using Equation (5) are drawn in Figure 3c. The initial probing force is 96 µN. The slopes with 30° and 70° inclinations after tip convolution are also depicted for comparison. At the traverse speed of 10 mm/s, the tip trajectory falls more rapidly than both slopes, which means that the probe can track both slopes with signal fidelity. When the traverse speed is further increased to 50 mm/s, the tip trajectory falls slower even than the convoluted 30° slope at the falling edge and the probe loses track, which means the microprobe has no capability to track slopes >30° at a traverse speed of 50 mm/s, applying a probing force of 96 µN.

The steepest slope that the microprobe can track, only considering the dynamic behavior, is presented in Figure 3d. The slope inclination increases rapidly from 0° at the start position. Then it keeps relatively constant and begins to decrease when the cantilever comes close to its deflection-free position. Since tip convolution smoothens sharp features, the moderate gradient at the starting point of the tip trajectory does not restrict mapping the surface feature if an appropriate initial probing force is selected. These curves help us to select the appropriate probing force if rough information about the surface is available. For example, with a traverse speed of 10 mm/s, an initial probing force of 96 µN is needed to track slope features with a height of 9 μm and an inclination of 70°. The initial probing force can be decreased to 32 μN for features of 3 μm height and an inclination of 30°.

From the above analysis, it can be concluded that the 5 mm long silicon microprobe has high dynamics and can track steep features with inclination variations up to 70° at traverse speeds up to 10 mm/s.

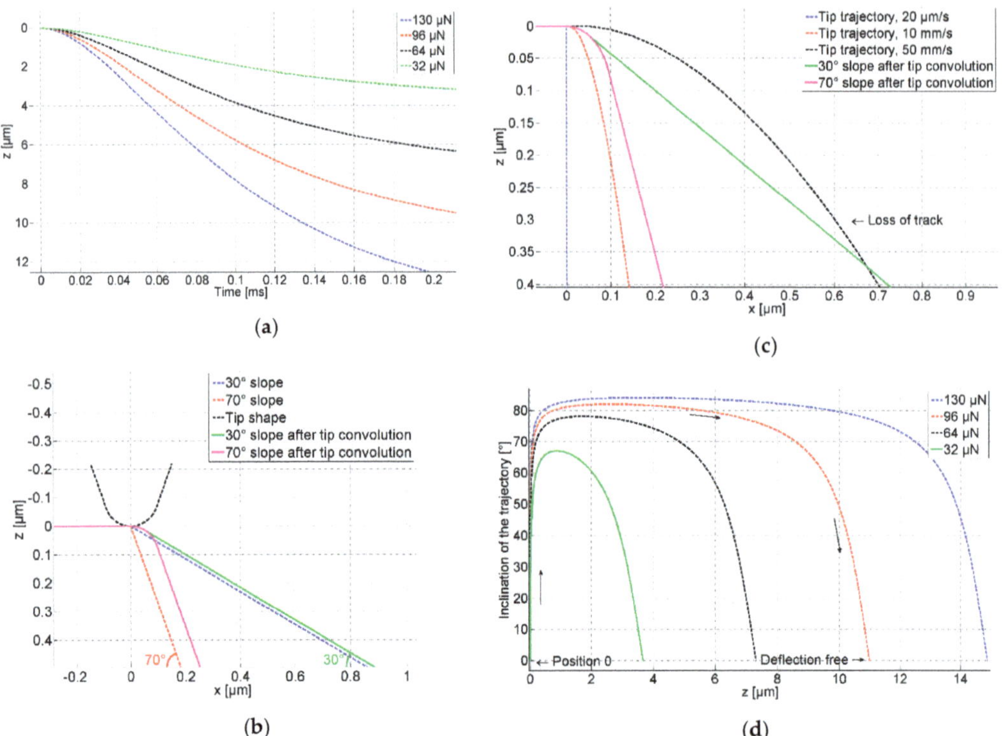

Figure 3. The dynamic behavior of the microprobe with a silicon tip. (**a**) The trajectory of the microprobe tip of the CAN50-2-5 drawing z with time under the condition of different initial probing forces; (**b**) The slope features before and after the tip convolution; (**c**) comparison of the tip trajectories and the slope features; (**d**) the steepest slope inclinations that the microprobe can track at the traverse speed of 10 mm/s.

2.4. The dynamic Behavior of the Microprobe with a Diamond Tip

Since undesired silicon tip abrasion exists in measurements, a diamond tip with a tip radius of 2 μm was glued to the microprobe to replace the silicon tip, as defined in DIN EN ISO 4288 [4]. The half opening angle of the diamond tip is about 45°, as shown in Figure 4. Limited by the half opening angle, the inclination of the slope feature that can be measured by the microprobe with the diamond tip is smaller than 45°.

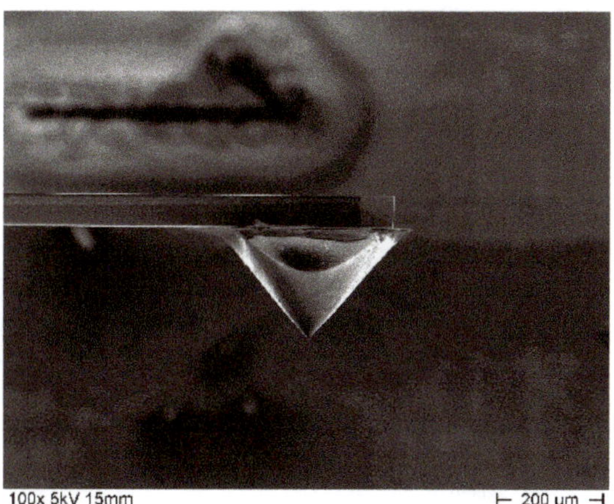

Figure 4. The microprobe with a glued 210 μm high diamond tip.

With a height of about 210 μm and the base area of about $5 \times 10^6\ \mu m^2$, the mass of the conical diamond tip is about 0.012 mg. The effective mass of the cantilever is increased to 0.036 mg (from 0.024 mg for the cantilever with silicon tip) and the dynamic performance of the microprobe is thus decreased. However, the surface variation after tip-sample convolution is moderated stronger by the 2 μm-radius tip than the 0.1 μm-radius silicon tip since the "cut-off" wavelength λ_t of the low-pass Gaussian filter formed by the 2 μm-radius diamond tip is longer according to

$$\lambda_t = 2\pi\sqrt{AR} \tag{12}$$

where R is the tip radius and A is the amplitude of the surface feature.

In consequence, the microprobe with the diamond tip can track the 45° slope at a traverse speed of 10 mm/s with an initial probing force of 96 μN, as calculated using Equation (5) and shown in Figure 5.

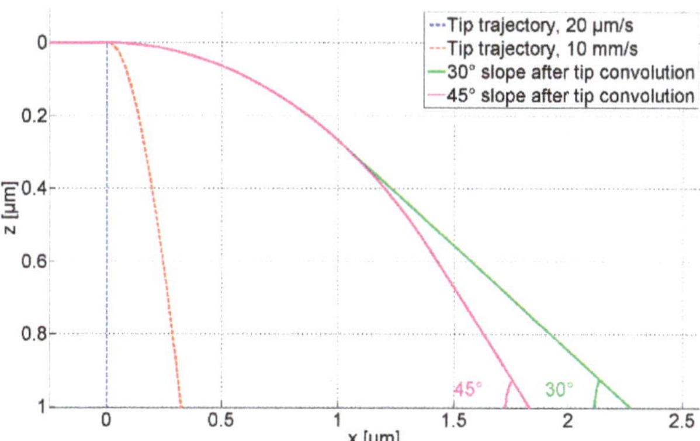

Figure 5. The trajectory of the microprobe with a diamond tip at different traverse speeds, with an initial probing force of 96 μN.

Table 1. Nominal parameters of the microprobe used to calculate the cantilever tip trajectories in Figures 3 and 5.

Nominal Parameters	Symbol	Value
Length	L	5 mm
Width	w	200 μm
Thickness	b	50 μm
cross-section area	$A = wb$	10^{-8} m^2
Density	ρ	2330 kg/m^3
Mass of the cantilever	$M = \rho Lwb$	0.1 mg
Mass of the tip	m_t	0.15 μg for the silicon tip of the CAN50-2-5 0.012 mg for the diamond tip
Half opening angle of the tip		20° for the silicon tip of the CAN50-2-5 45° for the diamond tip
effective mass of the cantilever	m	0.024 mg for the CAN50-2-5 0.036 mg for the cantilever with a diamond tip
Spring constant	k	8.45 N/m
Damping factor	d	0.001
Young's modulus	E	169 GPa
Area moment of inertia	$I_a = \frac{wb^3}{12}$	2.093×10^{-18} m^4
Tilt angle	θ	15°

3. Frequency Response of the Microprobes

The frequency response of a probe is important since it is a decisive factor in the measurement bandwidth of the probe. The demand on the measurement bandwidth of the probe increases linearly with the traverse speed because the frequency f exerted to the tip by the surface structure increases with the traverse speed v_x as given by

$$f = \frac{v_x}{\lambda} \tag{13}$$

where λ is the spatial wavelength of the structure.

Not only the frequency response of the microprobe without contact to a surface, but more important the frequency response of the probe in contact to a surface should be known.

Neglecting the mass of the probing tip the fundamental resonant frequency of a cantilever with an approximately rectangular shape and without a tip at its free end [23] can be calculated by

$$f_0 = \frac{1.758}{\pi L^2} \sqrt{\frac{E I_a}{\rho A}} \tag{14}$$

Using the parameters listed in Table 1, the resonant frequency of the CAN50-2-5 microprobe is calculated to be f_0 = 2.8 kHz. The frequency of the microprobe with an integrated silicon tip used in the measurement was measured to be about f_0 = 3.2 kHz, slightly higher than its nominal value, which is assigned to deviations from the nominal dimensions due to fabrication tolerances of the CAN50-2-5. Due to the effect of the added tip mass, a lower resonant frequency of f_0 = 2.2 kHz was measured with the microprobe with the diamond tip.

The contact resonant frequency of a probe is influenced by many factors, such as the tip radius, the artifact material, and the probing force. The first contact resonant frequency f_c of a probe can be roughly estimated to be about three to five times of its first free resonant frequency f_0.

If the "cut-off" spatial wavelength is λ_t = 1 μm, the contact resonant frequency of the probe is demanded to be above f_c = 10 kHz for the traverse speed of v_x = 10 mm/s according to Equation (13). This is just a rough estimation and not a hard criterion. The microprobe with a silicon tip (f_c = 9.6 to 16 kHz) may meet this requirement while the microprobe with a diamond tip (f_c = 6.6 to 11 kHz) possibly does not.

4. Experiment Results

4.1. "Tip Flight" Test

"Tip flight" tests (see Figure 6) were performed on flat smooth surfaces of a step artifact to prove the trackability of the microprobes. The idea was to measure with the microprobes across sloped features with different traverse speeds. To characterize tip flight, a normalized flight width was defined, which is the ratio of the flight width to the step structure height. It was used as a measure to evaluate the dynamics of the microprobes in comparison with the modeling described in Section 2 quantitatively.

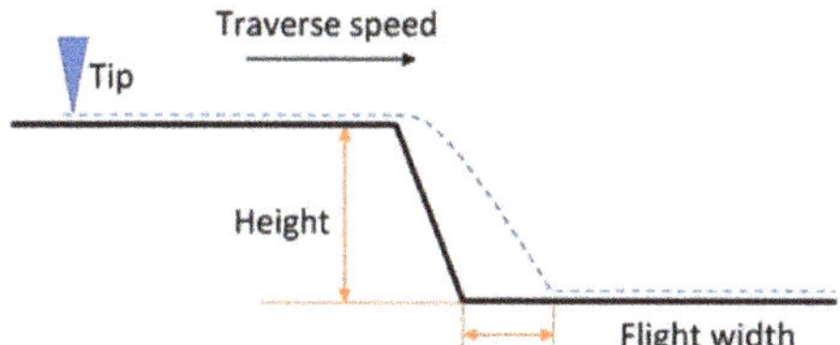

Figure 6. The tip flight test to measure the dynamic behavior of the microprobe.

The test was carried out with the instrument Profilscanner, a self-developed profiler at PTB [24–26]. The microprobe is glued and bonded on a microprobe holder and mounted on the tip scanning head of the Profilscanner which consists of an XYZ piezo stage (PI, model P-628.2CD for XY axes and P-622.ZCD for Z-axis) with a motion range of 800 μm $\times$ 800 μm $\times$ 250 μm (X $\times$ Y $\times$ Z).

Whether the XY stage can move steadily at the defined traverse speed influences the test results. The moving speeds of the XY piezo stage were measured before the 'flight test' with laser interferometers and the result proves that the XY piezo stage moves steadily at a speed up to 10 mm/s.

4.2. Artifacts and the Test Results

Two artifacts were used in the test. Artifact A is diamond turned from Cu and then coated with 15 μm Ni and 5 μm Cr on top (see Figure 7a). The nominal height is 10 μm and the slope inclination is about 30°. The nominal height of artifact B (see Figure 7b) is also 10 μm, but the slope inclination is 90°. It is made of steel. Because of the manufacturing tolerances, the edges of the top surface of artifact B are somewhat protruding.

Figure 8 shows the measured profiles by the CAN50-2-5 microprobe with the integrated silicon tip traversing across the artifacts with speeds of 20 μm/s, 5 mm/s, and 10 mm/s respectively. The initial probing forces are 96 μN for artifact A and 130 μN for artifact B. The measured heights of both artifacts significantly deviate from the nominal values given by the manufacturer. Since the height of artifact B is about 14 μm the static probing force needed to be increased to end up with the tip in contact with the ground surface of the artifact. Because of the influence of the mounting tilt angle ($\theta = 15°$) and the half opening angle of the tip (20°), the measured slope angle of the artifact B is about 55° instead of 90°. For both sloped artifacts, A and B, the profiles acquired at different traverse speeds agree well and flight widths are zero. This result proves the high fidelity of the measured profiles related to the fast-responding dynamics of the microprobe with a silicon tip.

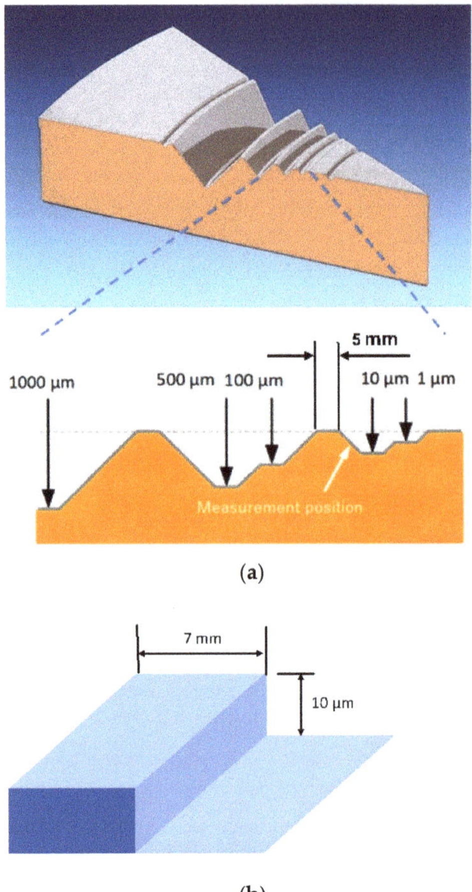

Figure 7. Artifacts used in "tip flight" test. (**a**) artifact A; (**b**) artifact B. The artifacts aren't drawn to scale.

The tip flight test was repeated with the microprobe with a diamond tip. Since the mounting tilt angle ($\theta = 15°$) and the half opening angle of the diamond tip ($45°$) leads to that the inclination that the diamond tip can measure is no more than $30°$, only artifact A was measured with the microprobe with a diamond tip. Again, an initial static probing force of 96 µN was used. As shown in Figure 9, the tip flight widths at different traverse speeds again were zero. This indicates the dynamics of the microprobe with the diamond tip is good enough for measuring the slope of $30°$ inclination up to traverse speeds of 10 mm/s, confirming the theoretical model developed above. However, the tip vibrated during the measurements on the artifact surface at the higher speeds of 5 mm/s and 10 mm/s. It means that the resonant response of the microprobe with the diamond tip is not high enough for such high-speed measurements. The bandwidth of the microprobe with the diamond tip will have to be improved by increasing the stiffness or decreasing the mass of the tip.

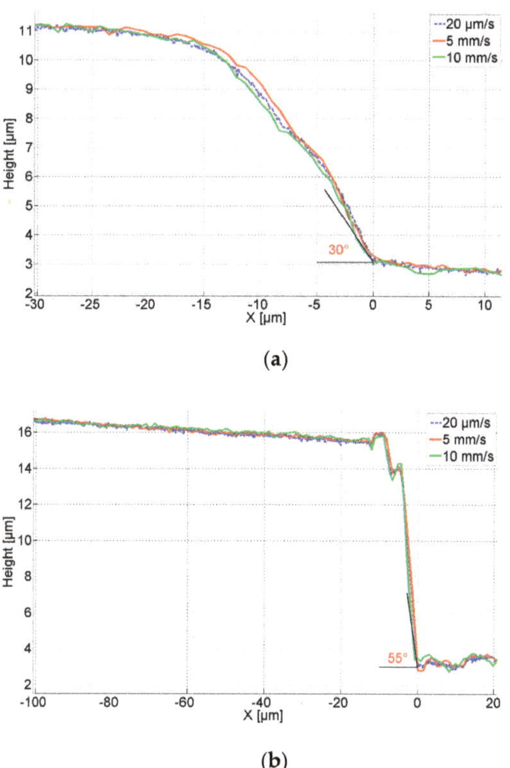

(a)

(b)

Figure 8. Measured profiles with the CAN50-2-5 microprobe with integrated silicon tip. (**a**) Measured profiles on artifact A (30° inclined sidewall) with an initial static probing force of 96 µN, at different traverse speeds; (**b**) Measured profiles on artifact B (90° inclined sidewall) with an initial static probing force of 130 µN, at different traverse speeds.

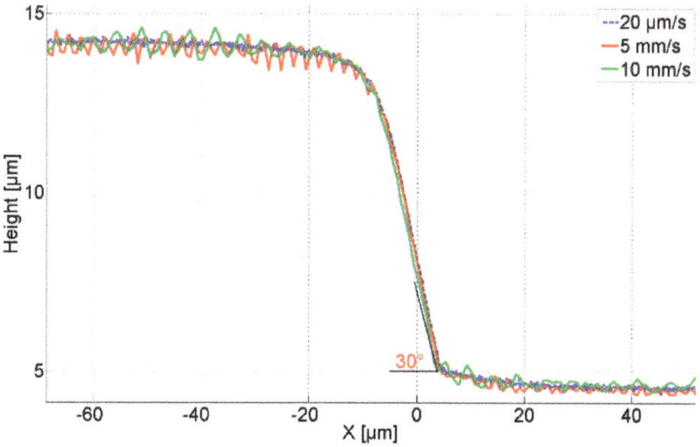

Figure 9. Measured profiles on artifact A using the microprobe with diamond tip with an initial static probing force of 96 µN, at different traverse speeds.

The "tip flight" test results demonstrate that the dynamics of both microprobes are high enough for the measurements at the speed of 10 mm/s, but the bandwidth of the microprobe with a diamond tip should be raised to meet the demand of trackability.

In the next step of our work, we will decrease the length of the microprobe to increase the resonance frequency and investigate the trackability further.

5. Summary

To investigate the trackability of piezoresistive silicon microprobes for high-speed surface roughness measurements, a theoretical dynamic model was derived to examine the dynamics of the microprobes. The resonant response of the microprobes was analyzed and tip-flight tests were performed. Both the theoretical analysis and the experimental results prove that the microprobes with integrated silicon tip have the capability of tracking surfaces with high fidelity up to traverse speeds of 10 mm/s. However, the resonant response of a microprobe with a glued diamond tip of 90° opening angle will have to be improved to fulfill the demands of high-speed roughness measurements.

Author Contributions: Conceptualization, M.X., Z.L. and U.B. Methodology, M.X., Z.L. and U.B. Formal Analysis, M.X. Investigation, M.X. and Z.L. Software, M.X. Writing-Original Draft Preparation, M.X. writing—review and editing, M.X., Z.L., M.F., E.P. and U.B. Funding acquisition, E.P. and U.B. All authors have read and agreed to the published version of the manuscript.

Funding: This research is funded by the EMPIR 17IND05 MicroProbes project. MicroProbes has received funding from the EMPIR program co-financed by the EU participating states and from the European Union's Horizon 2020 research and innovation program.

Institutional Review Board Statement: Not applicable.

Informed Consent Statement: Not applicable.

Data Availability Statement: All data and code will be made available on request to the correspondent author's email with appropriate justification.

Acknowledgments: We gratefully thank our colleagues, Helmut Wolff and Radovan Popadic, for preparing the microprobes and the artefacts.

Conflicts of Interest: The authors declare no conflict of interest.

References

1. Whitehouse, D.J. A revised philosophy of surface measuring systems. *Proc. Inst. Mech. Eng.* **1988**, *202*, 169. [CrossRef]
2. Whitehouse, D.J. Surface metrology. *Meas. Sci. Technol.* **1997**, *8*, 955. [CrossRef]
3. ISO. *ISO 4287. Geometrical Product Specifications (GPS)—Surface Texture: Profile Method—Terms, Definitions and Surface Texture Parameters*; International Organisation for Standardization: Geneva, Switzerland, 1998.
4. ISO. *ISO 4288. Geometric Product Specifications (GPS)—Surface Texture: Profile Method—Rules and Procedures for the Assessment of Surface Texture*; International Organisation for Standardization: Geneva, Switzerland, 1998.
5. Jaturunruangsri, S. Evaluation of Material Surface Profiling Methods: Contact Versus Non-Contact. Ph.D. Thesis, Brunel University, London, UK, 2014.
6. Kapłonek, W.; Ungureanu, M.; Nadolny, K.; Sutowski, P. Stylus profilometry in surface roughness measurements of the vertical conical mixing unit used in a food industry. *J. Mech. Eng.* **2018**, *47*, 1–8. [CrossRef]
7. Davinci, M.A.; Parthasarathi, N.L.; Borah, U.; Albert, S.K. Effect of the tracing speed and span on roughness parameters determined by stylus type equipment. *Measurement* **2014**, *48*, 368–377. [CrossRef]
8. Morrison, E. A prototype scanning stylus profilometer for rapid measurement of small surface areas. *Int. J. Mach. Tools Manufact.* **1995**, *35*, 325–331. [CrossRef]
9. Morrison, E. The development of a prototype high-speed stylus profilometer and its application to rapid 3D surface measurement. *Nanotechnology* **1996**, *7*, 37–42. [CrossRef]
10. Doering, L.; Brand, U.; Bütefisch, S.; Ahbe, T.; Weimann, T.; Peiner, E.; Frank, T. High-speed microprobe for roughness measurements in high-aspect-ratio microstructures. *Meas. Sci. Technol.* **2017**, *28*, 034009. [CrossRef]
11. Brand, U.; Xu, M.; Doering, L.; Langfahl-Klabes, J.; Behle, H.; Bütefisch, S.; Ahbe, T.; Peiner, E.; Völlmeke, S.; Frank, T.; et al. 2019 Long Slender Piezo-Resistive Silicon Microprobes for Fast Measurements of Roughness and Mechanical Properties inside Micro-Holes with Diameters below 100 μm. *Sensors* **2019**, *19*, 1410. [CrossRef] [PubMed]

12. Behrens, I.; Doering, L.; Peiner, E. Piezoresistive cantilever as portable micro force calibration standard. *J. Micromech. Microeng.* **2003**, *13*, 171–177. [CrossRef]

13. Peiner, E.; Balke, M.; Doering, L.; Brand, U.; Bartuch, H.; Völlmeke, S. Fabrication and test of piezoresistive cantilever sensors for high-aspect-ratio micro metrology. In Proceedings of the Mikrosystemtechnik Kongress 2007, Dresden, Germany, 15–17 October 2007; pp. 369–374.

14. Peiner, E.; Balke, M.; Doering, L. Slender tactile sensor for contour and roughness measurements within deep and narrow holes. *IEEE Sens. J.* **2008**, *8*, 1960–1967. [CrossRef]

15. Peiner, E.; Balke, M.; Doering, L.; Brand, U. Tactile probes for dimensional metrology with microcomponents at nanometer resolution. *Meas. Sci. Technol.* **2008**, *19*, 064001. [CrossRef]

16. Peiner, E.; Doering, L. Nondestructive evaluation of diesel spray holes using piezoresistive sensors. *IEEE Sens. J.* **2013**, *13*, 701–708. [CrossRef]

17. Bhushan, B. (Ed.) *Modern Tribology Handbook*; CRC Press: Boca Raton, FL, USA, 2001.

18. Yacoot, A.; Koenders, L. Aspects of scanning force microscope probes and their effects on dimensional measurement. *J. Phys. D Appl. Phys.* **2008**, *41*, 103001. [CrossRef]

19. Doll, J.C.; Pruitt, B.L. High-bandwidth piezoresistive force probes with integrated thermal actuation. *J. Micromech. Microeng.* **2012**, *22*, 095012. [CrossRef] [PubMed]

20. Leach, R.; Haitjema, H. Bandwidth characteristics and comparisons of surface texture measuring instruments. *Meas. Sci. Technol.* **2010**, *21*, 032001. [CrossRef]

21. Butt, H.-J.; Siedle, P.; Seifert, K.; Fender, K.; Seeger, T.; Bamberg, E.; Weisenhom, A.L.; Goldie, K.; Engel, A. Scan speed limit in atomic force microscopy. *J. Microsc.* **1993**, *169*, 75.

22. Villarrubia, J.S. Algorithms for scanned probe microscope image simulation, surface reconstruction, and tip estimation. *J. Res. Natl. Inst. Stand. Technol.* **1997**, *102*, 425–454. [CrossRef] [PubMed]

23. Rabe, U. Atomic force acoustic microscopy. In *Applied Scanning Probe Methods*; Bhushan, B., Fuchs, H., Eds.; Springer: New York, NY, USA, 2006; Volume II, pp. 37–90.

24. Xu, M.; Kirchhoff, J.; Brand, U. Development of a traceable profilometer for high-aspect-ratio micro-structures metrology. *Surf. Topogr. Metrol. Prop.* **2014**, *2*, 024002. [CrossRef]

25. Xu, M.; Kauth, F.; Mickan, B.; Brand, U. Traceable profile and roughness measurements inside micro sonic nozzles with the Profilscanner. In Proceedings of the 17th International Congress of Metrology, Paris, France, 21–24 September 2015; p. 13004.

26. Xu, M.; Kauth, F.; Mickan, B.; Brand, U. Traceable profile and roughness measurements inside micro sonic nozzles with the profilscanner to analyse the influence of inner topography on the flow rate characters. *Meas. Sci. Technol.* **2016**, *27*, 94001. [CrossRef]

Review

Microcantilever: Dynamical Response for Mass Sensing and Fluid Characterization

João Mouro [1,*], Rui Pinto [2], Paolo Paoletti [3] and Bruno Tiribilli [1]

1   Institute for Complex Systems (ISC-CNR), National Research Council, Via Madonna del Piano 10, 50019 Florence, Italy; bruno.tiribilli@isc.cnr.it
2   INESC—Microsystems e Nanotecnologies, Rua Alves Redol, 9, 1000-029 Lisbon, Portugal; ruimrpinto@yahoo.com
3   School of Engineering, University of Liverpool, Liverpool L69 3GH, UK; P.Paoletti@liverpool.ac.uk
*   Correspondence: joao.mouro@fi.isc.cnr.it

Abstract: A microcantilever is a suspended micro-scale beam structure supported at one end which can bend and/or vibrate when subjected to a load. Microcantilevers are one of the most fundamental miniaturized devices used in microelectromechanical systems and are ubiquitous in sensing, imaging, time reference, and biological/biomedical applications. They are typically built using micro and nanofabrication techniques derived from the microelectronics industry and can involve microelectronics-related materials, polymeric materials, and biological materials. This work presents a comprehensive review of the rich dynamical response of a microcantilever and how it has been used for measuring the mass and rheological properties of Newtonian/non-Newtonian fluids in real time, in ever-decreasing space and time scales, and with unprecedented resolution.

Keywords: microcantilever; mass sensing; rheology sensing; noise; non-Newtonian viscoelastic fluids

Citation: Mouro, J.; Pinto, R.; Paoletti, P.; Tiribilli, B. Microcantilever: Dynamical Response for Mass Sensing and Fluid Characterization. *Sensors* **2021**, *21*, 115. https://doi.org/10.3390/s21010115

Received: 9 December 2020
Accepted: 22 December 2020
Published: 27 December 2020

1. Introduction

Fluids play a key role in many sensing applications based on microelectromechanical systems (MEMS), being either the substance to be tested (when measuring rheological properties) or the support environment used to keep the substance of interest in its native or physiological state (when detecting proteins, DNA, or other analytes in a solution). The understanding of the interaction of the sensor with the surrounding medium is a key topic in the process of measuring the mass of analytes with extremely high—potentially single-molecule—accuracy, and when using MEMS sensors to study the rheology of simple and complex fluids. Such wide sensing applications span the fields of the food and process industry, environmental monitoring, healthcare, microfluidics, and others. Several of these problems still do not have an adequate solution, but huge progress has been made in the last two decades by exploiting microcantilevers, miniaturized beams supported at one end. Microcantilevers are a traditional but crucial MEMS design used for sensing, imaging, and time-keeping applications.

This paper presents a thorough review of how the complex dynamical response of the microcantilever excited by a periodic force and interacting with the surrounding environment can be used for mass and rheological sensing. Some examples of the latest and more significant results are provided, and the physical principles behind the applications are discussed. The work is organized as follows. In Section 2, the mechanics and dynamical response of the microcantilever are presented. The goal is to set the theoretical framework, consisting of some classical models, to determine the resonance frequencies and quality factors of each resonant mode in different media, which will be used throughout the rest of the work. Section 3 contains a discussion of the mechanisms used to excite the dynamical response of the probe, including open and closed-loop schemes typically found in sensing applications. Particular focus is dedicated to feedback loops, which have shown significant promise and exciting new results in recent years, by providing relevant examples that are currently being developed and studied to improve the performance of microcantilever-based sensors. A thorough discussion of noise measurements and mechanisms follows. This aspect is often overlooked in the literature, but it is a fundamental feature to consider when designing a sensor. Section 4 is dedicated to discussing the principle of mass sensing, sensitivities, and limits of detection and shows some examples of recent major achievements in this area. Finally, in Section 5 it is shown how the microcantilever can be used as a rheological sensor to measure the properties of Newtonian and non-Newtonian fluids in real time. Section 6 summarizes current open challenges and presents an outlook on future opportunities for microcantilever-based rheology and mass sensors with the aim of stimulating further research in this field.

2. Cantilever Mechanics and Dynamical Response

2.1. Euler–Bernoulli Beam

The flexural vibration of a thin uniform beam can be described by the well-known Euler–Bernoulli beam theory. This model is based on four main assumptions: (i) cantilever aspect ratios L/w and L/h are large ($L \gg w, h$), (ii) deflections are small when compared to the beam dimensions, (iii) the cantilever material is linear elastic and homogeneous, and (iv) no dissipation occurs during deformation. The schematic of such a cantilever subjected to an external force per unit length $q(x, t)$ is shown in Figure 1a. The external load is responsible for the existence of shear forces F_z and bending moments M_y that act on each element of the beam of infinitesimal length dx, as shown in Figure 1b.

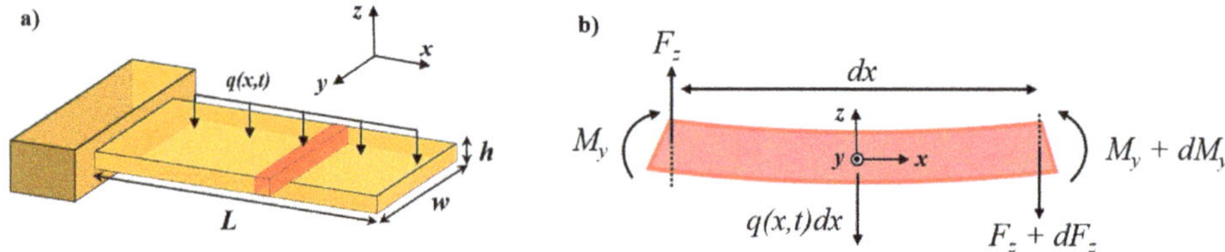

Figure 1. (**a**) Schematic of a cantilever beam of length L, width w, and thickness h. The cantilever is subjected to a distributed time-changing load per unit length, $q(x, t)$; (**b**) longitudinal cross section of an infinitesimal element dx of the same cantilever (red part highlighted in (**a**)), where the shear forces and bending moments act.

Balancing the forces (z-direction) and the bending moments (y-direction) acting on each infinitesimal element of the device and neglecting higher order terms leads to the following equations (see Figure 1b) [1,2]:

$$\begin{cases} F_z + dF_z - F_z + q(x,t)dx = \rho A dx \frac{\partial^2 W(x,t)}{\partial t^2} \\ M_y + dM_y - M_y - F_z dx + q(x,t)dx\frac{dx}{2} = 0 \end{cases} \Rightarrow \begin{cases} \frac{dF_z}{dx} + q(x,t) = \rho A \frac{\partial^2 W(x,t)}{\partial t^2} \\ \frac{dM_y}{dx} = F_z \end{cases}, \quad (1)$$

where $W(x,t)$ is the time-varying deflection at a distance x from the support; F_z and M_y are the shear forces and bending moments, respectively, acting on the element of the beam; ρ is the density of the structural material; $A = wh$ is the beam cross section; and $q(x,t)$ is a general distributed load per unit length. Merging the two equations yields:

$$\frac{d^2 M_y}{dx^2} + q(x,t) = \rho A \frac{\partial^2 W(x,t)}{\partial t^2}. \quad (2)$$

Upon the bending of the beam, the length of the neutral plane is given by $l = R d\theta$, with R and $d\theta$ the curvature radius and span angle of the bent beam, respectively. The strain at the planes above and below the neutral plane is given by $\varepsilon_x = \frac{\delta l}{l} = \frac{(R+z)d\theta - R d\theta}{R d\theta} = \frac{z}{R}$. Given that the material is elastic, the stress in the x-direction can then be calculated as $\sigma_x = E\varepsilon_x = E\frac{z}{R}$, with E indicating the Young's modulus of the material [2].

The bending moment around the y-axis, M_y, created by the tension forces in x-direction, $dF_x = \sigma_x dA = E\frac{z}{R}dydz$, applied to a distance z from the neutral plane, is given by:

$$M_y = \int dM_y = \int -z\vec{e_z} \times dF_x\vec{e_x} = \int -z^2\frac{E}{R}dydz = -\frac{E}{R}I_z, \tag{3}$$

where $I_z = \int_{-h/2}^{h/2} \int_{-w/2}^{w/2} z^2 dydz$ is the second moment of area of the cross section (for standard rectangular microcantilevers $I_z = \frac{wh^3}{12}$). Finally, for small curvatures the radius of the curvature can be approximated by $\frac{1}{R} = \frac{\partial^2 W(x,t)}{\partial x^2}$ [2], and therefore the bending moment can be expressed as:

$$M_y = -EI_z\frac{\partial^2 W(x,t)}{\partial x^2}. \tag{4}$$

Substituting Equation (4) into Equation (2) results in a differential equation usually known as the Euler–Bernoulli beam equation [1–3]:

$$EI_z\frac{\partial^4 W(x,t)}{\partial x^4} + \rho A\frac{\partial^2 W(x,t)}{\partial t^2} = q(x,t). \tag{5}$$

A general solution of this equation is obtained by performing a modal decomposition—i.e., by considering the microcantilever response as the linear superposition of simple vibrational modes. The first step in this process is to extract the natural resonance frequency and shape of each vibrational mode, assuming zero external forces—i.e., $q(x,t) = 0$.

It is assumed that the deflection at any point of the beam varies harmonically with time, so the general solution for each mode can be separated into a temporal term, $\psi(t)$, and a spatial term, $\Phi(x)$. The ansatz for the temporal term is a harmonic oscillation with natural angular frequency $\omega_{0,n}$, where the index n describes the resonant mode, such as $\psi_n(t) = e^{i\omega_{0,n}t}$. Therefore, the general solution of each individual mode can be written as [4]:

$$W_n(x,t) = \psi_n(t)\Phi_n(x). \tag{6}$$

Inserting the solution of Equation (6) into Equation (5) and rearranging the terms gives:

$$\frac{d^4\Phi_n(x)}{dx^4} = \beta_n^4\Phi_n(x), \tag{7}$$

with $\beta_n^4 = \left(\frac{\rho A\omega_{0,n}^2}{EI_z}\right)$. The solution to Equation (7) provides the spatial term of Equation (6):

$$\Phi_n(x) = c_1\cos(\beta_n x) + c_2\sin(\beta_n x) + c_3\cosh(\beta_n x) + c_4\sinh(\beta_n x). \tag{8}$$

Different boundary conditions are associated with each type of end constraints of the microresonator. For the particular case of the suspended cantilever considered in this paper, the boundary conditions reflect the fact that the displacement, $\Phi(x)$, and slope, $\frac{d\Phi(x)}{dx}$, must be zero at the clamped end ($x = 0$), while the bending moment, M_y, and shear force, F_z, are zero at the free-end ($x = L$) [1–4]—i.e.:

$$\Phi(0) = 0, \tag{9a}$$

$$\text{and } \frac{d\Phi(0)}{dx} = 0, \tag{9b}$$

$$M_y = EI_z\frac{\partial^2\Phi(L)}{\partial x^2} = 0, \tag{9c}$$

$$\text{and } F_z = EI_z\frac{\partial^3\Phi(L)}{\partial x^3} = 0. \tag{9d}$$

Imposing these four boundary conditions on Equation (8) results in a homogeneous system of four equations. Nontrivial solutions are obtained when the determinant of the matrix of the coefficients of these equations vanishes, which corresponds to the condition

$$\cos(\beta_n L)\cos h(\beta_n L) + 1 = 0. \tag{10}$$

Equation (10) can be numerically solved and the first consecutive roots $(\beta_n L)$ calculated, as shown in Figure 2a. Indicatively, $(\beta_1 L) = 1.875$, $(\beta_2 L) = 4.694$, $(\beta_3 L) = 7.855$, $(\beta_4 L) = 10.996$, etc. The condition expressed in Equation (10) can be asymptotically approximated by $\cos(\beta_n L) = 0$, with solutions given by $\beta_n L = \left(n - \frac{1}{2}\right)\pi$, with $n = 1, 2, 3, 4, \ldots$ also shown in Figure 2a [2,4].

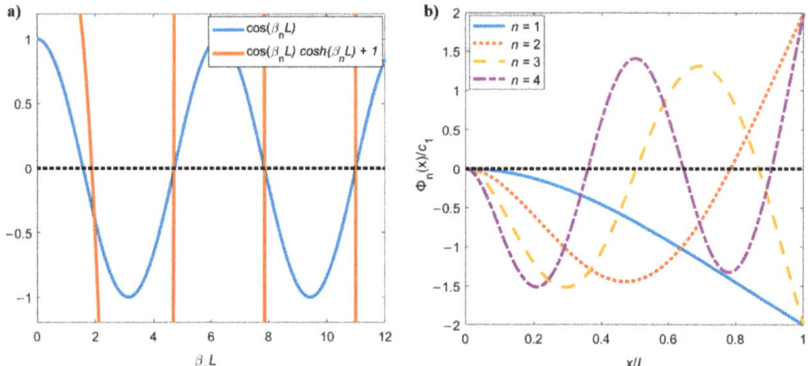

Figure 2. (**a**) Plot of Equation (10) (orange line), whose first four solutions $\beta_n L$ (indicated in the text) are the crossings with zero. The blue line shows the asymptotic approximation $\cos(\beta_n L) = 0$, which can be solved analytically, and that agrees with the numerical solution for $n \geq 2$; (**b**) mode shapes of the first four flexural modes of a cantilever.

The system of equations obtained from imposing the boundary conditions (Equation (9a)–(9d)) in Equation (8) is solved for the constants $c_{1,2,3,4}$. It is found that $c_1 = -c_3$, $c_2 = -c_4$, and $\frac{c_2}{c_1} = \frac{\sin(\beta_n L) - \sin h(\beta_n L)}{\cos(\beta_n L) + \cos h(\beta_n L)}$. For each particular value of $\beta_n L$, the spatial solution, referred to as the flexural mode shape, is obtained by substituting these results into Equation (8) [4]:

$$\Phi_n(x) = c_1\left(\cos(\beta_n x) - \cosh(\beta_n x) + \frac{\sin(\beta_n L) - \sin h(\beta_n L)}{\cos(\beta_n L) + \cos h(\beta_n L)}(\sin(\beta_n x) - \sin h(\beta_n x))\right), \tag{11}$$

where c_1 remains undefined until an external force is applied. These modal shapes are shown in Figure 2b for a generic c_1. Note how the number of nodes and the slope near the base of the cantilever increase with the mode number n.

The natural angular resonance frequency of each mode, $\omega_{0,n}$, can then be calculated using the respective β_n^4, thus obtaining:

$$\omega_{0,n} = \beta_n{}^2\sqrt{\frac{EI_z}{\rho A}} = \frac{(\beta_n L)^2 h}{L^2}\sqrt{\frac{E}{12\rho}}. \tag{12}$$

Finally, the complete solution of Equation (6) is [4]:

$$W_n(x,t) = a_1 e^{i\omega_{0,n}t}\left(\cos(\beta_n x) - \cosh(\beta_n x) + \frac{\sin(\beta_n L) - \sin h(\beta_n L)}{\cos(\beta_n L) + \cos h(\beta_n L)}(\sin(\beta_n x) - \sin h(\beta_n x))\right). \tag{13}$$

The Euler–Bernoulli beam model only accounts for free vibrations of undamped cantilevers. The free vibrations and mode shapes of clamped-clamped or supported beams can also be predicted using the Euler–Bernoulli model by replacing Equation (9) with the appropriate boundary conditions [1,4]. In addition, more complex continuous beam equations have been developed, which include the effects of rotary inertia and shear deformation [1] or the effect of axial tensile or compressive forces [1,2]. However, these models are seldomly used in sensing applications and therefore fall outside the scope of this review.

2.2. Harmonic Oscillations with a Single Degree of Freedom

Although a microcantilever is a continuous system with infinite degrees of freedom, its dynamical response can be accurately described by a single degree of freedom, given that in most applications a specific resonant mode dominates. In fact, these simplified models allow us to extract some useful information from the vibrations, such as the amplitude response and energy dissipation mechanisms, beyond what is provided by the more complex Euler–Bernoulli model. This section presents the theoretical background of the harmonic oscillator models.

2.2.1. Simple Harmonic Oscillator

The simplest possible model for describing the oscillatory motion of the cantilever is the undamped free oscillator. In this case, it is assumed that the cantilever is represented by an effective mass, m_{eff}, connected to a linear elastic spring with stiffness k_{eff}, whose potential energy is given by $U(z) = \frac{1}{2}k_{\text{eff}}z^2$, where z is the displacement from the equilibrium position $z = 0$. By using the restitutive force of this spring, $F = -\frac{dU(z)}{dz} = -k_{\text{eff}}z$ within Newton's law $F = m_{\text{eff}}\ddot{z}(t)$, one gets the following equation of motion [3,5]:

$$m_{\text{eff}}\ddot{z}(t) + k_{\text{eff}}\, z(t) = 0, \tag{14}$$

where the dots represent the time derivative. The solution of this second-order differential equation has the form of an oscillatory motion, described by $z(t) = A_0 e^{i(\omega t + \phi)}$, with A_0 and ϕ being the amplitude and phase of the motion, respectively, and ω being its frequency. Substituting this solution into Equation (14), one gets the expression for the (angular) natural resonance frequency (for each mode):

$$\omega_0 = \sqrt{\frac{k_{\text{eff}}}{m_{\text{eff}}}}. \tag{15}$$

This is the natural resonance frequency of the cantilever when it vibrates freely in vacuum, and it is equivalent to the resonance frequency calculated using the Euler–Bernoulli beam theory presented in Equation (12). A classical result from structural mechanics is that the effective stiffness of the cantilever is given by $k_{\text{eff}} = \frac{Ewh^3}{4L^3}$ [2]. This indicates that each resonant flexural mode predicted by the Euler–Bernoulli model can be described by an equivalent harmonic oscillator with a different m_{eff}. Therefore, by comparing Equation (15) with Equation (12), one can obtain the equivalent mass of each resonant mode in the limit of small damping (cantilevers vibrating in vacuum or air, for example):

$$m_{\text{eff},n} = \frac{3\rho h b L}{(\beta_n L)^4} = \frac{3m_c}{(\beta_n L)^4}, \tag{16}$$

where $m_c = \rho L h w$ is the total mass of the cantilever. Considering, for example, the fundamental resonance mode, where $\beta_1 L = 1.875$ (see Figure 2a), one gets $m_{\text{eff},1} = 0.24m_c$, as confirmed elsewhere [6,7].

2.2.2. Forced Damped Harmonic Oscillator

By introducing an intrinsic damping coefficient c and an external driving force $F_{drive}(t)$ in Equation (14), the equation of motion becomes [3,5]:

$$m_{\text{eff}}\ddot{z}(t) + c\dot{z}(t) + k_{\text{eff}}z(t) = F_{drive}(t). \tag{17}$$

Using Equation (15), defining a parameter $\gamma = \frac{c}{m_{\text{eff}}} = \frac{\omega_0}{Q}$, where $Q = \frac{m_{\text{eff}}\,\omega_0}{c}$ is called the quality factor and depends on the damping of the system, and assuming a periodic force $F_{drive}(t) = F_0 e^{i\omega t}$, Equation (17) is written as:

$$\ddot{z}(t) + \gamma \dot{z}(t) + \omega_0^2 z(t) = \frac{F_0}{m_{\text{eff}}} e^{i\omega t}. \tag{18}$$

A steady-state harmonic solution is given by $z(t) = A_0 e^{i(\omega t + \phi)}$, with A_0 being the amplitude of the motion and ϕ being the phase between the applied external force and the induced motion. Substituting such ansatz into Equation (18) yields:

$$\left(-\omega^2 + \omega_0^2 + i\gamma\omega\right) = \frac{F_0}{A_0 m_{\text{eff}}} e^{-i\phi}. \tag{19}$$

After separating the real and imaginary parts, the following system is obtained:

$$\begin{cases} \omega_0^2 - \omega^2 = \frac{F_0}{A_0 m_{\text{eff}}} \cos(\phi)\ \text{(a)} \\ -\gamma\omega = \frac{F_0}{A_0 m_{\text{eff}}} \sin(\phi)\ \text{(b)} \end{cases}, \tag{20}$$

which can be solved to get:

$$A_0 = \frac{F_0 / m_{\text{eff}}}{\sqrt{(\omega_0^2 - \omega^2)^2 + \left(\frac{\omega_0 \omega}{Q}\right)^2}}, \tag{21a}$$

$$\phi = \operatorname{atan}\left(-\frac{\omega_0 \omega}{Q(\omega_0^2 - \omega^2)}\right). \tag{21b}$$

Equation (21a) gives the amplitude response of the cantilever, whereas Equation (21b) represents the phase between the driving force and the motion of the cantilever. The amplitude and phase curves of microcantilevers with different values of Q are shown in Figure 3.

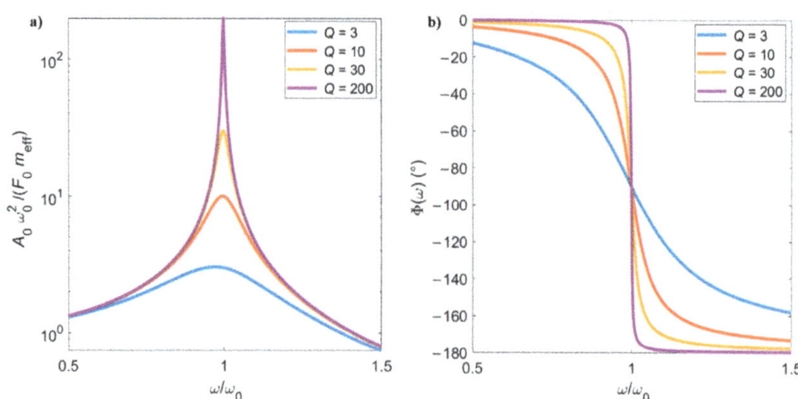

Figure 3. (**a**) Amplitude and (**b**) phase responses as function of the normalized excitation frequency of a forced and damped harmonic oscillator. Four different levels of damping (Q) are considered, typically encountered in microcantilevers vibrating in air or liquid mediums.

If the intrinsic damping coefficient c tends to zero, then the quality factor Q tends to infinite. In this case, the second term under the square root in Equation (21a) is negligible, and the maximum of the amplitude response will occur at the natural frequency—that is, at $\omega = \omega_0$. Conversely, when c increases, Q decreases and the amplitude response of the beam shifts to the left, with the peak of maximum amplitude occurring at a frequency lower than ω_0. The maximum amplitude can be obtained by solving $\frac{dA_0}{d\omega} = 0$, which gives the resonance frequency of intrinsically damped vibrations.

$$\omega_{res} = \omega_0 \sqrt{1 - \frac{1}{2Q^2}}. \tag{22}$$

As observed in Figure 3, low Qs decrease the peak amplitude, broadening the peak and shifting the maximum of the curve towards lower values. An even more pronounced shift occurs when the cantilever oscillates in damped environments, such as in the case of liquids, as will be discussed in the next section. The motion of the microcantilever is in phase with the drive force before the resonance, but lags $90°$ behind at the resonance point, and $180°$ after resonance. Low values of Q result in a wider linear region of phase change around the natural frequency. The phase of the resonator is of particular importance when closed-loop excitation systems are used, as will be discussed in Section 3.

The physical definition of Q is the ratio of the energies stored, E', and dissipated, $\Delta E''$, in the resonator, both averaged per oscillation cycle, at the resonance frequency [5], $Q = 2\pi \frac{E'}{\Delta E''}$. Q can be assessed experimentally from the amplitude response of the resonator using the bandwidth method by the expression:

$$Q \approx \frac{\omega_{res}}{\Delta \omega_{-3dB}} = \frac{\omega_{res}}{(\omega_2 - \omega_1)_{-3dB}}, \tag{23}$$

where ω_{res} corresponds to the maximum of the amplitude curve (Equation (22)), and $\Delta \omega_{-3dB}$ is the bandwidth at 3 dB below the maximum amplitude. In practice, measuring $\Delta \omega_{-3dB}$ corresponds to determining the frequencies of the two points, ω_2 and ω_1, at half the power (3 dB or 70.7%) of the maximum of the amplitude response.

2.2.3. General One-Degree-of-Freedom Equation of Motion for Microcantilevers

Despite the usefulness of the simple models discussed above, these are still of limited applicability to describe the complete dynamical response of a microcantilever in most applications. In fact, for sensing applications very often additional terms must be considered according to the excitation strategy used, the type of samples to be tested, or the properties of the surrounding fluid. A more general one-degree-of-freedom equation of motion can then be introduced as:

$$\ddot{z} + \frac{\omega_0}{Q}\dot{z} + \omega_0^2[1 + \lambda \cos(\Omega t + \Phi)]z + \frac{\alpha}{m_{\text{eff}}}z^3 + \frac{\eta}{m_{\text{eff}}}z^2\dot{z} + \frac{\delta}{m_{\text{eff}}}\left(z\dot{z}^2 + z^2\ddot{z}\right) = \frac{1}{m_{\text{eff}}}\left[F_{drive}(t) + F_{interaction}(z,\dot{z},\ddot{z},\ldots)\right], \tag{24}$$

where z represents the displacement of the microcantilever, m_{eff} is the effective mass, α is the nonlinear spring constant (also known as the Duffing parameter [8,9]), η is the coefficient of nonlinear damping, δ is the coefficient of nonlinear inertia [10,11], $F_{drive}(t)$ is a general external drive signal, $F_{interaction}(z,\dot{z},\ddot{z},\ldots)$ is a general interaction force between the cantilever and the surrounding environment, and $\lambda \cos(\Omega t + \Phi)$ is a function (proportional to the displacement of the beam) that is used to modulate the spring constant at frequency Ω, phase Φ (with respect to the drive force) and gain λ. Equation (24) is capable of describing distinct sources of non-linearities and different types of possible excitation mechanisms and external forces, and is applicable to any vibrational mode by using the respective effective mass and natural resonance frequency. This general equation can be used to describe rich and complex dynamics of the vibrating beam [4,12–17] across a wide variety of sensing applications, as will be discussed in more detail in the following sections.

2.3. Operation in Dissipative Fluids

To theoretically describe a cantilever beam moving in a fluid, Equation (5) is used and the general distributed load $q(x,t)$ is substituted with a driving force to excite the motion of the beam and a hydrodynamic load induced by the fluid [18]:

$$EI_z \frac{\partial^4 W(x,t)}{\partial x^4} + c_0 \frac{\partial W(x,t)}{\partial t} + m_0 \frac{\partial^2 W(x,t)}{\partial t^2} = F_{drive}(x,t) + F_{hydro}(x,t). \tag{25}$$

Compared to Equation (5), the dissipative intrinsic viscous term per unit length $c_0 = c/L$, was also introduced and $m_0 = \rho A = \rho wh$ is the mass of the cantilever per unit length.

Usually, the hydrodynamic load is decomposed into two terms: a pressure drag (inertial term, proportional to the acceleration) and a viscous drag (dissipative term, proportional to the velocity) [18]—i.e.,

$$F_{hydro}(x,t) = -m_A \frac{\partial^2 W(x,t)}{\partial t^2} - c_V \frac{\partial W(x,t)}{\partial t}, \tag{26}$$

where m_A is an added mass and c_V an added damping coefficient, both expressed per unit length. Equation (25) can then be re-written as:

$$EI_z \frac{\partial^4 W(x,t)}{\partial x^4} + (c_0 + c_V) \frac{\partial W(x,t)}{\partial t} + (m_0 + m_A) \frac{\partial^2 W(x,t)}{\partial t^2} = F_{drive}(x,t). \tag{27}$$

In [19], Sader solved Equation (27) analytically and showed that, in the limit of small dissipative effects (i.e., for $Q_n \gg 1$), the resonance frequency of the extrinsically damped vibrations in fluid with added mass and added damping (m_A and c_V), $\omega_{R,n}$, and the respective quality factor, Q_n, of each mode of the cantilever are given by [20]:

$$\frac{\omega_{R,n}}{\omega_{res,n}} = \left(1 + \frac{m_A}{m_0}\right)^{-\frac{1}{2}}, \tag{28a}$$

$$Q_n = \frac{\omega_{R,n}\,(m_0 + m_A)}{(c_0 + c_V)}, \tag{28b}$$

where $\omega_{res,n}$ is the resonance frequency of each mode for intrinsically damped vibrations calculated by Equation (22). Furthermore, Sader also showed that the amplitude response of each mode is given by [19]:

$$X \cong \left(\left(\omega_{R,n}^2 - \omega^2\right)^2 + \left(\frac{\omega_{R,n}\omega}{Q_n}\right)^2\right)^{-1/2}, \tag{29}$$

with X being a normalized amplitude. Equation (29) is readily identified as the amplitude response of the forced damped harmonic oscillator, as given by Equation (21a).

In summary, provided the dissipation is low; $Q_n \gg 1$; and, therefore, the resonant modes do not overlap, each resonant mode of a cantilever vibrating in liquid can be described by the harmonic oscillator model, with a resonance frequency $\omega_{R,n}$ and a quality factor Q_n.

The added mass, m_A, and added viscous damping coefficient, c_V, can be analytically calculated as functions of the properties of the liquid. This is done by solving the incompressible Navier–Stokes equations [21] and determining the velocity and pressure fields in the moving fluid that create a hydrodynamic load, $\Gamma(\omega)$, acting on the oscillating beam (see [22,23] for cantilevers with circular and rectangular cross sections, respectively).

The hydrodynamic load $\Gamma(\omega) = \Gamma'(\omega) + i\,\Gamma''(\omega)$ is a dimensionless function that contains inertial (real part, added mass) and dissipative (imaginary part, viscous damping) terms. For example, the added mass and viscous damping acting on a rectangular cantilever are described by [19,20]:

$$m_A = \frac{\pi}{4}\rho_f w^2 \Gamma_{rect}{}'(\omega), \tag{30a}$$

$$c_V = \frac{\pi}{4}\rho_f w^2 \omega \Gamma_{rect}{}''(\omega), \tag{30b}$$

where $\Gamma_{rect}{}'(\omega)$ and $\Gamma_{rect}{}''(\omega)$ are the real and imaginary parts of the hydrodynamic load acting on a cantilever with rectangular cross section, $\Gamma_{rect}(\omega)$, which is a function of the Reynolds number $Re = \frac{\rho_f \omega w^2}{4\eta}$, with ρ_f and η being the density and viscosity of the fluid, respectively.

Since it is not trivial to calculate $\Gamma_{rect}(\omega)$ analytically (see [23] for details), Sader used the strategy of correcting the hydrodynamic load calculated by Chen for a cylindrical cantilever, $\Gamma_{circ}(\omega)$ [22], with an empirical function $\Omega(\omega)$, in the form of $\Gamma_{rect}(\omega) = \Omega(\omega)\Gamma_{circ}(\omega)$, to match the analytical results for $\Gamma_{rect}(\omega)$ of Tuck [23]. Although treatable, Sader's correction function is still lengthy and numerical [19].

In a subsequent work, Maali et al. [20] further simplified the problem, by using simple polynomials to fit the real and imaginary parts of Sader's solution for $\Gamma_{rect}(\omega)$, thus obtaining the analytical expressions:

$$\Gamma_{rect}{}'(\omega) = a_1 + a_2 \frac{\delta}{w} = a_1 + \frac{a_2}{w}\sqrt{\frac{2\eta}{\rho_f \omega}}, \tag{31a}$$

$$\Gamma_{rect}{}''(\omega) = b_1 \frac{\delta}{w} + b_2 \left(\frac{\delta}{w}\right)^2 = \frac{b_1}{w}\sqrt{\frac{2\eta}{\rho_f \omega}} + \frac{2\eta}{\rho_f \omega}\left(\frac{b_2}{w}\right)^2, \tag{31b}$$

where $\delta = \sqrt{\frac{2\eta}{\rho_f \omega}}$ is the thickness of the layer surrounding the cantilever in which the velocity of the fluid drops by a factor of $1/e$ and $a_1 = 1.0553$, $a_2 = 3.7997$, $b_1 = 3.8018$, and $b_2 = 2.7364$. These expressions are valid for Reynolds numbers ranging between 1 and 1000 [20], which are the typical values for most of the microcantilever-based sensing applications, and can be applied for the first resonance modes. Equations (28)–(31) can be used in conjunction to obtain the dependence between the resonance frequency and quality factor of the resonant mode n and the rheological properties of the fluid, as will be shown in Section 5.

The results discussed in this section for the flexural oscillations of a microcantilever immersed in viscous fluids were experimentally validated in [24] and then extended to model torsional vibrations of the cantilever in [25], which can also be used in applications such as atomic force microscopy (AFM). The effect of a nearby wall on the frequency response of flexural and torsional oscillations of a cantilever immersed in viscous fluids was studied in [26]. Exact analytical solutions of the three-dimensional flow field and hydrodynamical load were presented, for the general case of a thin-blade (cantilever) performing flexural and torsional oscillations in viscous fluids, by Van Eysden and Sader [27]. These analytical calculations for the three-dimensional hydrodynamical load were subsequently incorporated in the initial analysis of flexural [19] and torsional [25] oscillations and used for the general case of arbitrary mode orders [28].

3. Excitation Schemes and Noise

3.1. Excitation Strategies

Dynamic sensing applications, especially when operating fluids, require an external force to induce vibrations of measurable amplitude and acceptable levels of signal-to-noise ratio. To maximize the efficiency of excitation, an external force must be applied at different frequencies corresponding to the microcantilevers resonant modes. This actuation is typically accomplished by using electrical, thermal, or acoustic actuation techniques [29]. Regardless of the physical mechanism used to generate the driving force (F_{drive} in Equation (24)), the excitation strategy can be broadly divided into two families, according to the way F_{drive} is applied and tuned. These two methodologies, namely "external excitation" and "feedback excitation" are briefly reviewed below.

3.1.1. External or Open-Loop Excitation Mechanisms

Traditional dynamic sensing applications are based on the assumption that the cantilever response can be well approximated by a single degree of freedom harmonic oscillator, as shown in Section 2.2.2. The adsorption of analytes on the sensor surface (for mass sensing application) and/or changes in the properties of the surrounding fluid (for rheology sensors) induce a shift in the natural frequency ω_0 and, potentially, a change in the quality factor Q as well. Such changes can be experimentally measured by performing frequency sweeps of the excitation force applied to the microcantilever and recording the corresponding amplitude response at each frequency, ideally recovering the theoretical curves shown in Figure 3. Such excitation mode is denoted as "open-loop" or "external", as the force applied to the cantilever is entirely controlled by the user, without feeding back any information from the cantilever response.

Most commonly, the excitation force is generated by a piezoelectric actuator embedded in the cantilever support which can set it into oscillation by creating an acoustic wave that propagates through the substrate materials. When the frequency of such wave matches (or is reasonably close to) the resonance frequency of the cantilever, the oscillation amplitude can be significant and detectable for measurements (as described in Section 2). This setup is the most commonly used for imaging and sensing in air or a vacuum, due to its simplicity and ease of operation. However, it presents major drawbacks for sensing applications in dissipative fluids. Indeed, most of the dynamic sensing applications are based on detecting small changes in the resonance frequency of the probe (see Sections 4 and 5), measured by performing frequency sweeps with the excitation force and detecting the maximum amplitude of the cantilever deflection. However, the forces acting on the cantilever by the surrounding fluid dramatically increase the amount of dissipation, as explained in Section 2.3, and therefore the Q factor of the resonant response is low (around 10 or below). As shown in Figure 3, in this condition the peak in the amplitude response is not very pronounced, making it difficult to: (i) tune the frequency of the excitation force to match the resonance and (ii) detect small changes in the resonance frequency itself induced by changes in the properties of the fluid. Even worse, a typical response of an acoustically excited cantilever suffers from the well-known "forest of peaks" phenomenon, where several spurious peaks appear in the amplitude response due to mode coupling with oscillations in the fluid and in the sample holder [30]. In these cases, identifying the right "peak"—corresponding to the cantilever resonance—and measuring small variations in its position becomes very challenging, as shown in Figure 4a for the case of a microcantilever oscillating in water and externally excited with a piezoelectric holder. The proper choice of material to be used as mechanical holder is key to obtaining reduced spurious peaks in the amplitude and phase curves [31].

To partially overcome these drawbacks, alternative physical sources have been considered to generate the driving force, with the aim of moving the point of excitation closer to the cantilever free end and minimizing mode coupling with the surrounding fluid. A common proposed technique is magnetic excitation [32], but nowadays thermal excitation, via an additional laser shining on the microcantilever [33] or by thermal effects in bi-layer cantilevers [34], and piezoelectric [35] or electrostatic [36] excitations are also commonly used for exciting the microdevices. With the point of application of the exciting force being co-located with the detection point (typically the cantilever tip), there is no need for a travelling wave to be formed in the probe and therefore the delay/phase-shift between excitation and deflection is minimized and there is very limited coupling with fluid vibratory modes. This results in a measured amplitude spectrum that is much closer to the theoretical one shown in Figure 3.

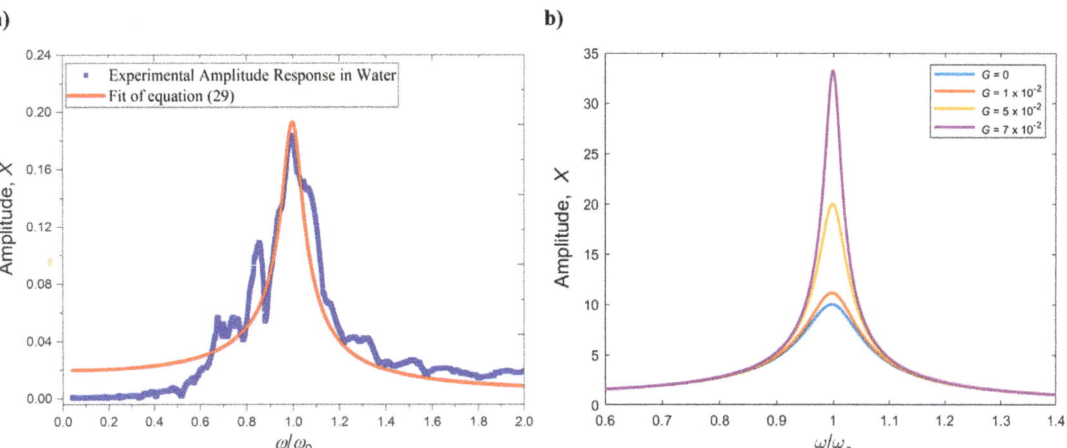

Figure 4. (**a**) Experimental amplitude response of an acoustically excited microcantilever oscillating in water, showing a noisy peak not accurately fitted by the theoretical amplitude curve; (**b**) dependence of the amplitude response of a microcantilever on the gain of the feedback loop in the Q-control method, for $\omega_0 = 1$ rad/s, original $Q = 10$ (for $G = 0$), $\frac{F_0}{m_{\text{eff}}} = 1$ N/kg and $\phi = \frac{\pi}{2}$ rad.

However, irrespective of the experimental implementation details of open-loop excitation, the fundamental problem of low Q factor and, correlated to this, a relatively inaccurate measurement of the frequency of the resonance peak remains.

3.1.2. Feedback or Closed-Loop Excitation Mechanisms

An alternative approach to improve the sensitivity and signal-to-noise ratio of the cantilever when used as a mass or rheology sensor is based on acting on the dynamic response of the cantilever itself, instead of on the physical actuation and detection methodology. Theoretically, if the Q factor of the amplitude response is significantly increased, then the issues related to poor signal-to-noise ratio in the identification of the resonance peak and the "forest of peaks" phenomenon become less pronounced. To achieve this goal, "feedback" or "closed-loop" approaches are needed, where the excitation force depends on the response of the microcantilever itself. Within this umbrella, several techniques have been proposed in the literature.

The original idea was proposed by Rodríguez et al. [37] and is commonly known as "Q-control", where the measured microcantilever deflection is amplified by a gain, delayed by an user-defined quantity and then added to the external harmonic excitation with a feedback loop. In its basic form, the equivalent single degree of freedom dynamics reads:

$$\ddot{z} + \frac{\omega_0}{Q}\dot{z} + \omega_0^2 z = \frac{1}{m_{\text{eff}}}\left[F_0 \cos \omega t - k_{\text{eff}}Gz\left(t - \frac{\phi}{\omega}\right)\right]. \tag{32}$$

When compared to Equation (24), $F_{drive} = F_0 \cos \omega t - k_{\text{eff}}Gz\left(t - \frac{\phi}{\omega}\right)$ consists of a harmonic external excitation with amplitude F_0 and frequency ω (same as in open-loop strategies), added up to a forcing component obtained by delaying the instantaneous detected deflection of the microcantilever $z(t)$ by a time $\tau_{delay} = \frac{\phi}{\omega}$ (with ϕ being a user-defined phase shift along the feedback loop) and then amplifying the delayed deflection by a feedback Q-control gain, $k_{\text{eff}}G$. It can be shown that by modulating the feedback gain G, the effective Q factor exhibited by the cantilever response is much larger than the intrinsic Q factor of the cantilever, as shown in Figure 4b [37].

An alternative and more effective approach, based on the concept of parametric resonance, has been proposed by Moreno et al. and Prakash et al. [38–40]. In these works Equation (24) turns into:

$$\ddot{z} + \frac{\omega_0}{Q}\dot{z} + \omega_0^2[1 - G\cos(2\omega_0 t)]z + \frac{\alpha}{m_{\text{eff}}}z^3 = \frac{F_{tip-sample}}{m_{\text{eff}}}, \tag{33}$$

where the cantilever is parametrically excited ($F_{drive}(t) = 0$) by a signal proportional to the displacement of the cantilever, amplified by G and at twice the natural frequency of the beam, $2\omega_0$, while the non-linear force acting on the tip of the cantilever, $F_{interaction} = F_{tip-sample}$, was used for imaging [41]. Once again, changes in the feedback gain G translate to variation in the effective Q factor, as shown in [14].

Similarly, Miller et al. [42] studied the possibility of using the nonlinear damping coefficient η to parametrically control the phase of a nonlinear resonator and implement a parametric phase-locked loop (PLL), turning Equation (24) into:

$$\ddot{z} + \frac{\omega_0}{Q}\dot{z} + \omega_0^2[1 + \lambda\cos(2(\omega_0 t + \Phi) + \delta)]z + \frac{\alpha}{m_{\text{eff}}}z^3 + \frac{\eta}{m_{\text{eff}}}z^2\dot{z} = 0, \tag{34}$$

where the parameter δ in Equation (34) is the phase setpoint of the PLL. Again, this technique does not require any external force F_{drive}, as the probe is self-oscillating.

It is worth noting that all the techniques discussed so far are capable of dramatically improving the cantilever response, but the feedback gains always have upper bounds corresponding to the probe oscillation becoming unstable and (theoretically) growing exponentially.

A different approach has been proposed by the authors in [43], where the intrinsic dynamics of the cantilever is first made unstable by amplifying the deflection signal and then stabilized by introducing a nonlinear saturation in the feedback loop. The resulting dynamics reads:

$$\ddot{z} + \frac{\omega_0}{Q}\dot{z} + \omega_0^2 z = \frac{1}{m_{\text{eff}}}\left[\text{sat}\left(Kz\left(t - \tau_{delay}\right)\right) - m_A\ddot{z} - c_A\dot{z}\right]. \tag{35}$$

In this work, the cantilever is subjected to a fluidic force given by $F_{interaction} = F_{fluidic}(\dot{z},\ddot{z}) = -m_A\ddot{z} - c_A\dot{z}$, where m_A and c_A describe the added mass and damping induced by the fluid. Additionally, the drive force of the cantilever has the form $F_{drive}(t) = \text{sat}\left(Kz\left(t - \tau_{delay}\right)\right)$, where the deflection z of the microcantilever is delayed by τ_{delay} (microseconds) along the feedback loop and amplified by a gain K, before being saturated and fed back as the driving force.

Although analytical solutions of Equation (35) are not available, accurate predictions of the overall response can be obtained by using Harmonic Balance methods [44]. These techniques are described in detail in [44,45], and it is assumed that the deflection in Equation (35) is approximately harmonic—i.e.:

$$z(t) \cong A_0\sin(\omega t). \tag{36}$$

If this solution is inserted into Equation (35), the response of the saturation can be expanded in harmonic terms (an approach known as Describing Function, see [44,45] for details) and all the terms of the same frequency are balanced together, resulting in analytical values for the critical gain K that initiates the self-oscillations, and amplitude and frequency of oscillation. According to classical Nyquist theory [44], the saturation term is required to stabilize the self-oscillations of the feedback loop for non-zero τ_{delay}. However no upper bound exists for K, with the self-sustained oscillation remaining always stable and therefore this excitation technique requires less tuning effort when compared with the other strategies described previously. This method was used to study a microcantilever used for imaging [46], for mass sensing [47], and for rheology sensor [48].

3.2. Detection Mechanisms

The mechanisms that can be employed for measuring the deflection and resonance share some of the physical mechanisms used for actuation, and include optical, capacitive, piezoelectric and piezoresistive sensing. Arguably the most common method to study the vibration of microstructures are the optical lever technique, where the motion of the cantilevers can be detected by reflecting a laser from the beam into a position-sensitive detector, and laser Doppler vibrometry, an interference-based optical technique. The two former methods depend on external equipment that is difficult to miniaturize or integrate, reducing the range of applications. However, this also stimulates an on-going effort to develop alternative optical detection methods to reduce the complexity of the detection configuration [49], using off-the-shelf components (DVD optical pickups, for example [50]), or implementing more recent paradigms, such as quantum sensing [51]. In the piezoresistive technique, the cantilever is fabricated with an integrated resistor having piezoresistive properties, and therefore the electrical resistance changes as a function of cantilever motion [52]. The use of piezoresistive detection is advantageous compared to optical detection because there is no need for laser or detector alignments. A disadvantage is that the current flowing in the piezoresistive layer causes temperature fluctuations in the cantilever, which may lead to parasitic cantilever deflection and to piezoresistive changes [53]. Electrostatic (capacitive) detection is also a widespread approach, which is possible to integrate with compact electronics. The capacitive detection is due to the changes in capacitance that arise from the displacement of the resonator relative to one or more fixed electrodes/gates (the resonator and the electrode/s are separated by a small gap, forming a capacitor). The capacitance method provides a high sensitivity and absolute displacement. However, the capacitance method is troublesome in an electrolyte solution due to the Faradic current between the resonator and electrodes, which obscures the desired signal. Another disadvantage is that capacitive sensing loses efficiency for nanoscale devices: capacitance scales as area/separation but practical limits on resonator–gate gaps limit reduction of their dimensions, and given the higher resonance frequencies of NEMS compared with MEMS, a large fraction of the drive and detection signals are lost through parasitic capacitances [54].

3.3. Noise

Many cantilever-based applications require detecting frequency shifts caused by small changes at the sensor surface or in the surrounding medium, as discussed in Sections 4 and 5 for the cases of mass and rheology sensing, respectively. Any real case resonator or oscillator shows fluctuations in its amplitude and frequency/phase responses, caused by the noise present in the mechanical system, surrounding environment and equipment, resulting in uncertainties in the measurements. Therefore, understanding how noise affects such measurements is key to assess the ultimate sensing performance. This section discusses how to experimentally measure noise, its physical origins, and how to use it to estimate the minimum detectable frequency shift and corresponding limit of detection.

3.3.1. Time Domain—Allan deviation

It is possible to quantify the frequency stability of an oscillator or resonator system in the time domain using the Allan variance (a measure of the frequency drift in a specific time window). The Allan deviation, $\sigma_y(\tau)$, is defined as the root mean square of the differences between consecutive relative frequency measurements taken in non-overlapping time windows of duration τ [55–57]:

$$\sigma_y(\tau) = \sqrt{\frac{1}{2(M-1)} \sum_{i=1}^{M-1} \left(\frac{f_{i+1} - f_i}{f_c}\right)^2}, \tag{37}$$

where M is the total number of frequency measurements taken, f_i is the ith frequency measurement (averaged in the time window with duration τ), and f_c is the nominal carrier frequency (typically the resonance frequency of the microcantilever, f_0). The Allan deviation depends on the time interval used to collect consecutive samples, commonly denoted as τ, or the integration time.

For short integration times, the Allan variance is dominated by the frequency/phase noise in the resonator and associated circuit. For long integration times, frequency drifts due to temperature variations, resonator ageing and other environmental conditions dominate. More information on the physical origin and manifestation of the noise mechanisms in MEMS/NEMS resonators can be found in [56–59] and will be shortly discussed later.

3.3.2. Frequency Domain—Spectral Densities

Another metric to study the frequency stability of the system is the noise density around the carrier (resonance) frequency. For continuous signals over time, such as the vibrations of a microresonator or oscillator, power spectral densities (PSD) can be defined. These show how the power of a signal is distributed over the frequency spectrum and are hence customarily called power spectra. For convenience, and to be able to apply the concept to any kind of signals (not only physical power), spectral densities are also defined by the variance of the signal over time or, in other words, by the squared deviation of the signal from its mean value over time. The variance of the signal at a certain frequency is then interpreted as the energy of the signal at that frequency (measured using a well-defined bandwidth in each frequency point).

Therefore, considering the frequency domain, frequency/phase instabilities can be measured by the spectral density of normalized frequency/phase fluctuations. These are given by:

$$S_y(f) = y_{rms}^2(f)\frac{1}{BW}, \tag{38a}$$

$$S_\Phi(f) = \Phi_{rms}^2(f)\frac{1}{BW}, \tag{38b}$$

where $S_y(f)$ and $S_\Phi(f)$ are, respectively, the spectral density of frequency and phase fluctuations; $y_{rms}(f)$ and $\Phi_{rms}(f)$ are the measured root mean squared (rms) value of normalized frequency and phase, respectively, in a band of Fourier frequencies containing the carrier frequency f_c; and BW is the width of the frequency band in Hz. The units of $S_y(f)$ are $1/\mathrm{Hz}$ and of $S_\Phi(f)$ are $\mathrm{rad}^2/\mathrm{Hz}$. $S_y(f)$ and $S_\Phi(f)$ are one-sided spectral densities, and apply over a Fourier (or sideband) frequency range f from 0 to ∞ [60,61]. The relation between these two quantities is given by [61]:

$$S_\Phi(f) = \left(\frac{f_c}{f}\right)^2 S_y(f). \tag{39}$$

$S_\Phi(f)$ has been historically utilized in metrology, but more recently the single-sideband phase noise, $\mathcal{L}(f)$, has become the prevailing quantity to measure phase noise among manufacturers and users of frequency standards [60]. $\mathcal{L}(f)$ is the noise power density normalized to the carrier power and can be defined as the ratio [60]:

$$\mathcal{L}(f) = \frac{P_{noise\ (1\ Hz)}(f)}{P_{signal}}, \tag{40}$$

where $P_{noise\ (1\ Hz)}(f)$ (units of dBm/ Hz) is the power density in one single sideband due to phase modulation (PM) by noise (for a 1 Hz bandwidth), and P_{signal} (units of dBm) is the power of the carrier. Usually, $\mathcal{L}(f)$ is expressed in decibels as $\log_{10} \mathcal{L}(f)$ and its units are dBc/Hz (dB below the carrier, where the power in each point was measured considering a 1 Hz bandwidth). Devices shall be characterized by a plot of $\mathcal{L}(f)$ versus Fourier (or offset) frequency f, as shown in [62], for example. The fact that Equation (40) is approximately half of $S_\Phi(f)$ (for small mean squared phase deviation [60]), led to a recent redefinition of $\mathcal{L}(f)$ as:

$$\mathcal{L}(f) = \frac{1}{2}S_\Phi(f). \tag{41}$$

3.3.3. Conversion between Frequency and Time Domain—Power Law Spectral Densities

It has been shown, from both theoretical considerations and experimental measurements, that spectral densities due to random noise can be accurately modeled by power-laws, where the spectral densities vary as a power of the Fourier frequency f [63]. More specifically, $S_y(f)$ and $S_\Phi(f)$ can be written as the sums:

$$S_y(f) = \sum_{\alpha=-2}^{+2} h_\alpha f^\alpha,\ 0 < f < f_h, \tag{42a}$$

$$S_\Phi(f) = \sum_{\alpha=-2}^{+2} f_c^2 h_\alpha f^{\alpha-2},\ 0 < f < f_h, \tag{42b}$$

with f_h as the high-frequency cut-off of an infinitely sharp low-pass filter (frequency after which the $S_\Phi(f)$ spectrum becomes flat) and h_α as the constants associated with each type of noise process. The two equations are related by Equation (39).

The random fluctuations are often represented by the sum of five independent noise processes: random walk frequency noise (constant h_{-2}, f^{-4} dependence on Fourier frequency), flicker of frequency (h_{-1}, f^{-3} dependence on frequency), white frequency noise (h_0, f^{-2} dependence on frequency), flicker of phase (h_1, f^{-1} dependence on frequency), and white phase noise (h_2 and f^0 dependence on frequency), as can be seen in Figure 5.

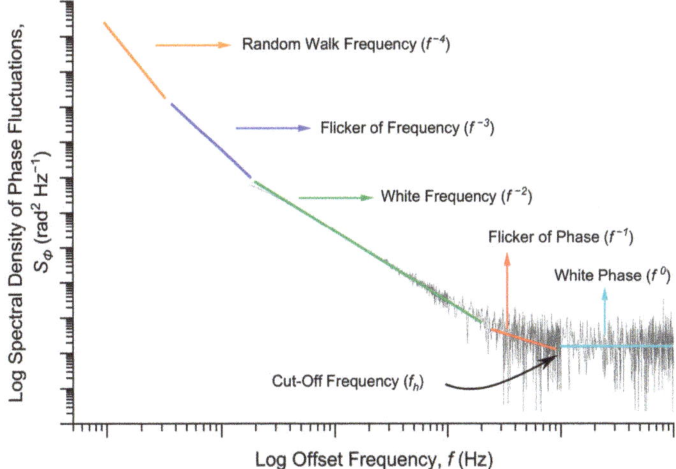

Figure 5. Power-laws of the spectral density of phase fluctuations, $S_\Phi(f)$, and corresponding noise mechanisms. A typical experimental curve of $S_\Phi(f)$ is plotted in light grey. The cut-off frequency (f_h) is also indicated.

Cutler and Searle derived the calculation of the Allan deviation, $\sigma_y(\tau)$, from the spectral density of the frequency fluctuations, $S_y(f)$, using the formula [60,63,64]:

$$\sigma_y(\tau) = \left[\int_0^\infty S_y(f)|H(f)|^2 df\right]^{\frac{1}{2}} = \left[2\int_0^{f_h} S_y(f)\frac{\sin^4(\pi\tau f)}{(\pi\tau f)^2}df\right]^{\frac{1}{2}}, \tag{43}$$

where $|H(f)|^2 = 2\frac{\sin^4(\pi\tau f)}{(\pi\tau f)^2}$ is the transfer function of an infinitely sharp low-pass filter, with $2\pi f_h\tau \gg 1$, used to count and average the frequency values for a time τ [60,63].

This integral can be analytically (for simple cases) or numerically solved as shown in [56,65–67] using the expressions for $S_y(f)$ given by Equation (42a). For the case of Equation (43), one obtains:

$$\sigma_y(\tau) = \sqrt{h_{-2}A\tau^1 + h_{-1}B\tau^0 + h_0 C\tau^{-1} + h_1 D\tau^{-2} + h_2 E\tau^{-2}}, \tag{44}$$

with $A = \frac{2\pi^2}{3}$, $B = 2\ln(2)$, $C = \frac{1}{2}$, $D = \frac{1.038 + 3\ln(2\pi f_h \tau)}{4\pi^2}$, and $E = \frac{3 f_h}{4\pi^2}$. Different dependencies of the Allan deviation with τ can be found. Additionally, the results depend on the type of filter chosen (other types of filters, such as a single-pole filter, can be used).

Experimentally, the slopes observed in the $S_\Phi(f)$ (or the equivalent $\mathcal{L}(f)$) spectrum can be fitted to power laws of the form $y = ax^b$, from where the constants h_α are extracted and the Allan deviation $\sigma_y(\tau)$ for a certain integration time τ is computed.

For micromechanical resonators in an oscillator configuration, the Allan deviation can be measured directly by probing the frequency of the signal at the oscillator output. For this purpose, an appropriate frequency counter can be used. Alternatively, the phase noise of the MEMS oscillator can be measured by using a high-stability reference oscillator and a phase detector, which can provide the instantaneous phase difference between the reference and the MEMS oscillator under test. Alternatively, a spectrum analyzer can be used to measure the single-sideband phase noise, provided that the resolution bandwidth of the instrument is narrower than the width of the resonance peak of the MEMS resonator. Furthermore, the reference oscillator for such measurements must exhibit significantly lower phase noise than the device under test. However, when using a sweeping spectrum analyzer, the amplitude noise and the phase noise of the cantilever oscillator are convoluted. If the amplitude noise of the device under test (MEMS oscillator or resonator) is comparable to its phase noise, then other techniques involving phase detectors, PLLs or delay lines must be employed to isolate and measure the phase noise (these can be found in many signal source analyzers).

3.3.4. Physical Origins of Noise

Readers are referred to the works [57–59,68] for a thorough and detailed analysis of some of the intrinsic noise mechanics present in MEMS resonators. In these works analytical expressions for the phase noise (frequency domain) and Allan deviation (time domain) are derived for the cases of thermomechanical fluctuations, temperature fluctuations, adsorption-desorption events, defect motion and moment exchange in gaseous environments [58,59,68]. In [56], the authors show an extensive experimental work to study the frequency fluctuations induced by the experimental setup and also by nonlinear phenomena, such as nonlinear damping, nonlinear mechanical properties (Duffing parameter) or nonlinear mode coupling in the resonators. Interestingly, they conclude that the frequency fluctuations that have been consistently observed within the MEMS community have no known physical origin, yet, or that current practices must be rethought.

As shown in [68], the Allan deviation can be calculated for various functional forms of the phase noise density. For example, considering a frequency noise with a $1/f$ component, the frequency fluctuations are given by $S_y(f) = Z\left(\frac{f_c}{f}\right)$ or, equivalently, $S_\Phi(f) = Z\left(\frac{f_c}{f}\right)^3$ where Z is a scale factor which encloses physical information about the noise process. Solving Equation (43) for the $S_y(f)$ of $1/f$ noise, one obtains an Allan deviation given by:

$$\sigma_y(\tau) = \sqrt{2\ln(4Z\pi f_c)\tau^0}. \tag{45a}$$

Note that the Allan deviation caused by the $1/f$ frequency noise mechanism is independent of the integration time τ. Additionally, the constant h_{-1} (shown in Equation (44)) can be related with the physical parameters of this type of noise. Another example from [68] is related to frequency drift, in which the frequency fluctuations are given by $S_y(f) = Z\left(\frac{f_c}{f}\right)^2$ or, equivalently, $S_\Phi(f) = Z\left(\frac{f_c}{f}\right)^4$, and the Allan deviation becomes:

$$\sigma_y(\tau) = \sqrt{\frac{2\pi^2}{3}Zf_c\tau^1},\tag{45b}$$

being now dependent on τ^1, with Z related to h_{-2} (see Equation (44)). Finally, for white frequency noise fluctuations, $S_y(f) = Z$ or, equivalently, $S_\Phi(f) = Z\left(\frac{f_c}{f}\right)^2$, the Allan deviation becomes:

$$\sigma_y(\tau) = \sqrt{\pi Z\tau^{-1}},\tag{45c}$$

dependent on τ^{-1} and with Z related to h_0 (see Equation (44)).

To conclude this section, an expression derived in [59,68,69] for the Allan deviation caused by events of adsorption-desorption of residual molecules around the resonator is discussed. The Allan deviation caused the adsorption-desorption events, $\sigma_{y_{AD}}(\tau)$, is given by:

$$\sigma_{y_{AD}}(\tau) = \sqrt{\frac{N_a\tau_r\sigma_{occ}^2}{2}\tau^{-1}}\frac{\delta m_{AD}}{m_c},\tag{46}$$

where δm_{AD} is the adsorbed-desorbed molecule mass, m_c is the total mass of the resonator, σ_{occ} is the variance in the occupation probability of any given site on the surface, N_a is the Avogadro number and τ_r is the correlation time for an adsorption-desorption event, with τ the integration time. This equation can be important for mass sensing applications and will be used later in this review, in Section 4.3.

3.3.5. Minimum Detectable Frequency Shift, $\delta f_{\min}$

A key question for any frequency-based sensing application is: what is the minimum measurable frequency shift, δf_{min}, that can be resolved in a realistic noisy system? As suggested in [58], in principle δf_{min} should be a shift comparable to the mean square noise in an average of series of frequency measurements, or in other words:

$$\delta f_{min} \approx \frac{1}{N}\sqrt{\sum_{i=1}^{N}(f_i - f_0)^2},\tag{47}$$

where f_i represents the consecutive ith frequency measurements and f_0 is the natural resonance frequency of a particular resonant mode vibrating in a dissipative medium.

Alternatively, to determine this minimum frequency shift, δf_{min}, one can also resort to the definition of the Allan deviation $\sigma_y(\tau)$ measured for a given integration time τ (Equation (37)), and think of:

$$\delta f_{min} \approx \sigma_y(\tau)f_0,\tag{48}$$

where $\sigma_y(\tau)$ is the Allan deviation caused by all the sources of noise in the system. In the next section, we show how this value of the minimum detectable frequency shift can be used to determine the ultimate limits of detection of the resonating sensors.

4. Mass Sensing

MEMS devices microfabricated with a wide variety of geometries and materials have been extensively used as mass sensors with different operating conditions. The minimum mass resolution ever reported in the literature is 1.7 yg (the mass of a proton), achieved by using a carbon nanotube resonator and demanding experimental conditions, such as a low-noise measurement setup, high-vacuum and cryogenic temperatures [70]. Nanoelectromechanical (NEM) resonators working in high-vacuum conditions have also been used as mass spectroscopes, to detect single biological molecules that adsorb in real time on the surface of the resonators, one by one, several hundred times [71–73]. Despite these remarkable demonstrations of mass sensing in vacuum, in some applications the cantilever must be able to operate in dissipative media, such as air and liquid, since these fluids are often the substance to be tested or the support for the analytes to be detected.

In particular, microcantilevers have been used to detect single virus, organic vapors, prostrate-specific antigen, and other analytes of interest in very small concentrations [74–78]. For some mass sensing applications, such as in the case of biosensing, the cantilever is typically functionalized with molecules that act as receptors and have a high affinity and selectivity for the analyte to be detected, as illustrated in Figure 6. The strategy used to immobilize the molecular probes on the surface of the microcantilever depends on the chemical nature of both the molecules and the surface. Depending on the target analyte to be detected, one can use antibodies, synthetic oligonucleotides, locked nucleic acids (LNA), or aptamers as recognition elements [79]. An illustrative comparison between the scale of common biological entities and MEMS sensors is also given in Figure 6.

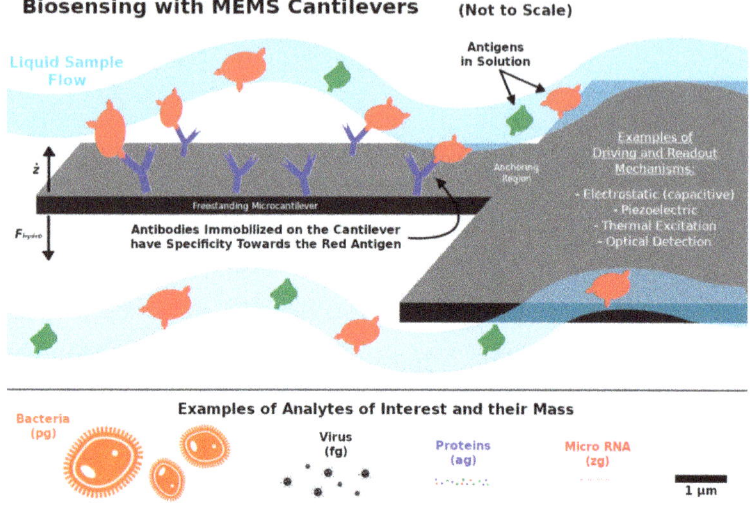

Figure 6. Illustration of a biosensing strategy with a microcantilever. The cantilever is functionalized with antibodies on the top surface (these are the biorecognition agents, which are chemically bound to the surface). Due to the specific binding of analytes (in this simple example, the red antigens are captured by the antibodies but the green ones are not), a resonance frequency shift or a static deflection will occur. On the bottom, the scale of the size and mass of bacteria, virus particles, proteins, and micro RNA strands is illustrated.

Some advantages of using microcantilevers as mass sensors include the possibility of probing the microscale, a high sensitivity due to their small dimensions, the ability to detect several analytes simultaneously, the fact that the detection does not require the analytes to be tagged (label-free detection), and the relative simplicity in interfacing the sensor with electronic readout and microfluidics [80].

4.1. Dynamic vs Static Sensing Modes

Two distinct mechanisms can be used for mass sensing with a microcantilever. In the first, called dynamic mode, the microcantilever oscillates with a constant frequency (as discussed in Section 2) and the binding of the target analyte induces a shift in this frequency (caused by changes in the effective mass and/or stiffness of the cantilever), which can be detected. In the second method, called static deflection mode, the cantilever is initially static and bends when the target analyte binds to the surface. This deflection, which can be detected, is caused by surface stress arising from the electrostatic repulsion between the molecules at the surface, or due to steric hindrance, for example. These two mass sensing modes are illustrated in Figure 7. In this work, only the dynamical mass sensing mode will be discussed. For a review of the static deflection mode, the reader is referred to [81–83].

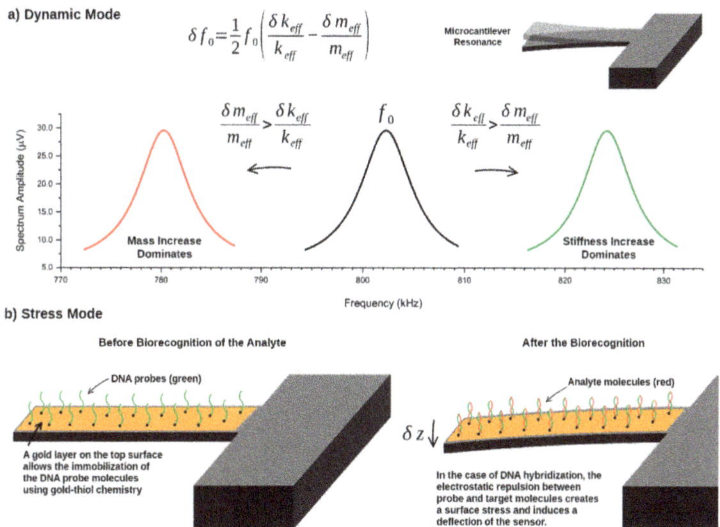

Figure 7. Mass sensing strategies. (**a**) Dynamic mode: the mass adsorbed on the surface of the cantilever causes changes in both its effective mass and stiffness, resulting in a shift in the resonance frequency, δf_0, to lower or higher values, depending on which effect dominates (red and green spectra). (**b**) Static deflection mode: the sensor deflects by a quantity δz due to intermolecular interactions at the surface of the sensor. In the case of a DNA biosensing experiment, the surface stress is due to the electrostatic repulsion between the molecules at the surface when complementary DNA (red strands) hybridizes with the initially immobilized DNA probes (green strands).

4.2. Mass Sensitivity

The dependence of the resonance frequency of the cantilever operating in vacuum with its effective mass and stiffness is obtained by differentiating Equation (15) with respect to these parameters, $\delta f_0 = \frac{\delta f_0}{\delta k_{\text{eff}}} \delta k_{\text{eff}} + \frac{\delta f_0}{\delta m_{\text{eff}}} \delta m_{\text{eff}}$. The final result is given by [84]:

$$\delta f_0 = \frac{1}{2} f_0 \left(\frac{\delta k_{\text{eff}}}{k_{\text{eff}}} - \frac{\delta m_{\text{eff}}}{m_{\text{eff}}} \right), \tag{49}$$

where k_{eff} and m_{eff} are the effective stiffness and mass of the resonant mode, respectively, and δk_{eff} and δm_{eff} are the corresponding infinitesimal changes. This equation can be applied to any flexural mode by using the appropriate effective mass (see Equation (16)). Equation (49) indicates that the resonance frequency f_0 of a mass sensor shifts when its effective mass and/or stiffness change due to the adsorption of an analyte on its surface. As discussed for the case of static mass sensing, the presence of analytes on the surface of the cantilever results in a surface stress. This surface stress also affects the resonance frequency of the cantilever and, therefore, the sensor behavior in the dynamical sensing mode. Several attempts have been made to relate the presence of surface stresses with changes in the effective stiffness of the cantilever [85,86], but a complete understanding of the physical mechanisms still remains elusive. This, in turn, impedes us from completely decoupling the simultaneous effects of the added mass and the surface stress (which impacts the effective stiffness) on the resonance frequencies [87].

However, when the adsorption of mass results in a distributed and low-density layer on the cantilever surface, or when the mass is adsorbed near its free end, the mass change

(δm_{eff}) effect dominates. Conversely, a high-density mass layer, or mass adsorbed near the support end of the microcantilever, causes a dominant stiffness-change effect (δk_{eff}) [88,89] These two competing effects are illustrated in Figure 7 and can be simultaneously measured by using two distinct vibrational modes of the resonator, provided that the adsorption position of the analyte is known [90]. In general, multimode excitation can be used to extract information about the mass, stiffness, and/or position of the analyte by measuring the resonance frequency shifts induced by the analyte on several flexural modes [91,92].

The mass sensitivity of a sensor operating in vacuum ($S_{mass,vac}$) is defined as the frequency shift induced by a mass added to the resonators (units of Hz kg^{-1}) and can be evaluated from Equation (49) as:

$$S_{mass,vac} = \frac{\delta f_0}{\delta m_{\text{eff}}} = -\frac{f_0}{2m_{\text{eff}}}, \tag{50}$$

where the stiffness change has been assumed to be negligible ($\delta k_{\text{eff}} \sim 0$). This equation shows that small cantilever effective mass and high natural resonance frequency are advantageous features to achieve high mass sensitivities in a vacuum.

To estimate the mass sensitivity for microcantilevers vibrating in fluids, one can at first combine Equations (15), (22), and (28a), thus expressing the resonance frequency with added mass and added damping, $\omega_{R,n}$, as:

$$\omega_{R,n} = \omega_{res,n}\left(1 + \frac{m_A}{m_0}\right)^{-\frac{1}{2}} = \left(\frac{k_{\text{eff}}}{m_{\text{eff},n}}\right)^{\frac{1}{2}}\left(1 - \frac{1}{2Q^2}\right)^{\frac{1}{2}}\left(1 + \frac{Lm_A}{\beta'_n m_{\text{eff},n}}\right)^{-\frac{1}{2}}, \tag{51}$$

where, as defined previously, $\omega_{res,n}$ is the resonance frequency of the nth mode of the intrinsically damped resonator, m_A and m_0 are the added mass by the fluid and the mass of the resonator, both per unit length, with $m_c = Lm_0 = \frac{m_{\text{eff},n}(\beta_n L)^4}{3} = \beta'_n m_{\text{eff},n}$, as shown in Equation (16).

The mass sensitivity in fluid, $S_{mass,fluid}$, for the nth mode, can then be obtained from Equation (51), by differentiating it with respect to these several parameters, $\delta f_R = \frac{\delta f_R}{\delta k_{\text{eff}}}\delta k_{\text{eff}} + \frac{\delta f_R}{\delta Q}\delta Q + \frac{\delta f_R}{\delta Lm_A}\delta Lm_A + \frac{\delta f_R}{\delta m_{\text{eff}}}\delta m_{\text{eff}}$. After some cumbersome calculations (see [93] for details) and admitting that the variations in the quality factor (defined in Equation (28b)) and stiffness are negligible ($\delta k_{\text{eff}} \sim 0$ and $\delta Q \sim 0$), one obtains:

$$\delta f_R = -\frac{f_R}{2}\left(\frac{L}{m_c + Lm_A}\right)\delta m_A - \frac{f_R}{2}\left(\frac{\beta'_n}{m_c + Lm_A}\right)\delta m_{\text{eff}}. \tag{52}$$

Finally, assuming that the properties of the fluid do not change, and therefore $\delta m_A \sim 0$ (see Equations (30a) and (31a) and Section 5 for cases where they do change), the mass sensitivity in fluids is given by:

$$S_{mass,fluid} = \frac{\delta f_R}{\delta m_{\text{eff}}} = -\frac{f_R}{2}\left(\frac{\beta'_n}{m_c + Lm_A}\right). \tag{53}$$

This expression indicates that a higher mass sensitivity in fluid is obtained with microcantilevers with reduced dimensions (eventually nanometric), which have small total mass m_c and high resonance frequency f_R, and for higher resonance modes (big values of $\beta'_n = \frac{(\beta_n L)^4}{3}$). Additionally, a smaller added mass m_A from the hydrodynamic load (see Equations (30) and (31)) is advantageous.

4.3. Limits of Detection (LoD)

The minimum mass that a resonator can detect is called limit of detection (*LoD*) and is defined as the mass of analyte that causes a frequency shift equal to three times the minimum detectable frequency shift δf_{min} [55], given a general sensitivity S:

$$LoD = 3\frac{\delta f_{min}}{S}.$$ (54)

As seen in Equations (47) and (48), δf_{min} depends on the stability of the resonance frequency, which, in its turn depends on the noise present in the system (Allan deviation). On the other hand, the mass sensitivity (S) depends on the operating conditions and can take the value of $S_{mass, vac}$ or $S_{mass,fluid}$ (Equations (50) and (53)).

In many applications, the sensor cannot operate under vacuum conditions. As seen, the operation of dynamic-mode mass sensors in gas or liquid media is challenging because the oscillation is damped by the fluid, decreasing S while increasing δf_{min}, and both factors contribute to the degradation of the mass *LoD*. The consequence is that MEMS flexural-mode resonators have typical limits of detection of hundreds of pg in air and even higher in liquids, rendering them ineffective for the detection of small analytes at low concentrations directly in liquid media.

In Section 3, the Allan deviation and phase noise were discussed in detail, and these formulas and methods allow the correct and complete characterization of the frequency stability of cantilever resonators and, in particular, mass sensors, as this stability has a critical effect on the *LoD*. However, in most of the mass sensing demonstrations found in literature, noise and/or frequency stability are very rarely mentioned. Instead, some different metrics, such as the quality factors or standard deviation, are arbitrarily used [94–109], making it difficult to compare the *LoDs* achieved in the different works. Works that show a detailed characterization of the frequency/phase stability can be found in [62,110–116]. Establishing coherent standards to be used when developing or assessing cantilever-based mass sensor should be encouraged [117].

To conclude this section, let us consider Equation (46) of Section 3 and combine it with Equation (50) (neglecting changes in stiffness). If it is assumed that: (i) in Equation (50), the shift in frequency caused by the adsorption-desorption events is the minimum detectable frequency shift ($\delta f_0 \sim \delta f_{min}$); (ii) the adsorbed-desorbed mass in Equation (46) contributes entirely for the change in effective mass of the cantilever in Equation (50), $\delta m_{AD} \sim \delta m_{\text{eff}}$; and (iii) $m_c = \beta'_n m_{\text{eff}}$ (Equation (16)), one gets for δf_{min}:

$$\delta f_{min} = \sigma_{y_{AD}}(\tau) f_0 Y \beta'_n,$$ (55)

with $Y = \left(2 N_a \tau_r \sigma_{occ}{}^2 \tau^{-1}\right)^{-\frac{1}{2}}$ as a numerical constant. This is a similar expression to that obtained in [113], and shows that the δf_{min} and consequently the *LoD* caused by mass adsorption-desorption events depend on the Allan deviation, the integration time, the natural frequency of the resonator, the resonance mode, and the physical constants in a non-trivial way.

As seen in Equation (54), the limit of detection can be decreased by increasing the sensitivity of the devices and/or reducing the minimum detectable frequency shift. However, as confirmed by Equations (53) and (55), both these parameters can have very complex dependences with geometry, vibration mode, physical processes of mass transfer or even the added mass by the surrounding dissipative fluid. This last term gains even more relevance when flexural vibration modes are used, since these modes displace a lot of fluid. This effect can be greatly reduced by using torsional or extensional modes in a mass sensor, for example, but also explored advantageously for rheological studies and the extraction of the fluid properties, as further discussed in Section 5. Flexural cantilevers can be useful for obtaining mass and rheology measurements simultaneously due to the close interaction between the resonator and the medium.

5. Viscosity Sensing

The vibrating microcantilever can be used for measuring the rheological properties of the fluid it is immersed in. In this section, some strategies to measure the viscosity and density of Newtonian fluids and the dynamic viscosity of non-Newtonian viscoelastic fluids are presented and discussed. The section starts with a brief revision on the response of viscoelastic materials to an applied shear stress.

5.1. Viscoelastic Materials

The viscosity, η, of a Newtonian incompressible fluid is defined as the proportionality constant between an applied shear stress to the fluid and the resulting shear strain rate. This is formally described by:

$$\tau_A = \eta \dot{\delta}_D, \tag{56}$$

where τ_A is the applied shear stress and δ_D is the shear strain, with $\dot{\delta}_D = \frac{d\delta_D}{dt}$ being the shear strain rate. This equation describes a purely viscous dashpot (hence the index "D"), where the shear force is proportional to the velocity. In a Newtonian fluid, the viscosity is constant and does not depend on the shear strain rate. Equation (56) is the simplest possible description of a viscous fluid and can be applied to some common liquids and gases, such as water or air.

However, most of the fluids of interest to biological applications do not follow Newton's law of viscosity (Equation (56)) and show a much more complex response to an applied shear stress. These fluids are termed non-Newtonian and their viscosity depends on the applied shear rate. A very common response of many solutions is shear-thinning, in which the viscosity decreases with an increasing shear rate. This arbitrarily complex behavior stems from the fact that these fluids have also an elastic response, in addition to the viscous response, and are hence also called viscoelastic fluids.

A Maxwell fluid is the simplest description of a viscoelastic fluid. This model considers that an elastic spring (obeying to Hooke's law) is added in series with the viscous dashpot, typical of purely viscous materials. The stress–strain relationship in an elastic spring is given by:

$$\dot{\tau}_A = G_0 \dot{\delta}_S, \tag{57}$$

where $\dot{\tau}_A$ is the applied shear stress rate, G_0 is the elasticity constant of the fluid, and δ_S is the shear strain of the spring (index "S") with $\dot{\delta}_S = \frac{d\delta_S}{dt}$ the shear strain rate. The total strain rate of the spring-dashpot series, $\dot{\delta}_{tot} = \dot{\delta}_D + \dot{\delta}_S$, when subjected to a shear stress is given by adding Equations (56) and (57):

$$\tau_A + \lambda \dot{\tau}_A = \eta \dot{\delta}_{tot}, \tag{58}$$

with $\lambda = \frac{\eta}{G_0}$ as a characteristic relaxation time. In the limit $G_0 \to \infty, \dot{\delta}_S \to 0$, Equation (58) reduces to Equation (56) and the fluid is purely viscous. In the case of $\eta \to \infty$, $\dot{\delta}_D \to 0$, Equation (58) reduces to Equation (57) and the fluid is purely elastic. Assuming that the applied shear stress and consequent total strain response are periodic with frequency ω and that the strain response of the material lags behind the applied stress by a phase φ, one gets:

$$\tau_A = \tau_0 e^{i(\omega t + \varphi)}, \tag{59a}$$

$$\delta_{tot} = \delta_0 e^{i\omega t}, \tag{59b}$$

$$G^* = \frac{\tau_0}{\delta_0} e^{i\varphi}, \tag{59c}$$

where G^* is a dynamic elastic modulus, defined by dividing the applied stress by the total strain of the system, and τ_0 and δ_0 are the amplitude of the shear stress and total strain, respectively. Equation (59c) reduces, respectively, to Hooke's and Newton's laws when the shear stress and strain are in phase ($\varphi = 0$, $G^* = G_0$) or in quadrature ($\varphi = \frac{\pi}{2}$, $G^* = \eta$). Viscoelasticity corresponds to any other value of φ. By substituting Equations (59a)–(59c) into Equation (58) and rearranging [32]:

$$\tau_0 e^{i\omega t}(1 + i\omega\lambda) = \delta_0 \eta i\omega \Rightarrow G^* = \frac{\omega^2\lambda^2 G_0}{1 + \omega^2\lambda^2} + i\frac{\omega\lambda G_0}{1 + \omega^2\lambda^2} \Rightarrow G^* = G' + iG'', \tag{60}$$

G^* is therefore defined as the sum of an elastic part, $G' = \frac{\omega^2\lambda^2 G_0}{1+\omega^2\lambda^2}$, and a viscous part, $G'' = \frac{\omega\lambda G_0}{1+\omega^2\lambda^2}$. The phase lag between the shear stress and the shear strain is given by $\varphi = \arctan\left(\frac{G''}{G'}\right)$.

The dynamic modulus G^* can be used to define a complex dynamic viscosity η^*, by equalling Newton's and Hooke's laws through the shear stress applied to the system:

$$\eta^* \dot{\delta}_{tot} = G^* \delta_{tot} \Rightarrow \eta^* = \frac{G^*}{i\omega} = \frac{G''}{\omega} - i\frac{G'}{\omega} \Rightarrow \eta^* = \eta' - i\eta''. \tag{61}$$

Therefore $\eta' = \frac{G''}{\omega} = \frac{\lambda G_0}{1+\omega^2\lambda^2}$ is the purely viscous part and $\eta'' = \frac{G'}{\omega} = \frac{\omega\lambda^2 G_0}{1+\omega^2\lambda^2}$ is the elastic viscosity. The components of both the dynamic modulus and dynamic viscosity are shown in Figure 8, as function of the frequency of the shear load.

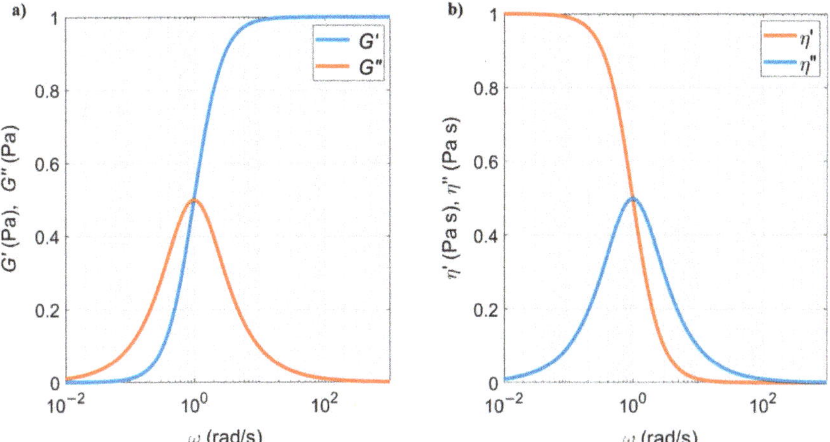

Figure 8. (**a**) Elastic (real) and viscous (imaginary) parts of the dynamic elastic modulus; (**b**) viscous (real) and elastic (imaginary) parts of the complex viscosity for $\lambda = 1$ s, $G_0 = 1$ Pa, and $\eta = 1$ Pa s in a Maxwell fluid.

At low excitation frequencies, the fluid is purely viscous ($\eta' = 1$ and $\eta'' = 0$), while at high frequencies it is purely elastic ($G' = 1$ and $G'' = 0$). When excited at intermediate frequencies, the Maxwell fluid is viscoelastic. The transition between viscous and elastic regimes occurs when the two curves cross each other.

5.2. Measuring Rheological Properties of Fluids Using Microcantilevers

5.2.1. Newtonian Fluids

Microcantilevers have been widely used to measure the rheological properties of Newtonian fluids. An early example of such a measurement is provided in [118], where the dependence of the resonance frequency of a cantilever on the viscosity of several aqueous solutions is used to monitor a chemical reaction. This section details how the rheological properties of Newtonian fluids can be measured using the theoretical framework described in Section 2. Two different strategies will be discussed.

The first strategy consists of measuring the amplitude response of the nth mode of the vibrating cantilever immersed in a viscus fluid. The experimentally measured amplitude response is fitted to Equation (29), allowing us to simultaneously extract the resonance frequency, $\omega_{R,n}$, and the quality factor, Q_n. Subsequently, the added inertial mass, m_A, and the added damping coefficient, c_V, are extracted from $\omega_{R,n}$ and Q_n using Equations (28a and (28b). Finally, the viscosity, η, and density of the fluid, ρ_f, can be simultaneously calculated using Maali's description of the hydrodynamic load by solving the system of equations given by (30) and (31). The sequence can be summarized as follows:

$$\text{fit } X \cong \left(\left(\omega_{R,n}{}^2 - \omega^2 \right)^2 + \left(\frac{\omega_{R,n}\omega}{Q_n} \right)^2 \right)^{-1/2} \Rightarrow \left\{ \begin{array}{c} \omega_{R,n} \\ Q_n \end{array} \right. \Rightarrow \left\{ \begin{array}{c} m_A \\ c_V \end{array} \right. \Rightarrow \left\{ \begin{array}{c} \eta \\ \rho_f \end{array} \right. .$$

This strategy has been described in [119,120], for example, and also used in [121] for the longitudinal modes of a microcantilever (not described here) and in [122,123] with a force applied at the free end of the cantilever. An interesting example of a self-excited microcantilever whose phase is used to measure the viscosity of different solutions is shown in [48].

Typically, iterative or numerical methods must be used to solve the system of equations that allow determining η and ρ_f. Additionally, an initial calibration is required in which the resonance frequency and quality factor of the cantilever vibrating in air ($\omega_{res,n}$) are measured. Furthermore, the fitted Equation (29) only describes the amplitude response around the resonance, and therefore only the rheological properties of the fluid at this frequency are measured.

To overcome this limitation, a second strategy has been developed in [124]. In this method, the amplitude and phase spectra are both experimentally measured for a wide range of excitation frequencies. Then, Equation (19), describing the complete transfer function of the nth mode of the forced damped harmonic oscillator oscillating in a fluid with resonance frequency $\omega_{R,n}$ and $\gamma = \frac{(c_0 + c_V)}{(m_0 + m_A)}$, is used:

$$\frac{A_0 m_{\text{eff}}}{F_0} e^{i\phi} = \frac{1}{1 - \left(\frac{\omega}{\omega_R} \right)^2 + i \frac{(c_0 + c_V)}{(m_0 + m_A)} \frac{\omega}{\omega_R{}^2}}. \tag{62}$$

Substituting Sader's result of $\omega_R = \omega_{res}\left(1 + \frac{m_A}{m_0} \right)^{-\frac{1}{2}}$ (Equation (28a)), using $e^{i\phi} = \cos(\phi) + i\sin(\phi)$, and considering a general ratio of amplitudes $\left| \frac{H(\omega)}{H_0} \right| = \frac{A_0 m_{\text{eff}}}{F_0}$, one gets the following complete transfer function:

$$\left| \frac{H(\omega)}{H_0} \right| (\cos(\phi) + i\sin(\phi)) = \frac{1}{1 - \left(\frac{\omega}{\omega_{res}} \right)^2 \left(1 + \frac{m_A}{m_0} \right) + i \frac{(c_0 + c_V)}{m_0} \frac{\omega}{\omega_{res}{}^2}}. \tag{63}$$

Finally, it is possible to use the experimentally measured amplitude $H(\omega)$ and phase $\phi(\omega)$ spectra to determine the added inertial mass and damping coefficient, m_A and c_V respectively, for all frequencies (not only limited to resonance), by equating the real and imaginary parts of the transfer function as:

$$\left\{ \begin{array}{l} Re\left(\frac{H(\omega)}{H_0} \right) = \left| \frac{H(\omega)}{H_0} \right| \cos(\phi(\omega)) = \left(1 - \left(\frac{\omega}{\omega_{res}} \right)^2 \left(1 + \frac{m_A}{m_0} \right) \right) \left| \frac{H(\omega)}{H_0} \right|^2 \\[2mm] Im\left(\frac{H(\omega)}{H_0} \right) = \left| \frac{H(\omega)}{H_0} \right| \sin(\phi(\omega)) = \left(-\frac{(c_0 + c_v)}{m_0} \frac{\omega}{\omega_{res}{}^2} \right) \left| \frac{H(\omega)}{H_0} \right|^2 \end{array} \right. \tag{64a}$$

$$\left\{ \begin{array}{l} m_A = \left[\left(1 - \left| \frac{H_0}{H(\omega)} \right| \cos(\phi(\omega)) \right) \left(\frac{\omega_{res}}{\omega} \right)^2 - 1 \right] m_0 \\[2mm] c_V = - \left| \frac{H_0}{H(\omega)} \right| \sin(\phi(\omega)) \frac{\omega_{res}{}^2}{\omega} m_0 - c_0 \end{array} \right. \tag{64b}$$

The rest of the procedure is identical to the first method: the viscosity, η, and density of the fluid, ρ_f, are determined from m_A and c_V using Equations (30) and (31). The fact that the rheological properties of the fluid can be measured for all range of frequencies is crucial for accurate measurements of viscoelastic fluids, as shown in Figure 8, since the properties of a viscoelastic fluid depend on the frequency of the shear load. This will be discussed in the next section.

To determine the limit of detection (LoD) of viscosity changes in a Newtonian fluid, Equation (54) can still be used, but where S is now the viscosity sensitivity ($S_{viscosity,fluid}$).

To determine the viscosity sensitivity of the microcantilever, the inertial added mass, given by Equations (30a) and (31a), can be differentiated with respect to the fluid viscosity and density, $\delta m_A = \frac{\delta m_A}{\delta \eta}\delta\eta + \frac{\delta m_A}{\delta \rho_f}\delta\rho_f$, to obtain:

$$\delta m_A = \frac{\pi}{4}wa_2\left(\frac{\rho_f}{2\eta\omega}\right)^{\frac{1}{2}}\delta\eta + \frac{\pi}{4}w^2a_1\delta\rho_f + \frac{\pi}{4}wa_2\left(\frac{\eta}{2\rho_f\omega}\right)^{\frac{1}{2}}\delta\rho_f. \tag{65}$$

This equation can then be substituted in Equation (52) to calculate the shift in resonance frequency caused by the added mass. Assuming that the variations in fluid density and effective mass of the cantilever are negligible ($\delta\rho_f \sim 0$ and $\delta m_{\text{eff}} \sim 0$), one obtains:

$$\delta f_R = -\frac{f_R}{2}\left(\frac{L}{m_c + Lm_A}\right)\left(\frac{\pi}{4}wa_2\left(\frac{\rho_f}{2\eta\omega}\right)^{\frac{1}{2}}\right)\delta\eta. \tag{66}$$

By rearranging this equation, substituting m_A from Equations (30a) and (31a) and considering the frequency of oscillation at resonance $\omega = 2\pi f_R$, the viscosity sensitivity in liquids is finally obtained:

$$S_{viscosity,fluid} = \frac{\delta f_R}{\delta\eta} = -\frac{f_R}{2}\left[\frac{\left(\frac{\rho_f}{4\pi f_R\eta}\right)^{\frac{1}{2}}}{\frac{4m_c}{a_2\pi Lw} + \rho_f w\frac{a_1}{a_2} + \left(\frac{\eta\rho_f}{\pi f_R}\right)^{\frac{1}{2}}}\right]. \tag{67}$$

The dependence of the sensitivity with the different geometrical parameters and fluid properties is complex and the design of the microcantilever can be optimized to reach an improved sensitivity and LoD. For example, these ideas have been applied by Dufour et al. for the development of a gas sensor in [125].

5.2.2. Viscoelastic Fluids

Measuring the viscoelastic properties of soft matter and fluids has become the focus of extensive research, given the key role these fluids play in biology and food manufacturing, just to cite two examples of relevant applications. Recently, several methods have emerged as powerful tools to investigate the dynamics and structure of soft matter or fluids at the micro or nanoscale [126,127]. One of these methods consists of using the microcantilever in a standard Atomic Force Microscopy (AFM) setup [128] and using the tip-sample interaction to probe the viscoelastic response of live cells [129,130] or soft surfaces [131].

In this section, the use of the microcantilever to measure the rheological properties of a viscoelastic fluid, in the context of the theoretical framework developed in the previous sections, is discussed.

This technique was initially proposed in [132,133], and the main idea is to incorporate the dynamic complex viscosity, $\eta^* = \eta' - i\eta''$, of Equation (61) into the hydrodynamic load (added inertial mass and viscous damping coefficient of Equations (30) and (31)), to get:

$$m_A = \frac{\pi}{4}\rho_f w^2\left(a_1 + \frac{a_2}{w}\sqrt{\frac{2(\eta' - i\eta'')}{\rho_f\omega}}\right), \tag{68}$$

$$c_V = \frac{\pi}{4}\rho_f w^2 \omega \left(\frac{b_1}{w}\sqrt{\frac{2(\eta' - i\eta'')}{\rho_f \omega}} + \frac{2(\eta' - i\eta'')}{\rho_f \omega}\left(\frac{b_2}{w}\right)^2 \right). \tag{69}$$

Using the identity of Equation (61), $\eta' - i\eta'' = \frac{G''}{\omega} - i\frac{G'}{\omega}$, and since $a_1 \sim 1$ and $a_2 \sim b_1$ [20] gives, after a cumbersome rearrangement [132],

$$m_A = \frac{\pi}{4}\rho_f w^2 + \frac{\pi}{2}b_2\frac{G'}{\omega^2} + \frac{b_1\pi}{2\sqrt{2}}\frac{\sqrt{\rho_f}w}{\omega}\left(\sqrt{\sqrt{G'^2 + G''^2} + G'}\right), \tag{70}$$

$$c_V = \frac{\pi}{2}b_2\frac{G''}{\omega} + \frac{b_1\pi}{2\sqrt{2}}\sqrt{\rho_f}w\left(\sqrt{\sqrt{G'^2 + G''^2} - G'}\right). \tag{71}$$

Figure 9 shows plots of the added mass and viscous damping per unit length as function of the shear load frequency (corresponding to the cantilever oscillation frequency, in the case of this work), as calculated by Equations (70) and (71), for some chosen values of the elastic and viscous components of the dynamic modulus. It can be observed that the added mass is more affected by variations in the elastic part of the dynamic modulus (G') (solid red and yellow lines), while the damping coefficient is more affected by variations in the viscous term of the dynamic modulus (purple and green dashed lines) (G'').

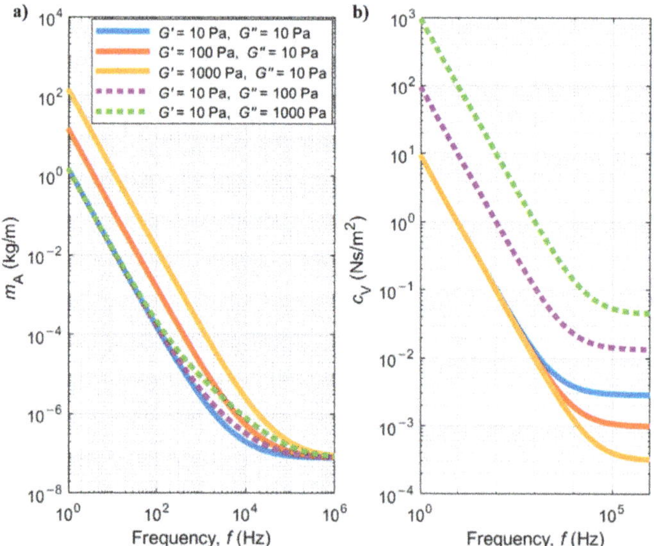

Figure 9. (**a**) Added mass and (**b**) viscous damping per unit length of a rectangular microcantilever (width $w = 10\ \mu$m) as a function of the frequency of oscillation in the fluid (water, $\rho_f = 1000\ \text{kg/m}^3$) and of the elastic and viscous parts of the dynamic modulus, calculated with Equations (70) and (71).

Experimentally, m_A and c_V are obtained from the measured amplitude, $H(\omega)$, and phase, $\phi(\omega)$, spectra using Equations (64a) and (64b), as described previously. Note that here, contrary to the method shown in the previous section, a prior knowledge of the fluid density is required, since the system of two equations is used to extract the two components of the dynamic modulus assuming a constant fluid density.

By defining the variables $B = \frac{b_1\pi}{2\sqrt{2}}\sqrt{\rho_f}w$, $C = \frac{\pi}{4}\rho_f w^2\omega$, and $D = \frac{\pi b_2}{2\omega}$, Equations (70) and (71) can be re-written as the following system of equations:

$$\begin{cases} m_A\omega = C + DG' + B\left(\sqrt{\sqrt{G'^2 + G''^2} + G'}\right) \text{ (a)} \\ c_V = DG'' + B\left(\sqrt{\sqrt{G'^2 + G''^2} - G'}\right) \text{ (b)} \end{cases} \qquad (72)$$

This system of equations can finally be solved to get the elastic and the viscous components of the dynamic modulus, G' and G'', respectively, as functions of the added mass, m_A, and damping coefficient, c_V [133,134]:

$$G''(\omega) = \frac{c_V}{D} - \frac{B}{D\sqrt{2D}}\sqrt{\sqrt{\left(\frac{B^2}{D} + 2(m_A\omega - C)\right)^2 + 4c_V^2} - \frac{B^2}{D} - 2(m_A\omega - C)}, \quad (73)$$

$$G'(\omega) = \frac{1}{D}\left(m_A\omega - C - \frac{B^2 G''}{c_V - DG''}\right). \qquad (74)$$

The two components of the dynamic modulus, calculated with Equations (73) and (74), are shown in Figure 10 as functions of the shear load frequency, and for some representative values of added mass and damping coefficient. In agreement with Figure 9, the elastic part of the dynamic modulus (G') mostly depends on the added mass (solid blue and yellow lines), while the viscous term of the dynamic modulus (G'') depends on the damping coefficient (purple and green dashed lines). It is also important to note that $G' \sim \omega^2$ and $G'' \sim \omega^1$, as predicted by the Maxwell model, and that $G' \to 0$ in the limit of low frequencies (viscous fluid). Therefore, the fluid behaves like a viscous liquid when G'' dominates and starts entering the viscoelastic regime when the values of G' approach those of G'' [133].

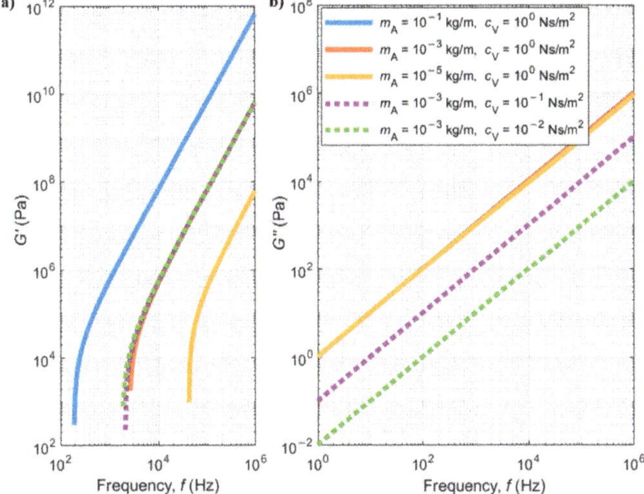

Figure 10. (**a**) Elastic and (**b**) viscous components of the dynamic modulus of a rectangular microcantilever (width w = 10 μm) as a function of the frequency of oscillation in the fluid (water, ρ_f = 1000 kg/m^3) and of the added mass and viscous coefficient, calculated with Equations (73) and (74).

The dynamic viscosity can be calculated from Equations (73) and (74) using $|\eta^*| = \left(\eta'^2 + \eta''^2\right)^{\frac{1}{2}} = \frac{1}{\omega}\left(G'^2 + G''^2\right)^{\frac{1}{2}}$. One of the greatest advantages of this method is the large range of frequencies that it can cover, since the microcantilever can be designed to operate and probe high-frequency viscoelastic fluid behavior that is not accessible by conventional rheometry [133].

6. Outlook and Further Challenges

This work presents a thorough analysis and discussion of the role of microcantilevers in mass and rheology sensing. Despite the great progress reported in the last two decades, many challenges, both theoretical and practical, remain to be tackled before unlocking the full foreseen potential for these microdevices.

Overcoming these challenges will likely require a complementary progress in both the fundamental understanding of the dynamical response of the microcantilever oscillating and changing mass in dissipative media and of the practical implementations of sensors based on microcantilevers.

Concerning the first set of challenges, a better understanding of the physical origins of noise, both intrinsic to the microcantilever or induced by the measuring apparatus, is required. As pointed out in Section 3.3.4, the noise mechanisms are yet not fully understood and the current practices for measuring noise may need to be rethought and standardized among the community. This would enable the development of benchmarks and metrics that can be used to better compare different techniques and tailor them to specific applications. In addition, the development of more complete models and the achievement of a deeper understanding of the dynamical non-linear response of the cantilever vibrating in a viscous medium are essential to fully exploit the behavior of non-linear systems and propose novel sensing approaches. Hysteresis, bifurcations, chaos, and energy transfer between coupled modes are only some of the effects that can arise from this highly non-linear system and that can be explored in applications. The dynamic interaction between the oscillating cantilever and a non-Newtonian viscoelastic fluid is still not well studied but has the potential of filling the gap between low-frequency characterization of these fluids (via bulk rheometry) and high-frequency measurements (via Brillouin or quartz resonators). Extract important information about the viscoelastic properties of the fluid may revolutionize point-of-care diagnostics and food processing. Extending our knowledge of the dynamic behavior in the presence of a non-zero flow velocity could also stimulate adoption in other emerging fields, such as in flow chemistry. Finally, fundamental knowledge of the kinetics of chemical reactions and optimal operation conditions that can be used for coating the beam or detecting new analytes is required.

Regarding the second set of challenges, one may mention the expected progress in microfabrication processes and materials, which can contribute to decrease costs and make high-volume industrial applications accessible, to develop new opto-electromechanical or biological functionalities to the microcantilever, or even allow the integration of sensors in curvilinear or complex three-dimensional (3D)-shaped surfaces. New geometries have started to emerge thanks to the enhanced fabrication techniques that are available today—see, for example, hollow cantilevers [135] or probes with integrated fibre optics for sensing [136]. Additionally, interfacing the beam with a proper circuitry, and integration with complementary metal-oxide-semiconductor (CMOS), is equally a crucial step for any commercial applications that will allow us to probe higher frequencies and ever-reducing time and space scales, in real-time, contributing actively to the promised next revolution of smart cities/homes and the Internet of Things (IoT), where common spaces can be filled with sensing devices that continuously monitor the environment and communicate with one another or with people.

Author Contributions: Conceptualization, J.M.; investigation, J.M., R.P. and P.P.; writing—original draft preparation, J.M., R.P. and P.P.; writing—review and editing, J.M., R.P., P.P. and B.T.; supervision, B.T.; project administration, B.T. All authors have read and agreed to the published version of the manuscript.

Funding: This research was funded by the European Union's Horizon 2020 research and innovation programme under the Marie Skłodowska-Curie grant agreement No. 842147.

Acknowledgments: The authors wish to acknowledge the support of the European Union to the MARS project funded by the European Union's Horizon 2020 research and innovation program under the Marie Skłodowska-Curie grant agreement No. 842147.

Conflicts of Interest: The authors declare no conflict of interest.

List of Symbols (in Order of Appearance in the Text)

L	cantilever length
w	cantilever width
h	cantilever thickness
t	time
x	space coordinate (distance from the cantilever support)
$q(x,t)$	time-varying distributed load acting on the beam at a distance x from the support, per unit length
$W(x,t)$	time-varying deflection of the beam at a distance x from the support
F_z	shear forces acting on the element of the beam
M_y	bending moment acting on the element of the beam
ρ	density of the structural material
A	area of rectangular beam cross section
I_z	second moment of area of the rectangular cross section beam
E	Young's modulus of the structural material
$\psi(t)$	temporal term solution of harmonic oscillation
$\Phi(x)$	spacial term solution of harmonic oscillation
$c_{1,2,3,4}$	constants of spacial term solution of harmonic oscillation
$f_{0,n}$	natural (undamped) resonance frequency of mode n
$\omega_{0,n}$	natural (undamped) radial resonance frequency of mode n
z	displacement of the one-degree-of-freedom microcantilever from the equilibrium position ($z = 0$)
$\dot{z}$	velocity of the one-degree-of-freedom microcantilever
$\ddot{z}$	acceleration of the one-degree-of-freedom microcantilever
k_{eff}	effective spring constant of the microcantilever
m_{eff}	effective mass of the nth resonant mode of the microcantilever
m_c	total mass of the microcantilever
c	intrinsic viscous damping coefficient
Q	quality factor
ω	excitation frequency
$F_0 e^{i\omega t}$	excitation harmonic force at ω, with amplitude F_0
A_0	amplitude of the motion at ω
ϕ	phase between the applied external force and the motion at ω
ω_{res}	resonance frequency of the nth mode of intrinsically damped resonators
m_0	mass of the cantilever per unit length
c_0	intrinsic viscous damping coefficient per unit length
$F_{hydro}(x,t)$	time-varying distributed hydrodynamic load, acting on the beam at a distance x, per unit length
m_A	added mass by interactions with the surrounding fluid, per unit length
c_V	added damping coefficient by interactions with the surrounding fluid, per unit length
$\omega_{R,n}$	resonance frequency of the nth mode of extrinsically damped resonators with added mass and damping
Q_n	quality factor of the nth mode

$\Gamma_{rect'}(\omega)$	real part of the hydrodynamic load acting on a microcantilever with rectangular cross section
$\Gamma_{rect''}(\omega)$	imaginary part of the hydrodynamic load acting on a microcantilever with rectangular cross section
ρ_f	density of the fluid
η	viscosity of the fluid
δ	thickness of the layer in which the velocity of the fluid drops by a factor of $1/e$
Re	Reynolds number
a_1, a_2, b_1, b_2	Maali's constants for $\Gamma_{rect}(\omega)$
τ	integration time
$\sigma_y(\tau)$	Allan deviation for time windows of duration τ
f_i	consecutive ith frequency measurements
f_c	nominal carrier frequency
$S_y(f)$	spectral density of frequency fluctuations
$S_\Phi(f)$	spectral density of phase fluctuations
$y_{rms}(f)$	measured root mean squared (rms) value of normalized frequency
$\Phi_{rms}(f)$	measured root mean squared (rms) value of normalized phase
BW	width of the frequency band in Hz
$P_{noise\ (1\ Hz)}(f)$	power density in one single sideband due to phase modulation by noise, for a 1 Hz bandwidth (dBm/Hz)
P_{signal}	total power of the carrier (dBm)
$\mathcal{L}(f)$	single-sideband phase noise, the ratio of $P_{noise\ (1\ Hz)}(f)$ to P_{signal} (dBc/Hz)
f_h	cut-off frequency of an infinitely sharp low-pass filter
$h_{-2}, h_{-1}, h_0,$ h_1, h_2	constants to fit power-laws to random walk frequency noise, flicker of frequency, white frequency noise, flicker of phase and white phase noise, respectively
A, B, C, D, E	numerical constants for conversion between frequency (spectral densities) and time (Allan deviation) domains
δf_{min}	minimum measurable frequency shift
LoD	limit of detection
δf_0	shift in the natural (undamped) resonance frequency
δf_R	shift in the damped resonance frequency of microcantilevers with added mass and damping;
δk_{eff}	infinitesimal change of the effective stiffness of the cantilever induced by the adsorbate
δm_{eff}	infinitesimal change of the effective mass of the cantilever induced by the adsorbate
δm_A	infinitesimal change of the added mass induced by the fluid
$\delta\eta$	infinitesimal change in the viscosity of the fluid
$\delta\rho_f$	infinitesimal change in the density of the fluid
S	sensitivity
$S_{mass,vac}$, $S_{mass,fluid}$	mass sensitivity in vacuum and in fluid
$S_{viscosity,fluid}$	viscosity sensitivity
$\tau_A, \dot{\tau}_A$	applied shear stress and shear stress rate
$\delta_D, \dot{\delta}_D$	shear strain and shear strain rate of a viscous dashpot
$\delta_S, \dot{\delta}_S$	shear strain and shear strain rate of an elastic spring
$\delta_{tot}, \dot{\delta}_{tot}$	shear strain and shear strain rate of the spring-dashpot series
G_0	elasticity constant of the fluid
λ	characteristic relaxation time of the fluid
ω	frequency of the applied shear stress and induced total strain response
φ	phase between applied stress and total strain response
τ_0	amplitude of the shear stress
δ_0	amplitude of the total strain response
G^*	dynamic elastic modulus
G', G''	elastic and viscous parts of the dynamic elastic modulus
η^*	complex dynamic viscosity
η', η''	viscous and elastic parts of the dynamic viscosity
$\left\|\frac{H(\omega)}{H_0}\right\|$	general ratio of amplitudes of the transfer function

References

1. Leissa, A.W.; Qatu, M.S. *Vibrations of Continuous Systems*; McGraw-Hill: New York, NY, USA, 2011.
2. Bao, M. *Analysis and Design Principles of MEMS Devices*; Elsevier Science: Amsterdam, The Netherlands, 2005.
3. Beards, C.F. *Structural Vibration: Analysis and Damping*; Arnold: London, UK, 1996.
4. Younis, M.I. *MEMS Linear and Nonlinear Statics and Dynamics*; Springer Science and Business Media LLC: Berlin/Heidelberg, Germany, 2011.
5. Meirovitch, L.; Parker, R. Fundamentals of Vibrations. *Appl. Mech. Rev.* **2001**, *54*, B100–B101. [CrossRef]
6. Kaajakari, V.; Mattila, T.; Oja, A.; Seppa, H. Nonlinear Limits for Single-Crystal Silicon Microresonators. *J. Microelectromech. Syst.* **2004**, *13*, 715–724. [CrossRef]
7. Dohn, S.; Svendsen, W.; Boisen, A.; Hansen, O. Mass and Position Determination of Attached Particles on Cantilever Based Mass Sensors. *Rev. Sci. Instruments* **2007**, *78*, 103303. [CrossRef] [PubMed]
8. Lifshitz, R.; Cross, M. Nonlinear Dynamics of Nanomechanical and Micromechanical Resonators. *Rev. Nonlinear Dyn. Complex.* **2009**, 1–52. [CrossRef]
9. Kovacic, I.; Brennan, M. *The Duffing Equation—Nonlinear Oscillators and Their Behaviour*; Wiley-VCH Verlag: Weinheim, Germany, 2011.
10. Villanueva, L.G.; Karabalin, R.B.; Matheny, M.H.; Chi, D.; Sader, J.E.; Roukes, M.L. Nonlinearity in Nanomechanical Cantilevers. *Phys. Rev. B* **2013**, *87*, 1–8. [CrossRef]
11. Venstra, W.J.; Westra, H.J.R.; Van Der Zant, H.S.J. Mechanical Stiffening, Bistability, and Bit Operations in A Microcantilever. *Appl. Phys. Lett.* **2010**, *97*, 193107. [CrossRef]
12. Zega, V.; Nitzan, S.; Li, M.; Ahn, C.H.; Ng, E.; Hong, V.; Yang, Y.; Kenny, T.; Corigliano, A.; Horsley, D.A. Predicting the Closed-Loop Stability and Oscillation Amplitude of Nonlinear Parametrically Amplified Oscillators. *Appl. Phys. Lett.* **2015**, *106*, 233111. [CrossRef]
13. Villanueva, L.G.; Karabalin, R.B.; Matheny, M.H.; Kenig, E.; Cross, M.C.; Roukes, M.L. A Nanoscale Parametric Feedback Oscillator. *Nano Lett.* **2011**, *11*, 5054–5059. [CrossRef]
14. Karabalin, R.B.; Masmanidis, S.C.; Roukes, M.L. Efficient Parametric Amplification in High and Very High Frequency Piezoelectric Nanoelectromechanical Systems. *Appl. Phys. Lett.* **2010**, *97*, 183101. [CrossRef]
15. Van Leeuwen, R.; Karabacak, D.M.; Van Der Zant, H.S.J.; Venstra, W.J. Nonlinear Dynamics of a Microelectromechanical Oscillator with Delayed Feedback. *Phys. Rev. B* **2013**, *88*, 1–5. [CrossRef]
16. Mestrom, R.M.C.; Fey, R.H.B.; Nijmeijer, H. Phase Feedback for Nonlinear MEM Resonators in Oscillator Circuits. *IEEE/ASME Trans. Mechatron.* **2009**, *14*, 423–433. [CrossRef]
17. Mouro, J.; Chu, V.; Conde, J.P. Dynamics of Hydrogenated Amorphous Silicon Flexural Resonators for Enhanced Performance. *J. Appl. Phys.* **2016**, *119*, 154501. [CrossRef]
18. Naik, T.; Longmire, E.K.; Mantell, S.C. Dynamic Response of a Cantilever in Liquid near a Solid Wall. *Sens. Actuators A Phys.* **2003**, *102*, 240–254. [CrossRef]
19. Sader, J.E. Frequency Response of Cantilever Beams Immersed in Viscous Fluids with Applications to the Atomic Force Microscope. *J. Appl. Phys.* **1998**, *84*, 64–76. [CrossRef]
20. Maali, A.; Hurth, C.; Boisgard, R.; Jai, C.; Cohen-Bouhacina, T.; Aimé, J.-P. Hydrodynamics of Oscillating Atomic Force Microscopy Cantilevers in Viscous Fluids. *J. Appl. Phys.* **2005**, *97*, 074907. [CrossRef]
21. Ekinci, K.L.; Yakhot, V.; Rajauria, S.; Colosqui, C.; Karabacak, D.M. High-Frequency Nanofluidics: A Universal Formulation of the Fluid Dynamics of MEMS and NEMS. *Lab Chip* **2010**, *10*, 3013–3025. [CrossRef]
22. Wambsganss, M.W.; Chen, S.S.; Jendrzejczyk, A.J. Added Mass and Damping of a Vibrating Rod in Confined Viscous Fluids. *J. Appl. Mech.* **1976**, *43*, 325–329. [CrossRef]
23. Tuck, E.O. Calculation of Unsteady Flows Due to Small Motions of Cylinders in a Viscous Fluid. *J. Eng. Math.* **1969**, *3*, 29–44. [CrossRef]
24. Chon, J.W.M.; Mulvaney, P.; Sader, J.E. Experimental Validation of Theoretical Models for the Frequency Response of Atomic Force Microscope Cantilever Beams Immersed in Fluids. *J. Appl. Phys.* **2000**, *87*, 3978–3988. [CrossRef]
25. Green, C.P.; Sader, J.E. Torsional Frequency Response of Cantilever Beams Immersed in Viscous Fluids with Applications to the Atomic Force Microscope. *J. Appl. Phys.* **2002**, *92*, 6262–6274. [CrossRef]
26. Green, C.P.; Sader, J.E. Frequency Response of Cantilever Beams Immersed in Viscous Fluids near a Solid Surface with Applications to the Atomic Force Microscope. *J. Appl. Phys.* **2005**, *98*, 114913. [CrossRef]
27. Van Eysden, C.A.; Sader, J.E. Small Amplitude Oscillations of a Flexible Thin Blade in a Viscous Fluid: Exact Analytical Solution. *Phys. Fluids* **2006**, *18*, 123102. [CrossRef]
28. Van Eysden, C.A.; Sader, J.E. Frequency Response of Cantilever Beams Immersed in Viscous Fluids with Applications to the Atomic Force Microscope: Arbitrary Mode Order. *J. Appl. Phys.* **2007**, *101*, 44908. [CrossRef]
29. Abdolvand, R.; Bahreyni, B.; Lee, J.E.-Y.; Nabki, F. Micromachined Resonators: A Review. *Micromachines* **2016**, *7*, 160. [CrossRef] [PubMed]
30. Labuda, A.; Kobayashi, K.; Kiracofe, D.; Suzuki, K.; Grütter, P.H.; Yamada, H. Comparison of Photothermal and Piezoacoustic Excitation Methods for Frequency and Phase Modulation Atomic Force Microscopy in Liquid Environments. *AIP Adv.* **2011**, *1*, 22136. [CrossRef]

31. Asakawa, H.; Fukuma, T. Spurious-Free Cantilever Excitation in Liquid by Piezoactuator with Flexure Drive Mechanism. *Rev. Sci. Instrum.* **2009**, *80*, 103703. [CrossRef]

32. Dufour, I.; Maali, A.; Amarouchene, Y.; Ayela, C.; Caillard, B.; Darwiche, A.; Guirardel, M.; Kellay, H.; Lemaire, E.; Mathieu, F. et al. The Microcantilever: A Versatile Tool for Measuring the Rheological Properties of Complex Fluids. *J. Sens.* **2012**, *2012*, 1–9. [CrossRef]

33. Pini, V.; Tiribilli, B.; Gambi, C.M.C.; Vassalli, M. Dynamical Characterization of Vibrating AFM Cantilevers Forced by Photothermal Excitation. *Phys. Rev. B* **2010**, *81*, 2–6. [CrossRef]

34. Lee, J.O.; Choi, K.; Choi, S.-J.; Kang, M.-H.; Seo, M.-H.; Kim, I.-D.; Yu, K.; Yoon, J.-B. Nanomechanical Encoding Method Using Enhanced Thermal Concentration on a Metallic Nanobridge. *ACS Nano* **2017**, *11*, 7781–7789. [CrossRef]

35. Zaghloul, U.; Piazza, G. Sub-1-Volt Piezoelectric Nanoelectromechanical Relays with Millivolt Switching Capability. *IEEE Electron Device Lett.* **2014**, *35*, 669–671. [CrossRef]

36. Rana, S.; Mouro, J.; Bleiker, S.J.; Reynolds, J.D.; Chong, H.M.; Niklaus, F.; Pamunuwa, D. Nanoelectromechanical Relay without Pull-in Instability for High-Temperature Non-Volatile Memory. *Nat. Commun.* **2020**, *11*, 1181–1910. [CrossRef] [PubMed]

37. Rodriguez, T.R.; Garcia, R. Theory of Q Control in Atomic Force Microscopy. *Appl. Phys. Lett.* **2003**, *82*, 4821–4823. [CrossRef]

38. Moreno-Moreno, M.; Raman, A.; Gómez-Herrero, J.; Reifenberger, R. Parametric Resonance Based Scanning Probe Microscopy. *Appl. Phys. Lett.* **2006**, *88*, 193108. [CrossRef]

39. Prakash, G.; Raman, A.; Rhoads, J.; Reifenberger, R.G. Parametric Noise Squeezing and Parametric Resonance of Microcantilevers in Air and Liquid Environments. *Rev. Sci. Instrum.* **2012**, *83*, 065109. [CrossRef]

40. Prakash, G.; Hu, S.; Raman, A.; Reifenberger, R. Theoretical Basis of Parametric-Resonance-Based Atomic Force Microscopy. *Phys. Rev. B* **2009**, *79*, 094304. [CrossRef]

41. Garcia, R. Dynamic Atomic Force Microscopy Methods. *Surf. Sci. Rep.* **2002**, *47*, 197–301. [CrossRef]

42. Miller, J.M.; Shin, D.D.; Kwon, H.-K.; Shaw, S.W.; Kenny, T.W. Phase Control of Self-Excited Parametric Resonators. *Phys. Rev. Appl.* **2019**, *12*, 044053. [CrossRef]

43. Mouro, J.; Tiribilli, B.; Paoletti, P. Nonlinear Behaviour of Self-Excited Microcantilevers in Viscous Fluids. *J. Micromech. Microeng.* **2017**, *27*, 095008. [CrossRef]

44. Khalil, H.K. *Nonlinear Systems, New International Edition*; Pearson: London, UK, 2014; p. XVIII.

45. Basso, M.; Paoletti, P.; Tiribilli, B.; Vassalli, M. Modelling and Analysis of Autonomous Micro-Cantilever Oscillations. *Nanotechnology* **2008**, *19*, 475501. [CrossRef]

46. Basso, M.; Paoletti, P.; Tiribilli, B.; Vassalli, M. AFM Imaging via Nonlinear Control of Self-Driven Cantilever Oscillations. *IEEE Trans. Nanotechnol.* **2010**, *10*, 560–565. [CrossRef]

47. Mouro, J.; Tiribilli, B.; Paoletti, P. A Versatile Mass-Sensing Platform with Tunable Nonlinear Self-Excited Microcantilevers. *IEEE Trans. Nanotechnol.* **2018**, *17*, 751–762. [CrossRef]

48. Mouro, J.; Tiribilli, B.; Paoletti, P. Measuring Viscosity with Nonlinear Self-Excited Microcantilevers. *Appl. Phys. Lett.* **2017**, *111*, 144101. [CrossRef]

49. Kim, W.; Kouh, T. Simple Optical Knife-Edge Effect Based Motion Detection Approach for a Microcantilever. *Appl. Phys. Lett.* **2020**, *116*, 163104. [CrossRef]

50. Ulčinas, A.; Picco, L.M.; Berry, M.; Hörber, J.H.; Miles, M.J. Detection and Photothermal Actuation of Microcantilever Oscillations in Air and Liquid Using a Modified DVD Optical Pickup. *Sens. Actuators A Phys.* **2016**, *248*, 6–9. [CrossRef]

51. Pooser, R.C.; Lawrie, B. Ultrasensitive Measurement of Microcantilever Displacement below the Shot-Noise Limit. *Optica* **2015**, *2*, 393–399. [CrossRef]

52. Boisen, A.; Thundat, T. Design & Fabrication of Cantilever Array Biosensors. *Mater. Today* **2009**, *12*, 32–38. [CrossRef]

53. Hwang, K.S.; Lee, S.-M.; Kim, S.K.; Lee, J.H.; Kim, T.S. Micro-and Nanocantilever Devices and Systems for Biomolecule Detection. *Annu. Rev. Anal. Chem.* **2009**, *2*, 77–98. [CrossRef]

54. Arlett, J.; Myers, E.; Roukes, M.L. Comparative Advantages of Mechanical Biosensors. *Nat. Nanotechnol.* **2011**, *6*, 203–215. [CrossRef]

55. Braind, O.; Dufour, I.; Heinrich, S.M.; Josse, F. *Resonant MEMS—Fundamentals, Implementation and Application*; Wiley-VCH Verlag: Weinheim, Germany, 2015.

56. Sansa, M.; Sage, E.; Bullard, E.C.; Gély, M.; Alava, T.; Colinet, E.; Naik, A.K.; Villanueva, G.; Duraffourg, L.; Roukes, M.L.; et al. Frequency Fluctuations in Silicon Nanoresonators. *Nat. Nanotechnol.* **2016**, *11*, 552–558. [CrossRef]

57. Feng, X.L.; White, C.J.; Hajimiri, A.; Roukes, M.L. A Self-Sustaining Ultrahigh-Frequency Nanoelectromechanical Oscillator. *Nat. Nanotechnol.* **2008**, *3*, 342–346. [CrossRef]

58. Ekinci, K.L.; Yang, Y.T.; Roukes, M.L. Ultimate Limits to Inertial Mass Sensing Based upon Nanoelectromechanical Systems. *J. Appl. Phys.* **2004**, *95*, 2682–2689. [CrossRef]

59. Vig, J.; Kim, Y. Noise in Microelectromechanical System Resonators. *IEEE Trans. Ultrason. Ferroelectr. Freq. Control.* **1999**, *46*, 1558–1565. [CrossRef] [PubMed]

60. *IEEE Standard Definitions of Physical Quantities for Fundamental Frequency and Time Metrology-Random Instabilities in IEEE Std Std 1139-2008*; IEEE: Piscataway, NJ, USA, 2009. [CrossRef]

61. Riley, W.J. *Handbook of Frequency Stability Analysis*; NIST: Gaithersburg, MD, USA, 2008; p. 31.

62. Pinto, R.M.R.; Brito, P.; Chu, V.; Conde, J.P. Thin-Film Silicon MEMS for Dynamic Mass Sensing in Vacuum and Air: Phase Noise, Allan Deviation, Mass Sensitivity and Limits of Detection. *J. Microelectromech. Syst.* **2019**, *28*, 390–400. [CrossRef]
63. Rutman, J.; Walls, F. Characterization of Frequency Stability in Precision Frequency Sources. *Proc. IEEE* **1991**, *79*, 952–960. [CrossRef]
64. Cutler, L.; Searle, C. Some Aspects of the Theory and Measurement of Frequency Fluctuations in Frequency Standards. *Proc. IEEE* **1966**, *54*, 136–154. [CrossRef]
65. Rutman, J. ; Characterization of Frequency Stability: A Transfer Function Approach and Its Application to Measurements via Filtering of Phase Noise. *IEEE Trans. Instrum. Meas.* **1974**, *23*, 40–48. [CrossRef]
66. Comment on "Characterization of Frequency Stability". *IEEE Trans. Instrum. Meas.* **1972**, *21*, 85. [CrossRef]
67. Allan, D.W. Time and Frequency Characterization Estimation and Prediction of Precision Clocks and Oscillators Allan IEEE Transaction on Ultrasonics Ferroelectrics. *IEEE Trans. Ultrason. Ferroelectr. Freq. Control.* **1987**, *34*, 647–654. [CrossRef]
68. Cleland, A.N.; Roukes, M.L. Noise Processes in Nanomechanical Resonators. *J. Appl. Phys.* **2002**, *92*, 2758–2769. [CrossRef]
69. Yang, Y.T.; Callegari, C.; Feng, X.L.; Roukes, M.L. Surface Adsorbate Fluctuations and Noise in Nanoelectromechanical Systems. *Nano Lett.* **2011**, *11*, 1753–1759. [CrossRef]
70. Chaste, J.; Eichler, A.; Moser, E.J.; Ceballos, G.; Rurali, R.; Bachtold, A. A Nanomechanical Mass Sensor with Yoctogram Resolution. *Nat. Nanotechnol.* **2012**, *7*, 301–304. [CrossRef] [PubMed]
71. Naik, A.K.; Hanay, M.S.; Hiebert, W.K.; Feng, X.L.; Roukes, M.L. Towards Single-Molecule Nanomechanical Mass Spectrometry. *Nat. Nanotechnol.* **2009**, *4*, 445–450. [CrossRef] [PubMed]
72. Hanay, M.S.; Kelber, S.; Naik, A.K.; Chi, D.; Hentz, S.; Bullard, E.C.; Colinet, E.; Duraffourg, L.; Roukes, M.L. Single-Protein Nanomechanical Mass Spectrometry in Real Time. *Nat. Nanotechnol.* **2012**, *7*, 602–608. [CrossRef] [PubMed]
73. Domínguez-Medina, S.; Fostner, S.; Defoort, M.; Sansa, M.; Stark, A.-K.; Halim, M.A.; Vernhes, E.; Gely, M.; Jourdan, G.; Alava, T.; et al. Neutral Mass Spectrometry of Virus Capsids above 100 Megadaltons with Nanomechanical Resonators. *Science* **2018**, *362*, 918–922. [CrossRef]
74. Gupta, A.; Akin, D.; Bashir, R. Single Virus Particle Mass Detection Using Microresonators with Nanoscale Thickness. *Appl. Phys. Lett.* **2004**, *84*, 1976–1978. [CrossRef]
75. Then, D.; Vidic, A.; Ziegler, C. A Highly Sensitive Self-Oscillating Cantilever Array for the Quantitative and Qualitative Analysis of Organic Vapor Mixtures. *Sens. Actuators B Chem.* **2006**, *117*, 1–9. [CrossRef]
76. Vančura, C.; Li, Y.; Lichtenberg, J.; Kirstein, K.-U.; Hierlemannb, A.; Josse, F. Liquid-Phase Chemical and Biochemical Detection Using Fully Integrated Magnetically Actuated Complementary Metal Oxide Semiconductor Resonant Cantilever Sensor Systems. *Anal. Chem.* **2007**, *79*, 1646–1654. [CrossRef]
77. Li, M.; Tang, H.X.; Roukes, M.L. Ultra-Sensitive NEMS-Based Cantilevers for Sensing, Scanned Probe and Very High-Frequency Applications. *Nat. Nanotechnol.* **2007**, *2*, 114–120. [CrossRef]
78. Zhou, X.; Wu, S.; Liu, H.; Wu, X.; Zhang, Q. Nanomechanical Label-Free Detection of Aflatoxin B1 Using a Microcantilever. *Sens. Actuators B Chem.* **2016**, *226*, 24–29. [CrossRef]
79. Drummond, T.G.; Hill, M.G.; Barton, J.K. Electrochemical DNA Sensors. *Nat. Biotechnol.* **2003**, *21*, 1192–1199. [CrossRef]
80. Conde, J.P.; Madaboosi, N.; Soares, R.R.G.; Fernandes, J.T.S.; Novo, P.; Moulas, G.; Chu, V. Lab-on-Chip Systems for Integrated Bioanalyses. *Essays Biochem.* **2016**, *60*, 121–131. [CrossRef] [PubMed]
81. Alvarez, M.; Lechuga, L.M. Microcantilever-Based Platforms as Biosensing Tools. *Analyst* **2010**, *135*, 827–836. [CrossRef] [PubMed]
82. Mathew, R.; Sankar, A.R. A Review on Surface Stress-Based Miniaturized Piezoresistive SU-8 Polymeric Cantilever Sensors. *Nano. Micro Lett.* **2018**, *10*, 1–41. [CrossRef] [PubMed]
83. Pinto, R.M.R.; Chu, V.; Conde, J.P. Label-Free Biosensing of DNA in Microfluidics using Amorphous Silicon Capacitive Micro-Cantilevers. *IEEE Sens. J.* **2020**, *20*, 1. [CrossRef]
84. Johnson, B.N.; Mutharasan, R. Biosensing Using Dynamic-Mode Cantilever Sensors: A Review. *Biosens. Bioelectron.* **2012**, *32*, 1–18. [CrossRef]
85. Lachut, M.J.; Sader, J.E. Effect of Surface Stress on the Stiffness of Cantilever Plates. *Phys. Rev. Lett.* **2007**, *99*, 206102. [CrossRef]
86. Karabalin, R.B.; Villanueva, G.; Matheny, M.H.; Sader, J.E.; Roukes, M.L. Stress-Induced Variations in the Stiffness of Micro- and Nanocantilever Beams. *Phys. Rev. Lett.* **2012**, *108*, 236101. [CrossRef]
87. Sohi, A.N.; Nieva, P.M. Size-Dependent Effects of Surface Stress on Resonance Behavior of Microcantilever-Based Sensors. *Sens. Actuators A Phys.* **2018**, *269*, 505–514. [CrossRef]
88. Tamayo, J.; Kosaka, P.M.; Ruz, J.J.; Paulo, A.S.; Calleja, M. Biosensors Based on Nanomechanical Systems. *Chem. Soc. Rev.* **2013**, *42*, 1287–1311. [CrossRef]
89. Tamayo, J.; Ramos, D.; Mertens, J.; Calleja, M. Effect of the Adsorbate Stiffness on the Resonance Response of Microcantilever Sensors. *Appl. Phys. Lett.* **2006**, *89*, 224104. [CrossRef]
90. Gil-Santos, E.; Ramos, D.; Martínez, J.; Fernández-Regúlez, M.; García, R.; San Paulo, Á.; Calleja, M.; Tamayo, J. Nanomechanical Mass Sensing and Stiffness Spectrometry Based on Two-Dimensional Vibrations of Resonant Nanowires. *Nat. Nanotechnol.* **2010**, *5*, 641–645. [CrossRef] [PubMed]
91. Olcum, S.; Cermak, N.; Wasserman, C.; Manalis, S.R. High-speed multiple-mode mass-sensing resolves dynamic nanoscale mass distributions. *Nat. Commun.* **2015**, *6*, 1–8. [CrossRef] [PubMed]

92. Malvar, O.; Ruz, J.J.; Kosaka, P.M.; Domínguez, C.M.; Gil-Santos, E.; Calleja, M.; Tamayo, J. Mass and Stiffness Spectrometry of Nanoparticles and Whole Intact Bacteria by Multimode Nanomechanical Resonators. *Nat. Commun.* **2016**, *7*, 13452. [CrossRef] [PubMed]

93. Dufour, I.; Heinrich, S.M.; Josse, F. Strong-Axis Bending Mode Vibrations for Resonant Microcantilever (Bio)Chemical Sensors in Gas or Liquid Phase. In Proceedings of the IEEE International Frequency Control Symposium and Exposition, Montreal, QC, Canada, 23–27 August 2004; pp. 193–199. [CrossRef]

94. Thundat, T.; Warmack, R.J.; Chen, G.Y.; Allison, D.P. Thermal and Ambient-Induced Deflections of Scanning Force Microscope Cantilevers. *Appl. Phys. Lett.* **1994**, *64*, 2894–2896. [CrossRef]

95. Gupta, A.; Akin, D.; Bashir, R. Detection of Bacterial Cells and Antibodies Using Surface Micromachined Thin Silicon Cantilever Resonators. *J. Vac. Sci. Technol. B Microelectron. Nanometer Struct.* **2004**, *22*, 2785. [CrossRef]

96. Xie, H.; Vitard, J.; Haliyo, S.; Régnier, S. Enhanced Sensitivity of Mass Detection Using the First Torsional Mode of Microcantilevers. *Meas. Sci. Technol.* **2008**, *19*. [CrossRef]

97. Ghatkesar, M.K.; Braun, T.; Barwich, V.; Ramseyer, J.-P.; Gerber, C.; Hegner, M.; Lang, H.P. Resonating Modes of Vibrating Microcantilevers in Liquid. *Appl. Phys. Lett.* **2008**, *92*, 043106. [CrossRef]

98. Parkin, J.D.; Hähner, G. Mass Determination and Sensitivity Based on Resonance Frequency Changes of the Higher Flexural Modes of Cantilever Sensors. *Rev. Sci. Instrum.* **2011**, *82*, 35108. [CrossRef]

99. Jiang, Y.; Zhang, M.; Duan, X.; Zhang, H.; Pang, W. A Flexible, Gigahertz, and Free-Standing Thin Film Piezoelectric MEMS Resonator with High Figure of Merit. *Appl. Phys. Lett.* **2017**, *111*, 023505. [CrossRef]

100. Yu, H.; Chen, Y.; Xu, P.; Xu, T.; Bao, Y.; Li, X. μ-'Diving Suit' for Liquid-Phase High-Q Resonant Detection. *Lab Chip* **2016**, *16*, 902–910. [CrossRef]

101. Lassagne, B.; García-Sánchez, D.; Aguasca, A.; Bachtold, A. Ultrasensitive Mass Sensing with a Nanotube Electromechanical Resonator. *Nano Lett.* **2008**, *8*, 3735–3738. [CrossRef] [PubMed]

102. Jensen, K.H.; Kim, K.; Zettl, A. An Atomic-Resolution Nanomechanical Mass Sensor. *Nat. Nanotechnol.* **2008**, *3*, 533–537. [CrossRef] [PubMed]

103. Davis, Z.; Abadal, G.; Helbo, B.; Hansen, O.; Campabadal, F.; Pérez-Murano, F.; Esteve, J.; Figueras, E.; Verd, J.; Barniol, N.; et al. Monolithic Integration of Mass Sensing Nano-Cantilevers with CMOS Circuitry. *Sens. Actuators A Phys.* **2003**, *105*, 311–319. [CrossRef]

104. Teva, J.; Abadal, G.; Torres, F.; Verd, J.; Pérez-Murano, F.; Barniol, N. A Femtogram Resolution Mass Sensor Platform, Based on SOI Electrostatically Driven Resonant Cantilever. Part I: Electromechanical Model and Parameter Extraction. *Ultramicroscopy* **2006**, *106*, 800–807. [CrossRef]

105. Lu, J.; Ikehara, T.; Zhang, Y.; Mihara, T.; Itoh, T.; Maeda, R. Characterization and Improvement on Quality Factor of Microcantilevers with Self-Actuation and Self-Sensing Capability. *Microelectron. Eng.* **2009**, *86*, 1208–1211. [CrossRef]

106. Narducci, M.; Figueras, E.; Lopez, M.J.; Gràcia, I.; Santander, J.; Ivanov, P.; Fonseca, L.; Cané, C. Sensitivity Improvement of a Microcantilever Based Mass Sensor. *Microelectron. Eng.* **2009**, *86*, 1187–1189. [CrossRef]

107. Linden, J.; Thyssen, A.; Oesterschulze, E. Suspended Plate Microresonators with High Quality Factor for the Operation in Liquids. *Appl. Phys. Lett.* **2014**, *104*, 191906. [CrossRef]

108. Badarlis, A.; Pfau, A.; Kalfas, A. Measurement and Evaluation of the Gas Density and Viscosity of Pure Gases and Mixtures Using a Micro-Cantilever Beam. *Sensors* **2015**, *15*, 24318–24342. [CrossRef]

109. Gfeller, K.Y.; Nugaeva, N.; Hegner, M. Micromechanical Oscillators as Rapid Biosensor for the Detection of Active Growth of Escherichia Coli. *Biosens. Bioelectron.* **2005**, *21*, 528–533. [CrossRef]

110. Ramos, D.; Mertens, J.; Calleja, M.; Tamayo, J. Phototermal Self-Excitation of Nanomechanical Resonators in Liquids. *Appl. Phys. Lett.* **2008**, *92*, 173108. [CrossRef]

111. Villarroya, M.; Verd, J.; Teva, J.; Abadal, G.; Forsen, E.; Murano, F.P.; Uranga, A.; Figueras, E.; Montserrat, J.; Esteve, J.; et al. System on Chip Mass Sensor Based on Polysilicon Cantilevers Arrays for Multiple Detection. *Sens. Actuators A Phys.* **2006**, *132*, 154–164. [CrossRef]

112. Verd, J.; Uranga, A.; Abadal, G.; Teva, J.L.; Torres, F.; López, J.; Pérez-Murano, F.; Esteve, J.; Barniol, N. Monolithic CMOS MEMS Oscillator Circuit for Sensing in the Attogram Range. *IEEE Electron. Device Lett.* **2008**, *29*, 146–148. [CrossRef]

113. Jin, D.; Li, X.; Liu, J.; Zuo, G.; Wang, Y.; Liu, M.; Yu, H. High-Mode Resonant Piezoresistive Cantilever Sensors for Tens-Femtogram Resoluble Mass Sensing in Air. *J. Micromech. Microeng.* **2006**, *16*, 1017–1023. [CrossRef]

114. Lochon, F.; Fadel, L.; Dufour, I.; Rebière, D.; Pistré, J. Silicon Made Resonant Microcantilever: Dependence of the Chemical Sensing Performances on the Sensitive Coating Thickness. *Mater. Sci. Eng. C* **2006**, *26*, 348–353. [CrossRef]

115. Burg, T.P.; Godin, M.; Knudsen, S.M.; Shen, W.; Carlson, G.; Foster, J.S.; Babcock, K.; Manalis, S.R. Weighing of Biomolecules, Single Cells and Single Nanoparticles in Fluid. *Nat. Cell Biol.* **2007**, *446*, 1066–1069. [CrossRef]

116. Ivaldi, P.; Abergel, J.; Matheny, M.H.; Villanueva, G.; Karabalin, R.B.; Roukes, M.L.; Andreucci, P.; Hentz, S.; Defaÿ, E. 50 nm Thick AlN Film-Based Piezoelectric Cantilevers for Gravimetric Detection. *J. Micromech. Microeng.* **2011**, *21*, 085023. [CrossRef]

117. Vig, J.R.; Walls, F. A Review of Sensor Sensitivity and Stability. In Proceedings of the IEEE/EIA International Frequency Control Symposium and Exhibition. Institute of Electrical and Electronics Engineers (IEEE), Kansas City, MO, USA, 7–9 June 2000; pp. 30–33.

18. Ahmed, N.; Nino, D.F.; Moy, V.T. Measurement of Solution Viscosity by Atomic Force Microscopy. *Rev. Sci. Instrum.* **2001**, *72*, 2731–2734. [CrossRef]
19. Boskovic, S.; Chon, J.W.M.; Mulvaney, P.; Sader, J.E. Rheological Measurements Using Microcantilevers. *J. Rheol.* **2002**, *46*, 891. [CrossRef]
20. Vančura, C.; Dufour, I.; Heinrich, S.M.; Josse, F.; Hierlemann, A. Analysis of Resonating Microcantilevers Operating in a Viscous Liquid Environment. *Sens. Actuators A Phys.* **2008**, *141*, 43–51. [CrossRef]
21. Castille, C.; Dufour, I.; Lucat, C. Longitudinal Vibration Mode of Piezoelectric Thick-Film Cantilever-Based Sensors in Liquid Media. *Appl. Phys. Lett.* **2010**, *96*, 154102. [CrossRef]
22. Youssry, M.; Belmiloud, N.; Caillard, B.; Ayela, C.; Pellet, C.; Dufour, I. A Straightforward Determination of Fluid Viscosity and Density Using Microcantilevers: From Experimental Data to Analytical Expressions. *Sens. Actuators A Phys.* **2011**, *172*, 40–46. [CrossRef]
23. Dufour, I.; Lemaire, E.; Caillard, B.; Debéda, H.; Lucat, C.; Heinrich, S.; Josse, F.; Brand, O. Effect of Hydrodynamic Force on Microcantilever Vibrations: Applications to Liquid-Phase Chemical Sensing. *Sens. Actuators B Chem.* **2014**, *192*, 664–672. [CrossRef]
24. Belmiloud, N.; Dufour, I.; Colin, A.; Nicu, L. Rheological Behavior Probed by Vibrating Microcantilevers. *Appl. Phys. Lett.* **2008**, *92*, 041907. [CrossRef]
25. Iglesias, L.; Boudjiet, M.; Dufour, I. Discrimination and Concentration Measurement of Different Binary Gas Mixtures with a Simple Resonator through Viscosity and Mass Density Measurements. *Sens. Actuators B Chem.* **2019**, *285*, 487–494. [CrossRef]
26. Liu, W.; Wu, C. Rheological Study of Soft Matters: A Review of Microrheology and Microrheometers. *Macromol. Chem. Phys.* **2017**, *219*, 1–10. [CrossRef]
27. Waigh, T.A. Advances in the Microrheology of Complex Fluids. *Rep. Prog. Phys.* **2016**, *79*, 074601. [CrossRef]
28. Garcia, R. Nanomechanical Mapping of Soft Materials with the Atomic Force Microscope: Methods, Theory and Applications. *Chem. Soc. Rev.* **2020**, *49*, 5850–5884. [CrossRef]
29. Hecht, F.M.; Rheinlaender, J.; Schierbaum, N.; Goldmann, W.H.; Fabry, B.; Schäffer, T.E. Imaging Viscoelastic Properties of Live Cells by AFM: Power-Law Rheology on the Nanoscale. *Soft Matter.* **2015**, *11*, 4584–4591. [CrossRef]
30. Efremov, Y.M.; Okajima, T.; Raman, A. Measuring Viscoelasticity of Soft Biological Samples Using Atomic Force Microscopy. *Soft Matter.* **2020**, *16*, 64–81. [CrossRef]
31. Haviland, D.B.; Van Eysden, C.A.; Forchheimer, D.; Platz, D.; Kassa, H.G.; Leclère, P. Probing Viscoelastic Response of Soft Material Surfaces at the Nanoscale. *Soft Matter.* **2016**, *12*, 619–624. [CrossRef]
32. Belmiloud, N.; Dufour, I.; Nicu, L.; Colin, A.; Pistre, J. Vibrating Microcantilever used as Viscometer and Microrheometer. In Proceedings of the 5th IEEE Conference on Sensors, Daegu, Korea, 22–25 October 2006; Volume 4, pp. 753–756. [CrossRef]
33. Youssry, M.; Lemaire, E.; Caillard, B.; Colin, A.; Dufour, I. On-Chip Characterization of the Viscoelasticity of Complex Fluids Using Microcantilevers. *Meas. Sci. Technol.* **2012**, *23*, 125306. [CrossRef]
34. Lemaire, E.; Heinisch, M.; Caillard, B.; Jakoby, B.; Dufour, I. Comparison and Experimental Validation of Two Potential Resonant Viscosity Sensors in the Kilohertz Range. *Meas. Sci. Technol.* **2013**, *24*, 084005. [CrossRef]
35. Meister, A.; Gabi, M.; Behr, P.; Studer, P.; Vörös, J.; Niedermann, P.; Bitterli, J.; Polesel-Maris, J.; Liley, M.; Heinzelmann, H.; et al. FluidFM: Combining Atomic Force Microscopy and Nanofluidics in a Universal Liquid Delivery System for Single Cell Applications and Beyond. *Nano Lett.* **2009**, *9*, 2501–2507. [CrossRef] [PubMed]
36. Van Hoorn, C.H.; Chavan, D.C.; Tiribilli, B.; Margheri, G.; Mank, A.J.G.; Ariese, F.; Iannuzzi, D. Opto-Mechanical Probe for Combining Atomic Force Microscopy and Optical Near-Field Surface Analysis. *Opt. Lett.* **2014**, *39*, 4800–4803. [CrossRef]

Review

Review: Cantilever-Based Sensors for High Speed Atomic Force Microscopy

Bernard Ouma Alunda [1] and **Yong Joong Lee** [2,*]

[1] School of Mines and Engineering, Taita Taveta University, P.O. Box 635-80300 Voi, Kenya; benalunda10@gmail.com
[2] School of Mechanical Engineering, Kyungpook National University, Daegu 41566, Korea
* Correspondence: yjlee76@knu.ac.kr

Received: 23 July 2020; Accepted: 12 August 2020; Published: 25 August 2020

Abstract: This review critically summarizes the recent advances of the microcantilever-based force sensors for atomic force microscope (AFM) applications. They are one the most common mechanical spring–mass systems and are extremely sensitive to changes in the resonant frequency, thus finding numerous applications especially for molecular sensing. Specifically, we comment on the latest progress in research on the deflection detection systems, fabrication, coating and functionalization of the microcantilevers and their application as bio- and chemical sensors. A trend on the recent breakthroughs on the study of biological samples using high-speed atomic force microscope is also reported in this review.

Keywords: microcantilever; atomic force microscope; ultra-short cantilevers; high-speed atomic force microscope; biosensors

---

1. Introduction

Since its debut in 1986, atomic force microscopy has evolved from a wobbly method to one of the most utilized tool for nanoscale characterizations. The early years of atomic force microscope (AFM) use were devoted to pushing the resolution boundary, to some unimaginable extent. However, during the last decade, research has been dedicated more to force measurements, identification and characterization of processes at the molecular level, force spectroscopy, and chemical force microscopy. The atomic force microscope has the ability to detect pico-newton scale intermolecular forces using a microcantilever as a force sensor thus aiding in the investigation of intermolecular interactions between receptors and ligands in biological systems in addition to mechanics of the single living cells and biomolecules [1–3].

The AFM microcantilevers are not restricted to the measurement of forces and displacements accurately and precisely, but owing to their ability to be used as a spring, they can be used as a motion sensor to detect nanoscale vibrations of various prokaryotic and eukaryotic cells [4]. From single-molecule to single-cell manipulation, the AFM became a multifunctional toolbox for observing and measuring various biophysical parameters of cellular and subcellular assemblies and machineries [5,6].

In addition to microcantilevers, microfabricated devices of different geometries such as flat pattern, micro-fluidic devices, micropillars, and microwells have been developed, and they are used to study the forces generated by various cells. The beauty of the micropillars, for example, is that they can be fabricated to the submicron range by using several nanobased techniques that include molding and lithography. In addition, the methods provide an easy way to vary the geometric parameters. Thus, the micropillars provide a versatile platform on which various cells can be examined [7,8]. Furthermore, the micropillars can be easily modified like the microcantilevers so that the cells are only attracted to the top surface only, while the remaining parts are covered with a repelling hydrogel layer. The micropillar-based sensors are promising tools for the measurements of biological samples in addition to other sensing applications including the flow of fluids and shear stresses [9]. The microchannel devices have also been developed as sensors for biological and chemical applications. These sensors are used for various biosensing applications because they can allow parallel processing of numerous samples within the same chip. It has been reported that microchannel biosensors can not only increase the detection sensitivity but also decrease the cost when compared to the conventional detection methods [10]. Other advantages of the microchannels-based sensors include real-time detection, high throughput, enhanced analytical performance, and portability. Moreover, it is possible to analyze most biomolecules in their solutions so as to imitate the natural environment close to in vivo. Thus, microchannel devices have become the most suitable methods for the development of some specific biosensors. Microchannels have critical length dimensions in the range of 1 to 100 mm and are characterized by a high surface area to volume ratio. The main limitation of microchannel biosensors is that most of the techniques involved in the fabrication of these devices are only capable of creating features with specific geometries. It is not possible to mimic those in their natural vasculature. The sensors, however, are reported to exhibit numerous excellent characteristics including low cost, portability, high sensitivity, and simple instrumentation [11]. Mi et al. [12] reported an amperometric lactate biosensor based on electrodes modified by Prussian blue. They immobilized the lactate oxidase enzyme using chitosan–carbon nanotubes. The biosensor was integrated with flow microchannels, and they were able to achieve a high sensitivity of 567 nA mM^{-1} mm^{-2}.

The atomic force microscope has several advantages when compared to other microscopic surface characterization techniques, such as optical fluorescent microscopy (OFM), optical confocal laser scanning microscopy (OCLSM), transmission electron microscopy (TEM), and scanning electron microscopy (SEM). Quantifiable and accurate surface height information, down to the sub-nanometer scale level, is attainable by using the atomic force microscope. On the contrary, OFM, OCLSL, SEM, or TEM cannot provide three-dimensional topographies. In addition, the AFM can permit imaging in air, aqueous, or even under vacuum conditions over a wide range of temperatures. The feasibility of observing the samples in liquid media at room temperature [13] and the capability of scanning an area of interest from the nanometer to the sub-millimeter scale open the possibility of studying many systems under physiological conditions from the macro level to cells and tissues [14] at an unrivaled resolution. Furthermore, the sample preparations are considerably easier compared to the TEM or SEM. Researchers can take advantage of the simple sample preparation for AFM, which allows studying living samples through surface imaging and mechanical mapping at the same time. For example, in cancerology, the AFM has been extensively used as an innovative diagnostic tool to explore the effects of cytotoxic drugs [15]. With a relatively simple setup and principle, AFM can probe the tissue dynamics at the nano-scale. After image acquisitions using the AFM, other surface mechanical/electrical/magnetic property characterizations can be performed in both quantitative and qualitative manners [16].

Compared to AFM capable of observing high-resolution of cellular processes in their native environments, the electron microscopy methods such as scanning electron microscope (SEM) and transmission electron microscope (TEM) can equally achieve nanometer level resolution but are limited in several aspects [17]. They require extensive sample preparations owing to the high-vacuum conditions required for the operations and limited sampling speed for possible real-time observations [18]. Furthermore, the traditional electron microscopes only allow imaging of samples in the unhydrated state. Even with the development of the environmental electron microscope, it is still not possible to image in a perfect liquid environment [19]. Other researchers have also reported potential damaging of cells by the electron irradiations [20]. The damages may include breakage of the molecular bonds, death of the cells, and generation of the reactive solvate electrons [21]. However, it has been recently reported that there is a possibility of employing electron microscope for the study of live bacterial cells if the radiation dosage is a few orders above the lethal dose needed to cause reproductive cell death [22].

AFM provides a technology that can also be integrated with other microscopic and spectroscopic techniques such as laser scanning confocal microscopy (LSCM) [23,24], total internal reflection fluorescence microscopy (TIRFM) [25–27], aperture correction microscopy (ACM) [28], correlative stimulated emission depletion microscopy (STEDM) [29–31], fluorescence lifetime imaging microscopy (FLIM) [32,33], stochastic optical reconstruction microscopy (STORM) [34,35], super-resolution fluorescence microscopy (SRFM) [36], tip-enhanced raman spectroscopy (TERS) [37–39], scanning near-field optical microscopy (SNOM) [40–42], and Förster resonance energy transfer (FRET) [43]. These correlative approaches offer a good spatial (nm) and high temporal (ms) resolution to study cellular and molecular biophysics. For example, Newton and his co-workers [24] developed a novel approach for quantifying the binding events of a single virus onto the surface receptors of a mammalian cell surface. They integrated a force–distance-based atomic force microscopy and fluorescence microscopy operated from below. By using the functionalized microcantilevers, they were able to gain an insight into the initial binding effects of a single pseudo-typed rabies virus to the avian tumor virus cells receptors artificially expressed in MDCK cells. A schematic of their set-up in Figure 1 shows the functionalized microcantilever using amine groups probing the MDCK cells with TVA950 receptors.

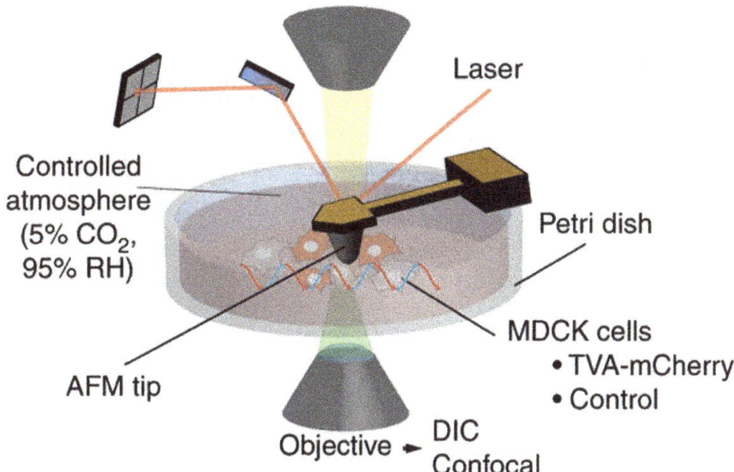

Figure 1. Combined force distance-based atomic force microscope and confocal microscopy for life science used to study the binding effects of a single virus onto the surface of a mammalian cell. Reprinted from the work in [24].

The microcantilever is the heart of the atomic force microscope and its existence can be associated with the invention of the atomic force microscope over three (3) decades ago [44]. The microcantilever sensor has a sharp probe attached to the free end. With the presence of a cantilever probe, the amplitudes, phases, and frequencies of various modes of resonance can be utilized. Moreover, the AFM can create nanoscale patterns with lithographic techniques using a conductive probe or obtain mapping of sample chemical identity when combined with optical spectroscopy techniques. They have attracted tremendous attention, and numerous biological [45], biomedical [46,47], physical, and chemical [48–51] applications have been demonstrated.

For biological applications, the microcantilever biosensors should be sensitive, fast, and flexible for identification of biomolecules and high-throughput screening in the pharmaceutical industries [52].

They have also been applied in the study of biosample stiffness measurements [53], surface morphological and mechanical analysis [54], and viscosity–density sensing in liquid media.

For physical detection, Markidou and his co-workers [55] developed a sensor that can measure the shear and elastic modulus of a soft material using a piezoelectric microcantilever. The microcantilever consisted of a highly piezoelectric material with a stainless steel material coating. Thus, in response to an applied voltage, the material bends creating an axial force capable of generating stress on the soft material. Other physical parameter detection measurements that have been conducted using microcantilevers include viscosity measurements, thermal analysis in picoliter solid samples, and detection and identification of trace amounts of biological species using a combination of micro-calorimetric spectroscopy and microcantilever thermal detectors [56–59].

The ability of microcantilevers to change their vibrational frequencies or levels of deflection upon adsorbing molecules on their surface makes them excellent probes that can act as chemical, physical, or biological sensors at the nanoscale. Changes in the vibrational frequency of micromechanical devices can be used to measure viscosity, density, and flow rates in various systems. Deflections of the cantilever are due to the stress from the molecular adsorption, which can be upward or downward depending on the type of chemical bonding of the molecule. In these systems, the change in the frequency of a microcantilever has been reported to be proportional to the magnitude of the adsorbed mass [48,60–62]. By using this phenomenon, the microcantilevers have been employed in the measurement of various physical phenomena such as humidity [63–67], temperature [68,69], and pressure [70,71]. For temperature sensing, the deflection of a silicon nitride microcantilever coated with a different material such as gold or aluminum is monitored. When there is a temperature change, the microcantilever bends due to the differences in the coefficients of thermal expansions of two materials [72,73]. Such sensors are found to exhibit a linear relationship between the deflection of the microcantilever and the changes in temperature as opposed to the thermal resistance or bi-metallic temperature sensors. The thermal resistance temperature sensors possess high thermal nonlinearity whereas the bi-metallic temperature sensors have a slow response and low resolution.

For humidity sensing, the microcantilevers can be coated with a water absorbing polymer [74]. For example, Singamaneni and his co-workers achieved a sensitive humidity sensor by coating a flexible silicon nitride microcantilever with a plasma-polymerized methacrylonitrile monolayer [75]. A linear relationship was observed as a function of time for both humidification and desiccation and a fast response time. Often, the microcantilever chemically or physically interacts with the environment resulting in an increase in the mass of the cantilever thus decreasing the resonant frequency. Sometimes the chemical reaction can cause an increase in the stiffness of the material. Their beauty, however, lies in their miniature structures, simplicity, and the possibility of mass production with good reproducibility [49].

In gas detection, for example, the gas molecules in the area surrounding the sensor are selectively adsorbed leading to an increase in the mass of the microcantilever that causes a proportional shift in the vibrational frequency according to the mass of the adsorbed gases. Another variable that can be used to quantify the amount of the adsorbed gases in the microcantilever is the change in the resonant frequency caused by the surface stresses. However, when compared to the mass loading, the effects of such stresses on the resonant frequency are found to be insignificant [76,77].

In this review, we look at the recent developments in microcantilever-based sensors in atomic force microscopy, latest improvements in various methods of microcantilever excitations for atomic force microscopes, progress in microcantilever fabrication, and modification suitable for biosensors or chemical sensors. We also address the progress in the development of ultrashort-microcantilevers and high-speed atomic force microscopy and their application to the study of life sciences.

2. Generic Operations of the Microcantilever

There are basically two modes of operations of the microcantilevers: static mode and dynamic mode. In the static mode, the microcantilever remains stationary, and its deflections only depend on surface stress variation. However in the dynamic mode, the microcantilever is externally actuated to oscillate about its natural resonant frequency. Thus, it is possible to accurately determine any mass change caused by adsorption of molecular layers. Typically, either the upper or both the upper and the bottom surfaces of the cantilever are coated with an active film followed by the close observations of the changes in the resonant frequencies or the quality factor caused by an addition of mass on the sensor [78,79]. The general overview of the operations of the microcantilevers is shown in Figure 2.

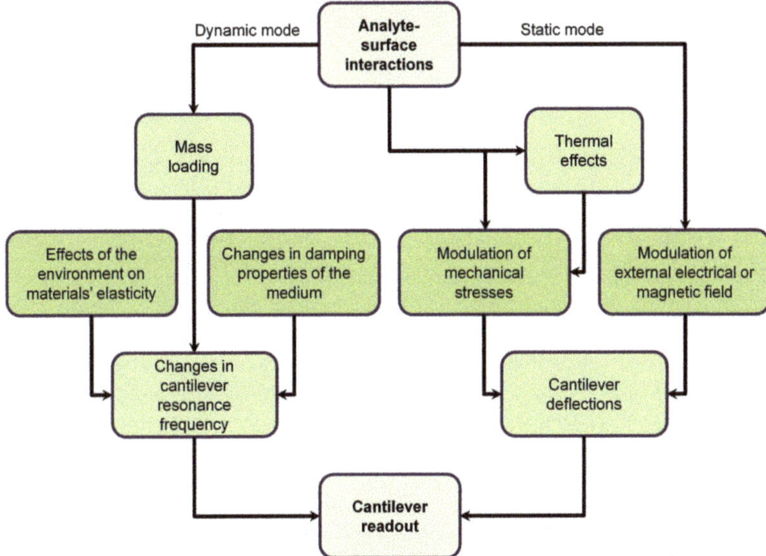

Figure 2. A flowchart describing the modes of operation and the principle of transduction for the microcantilevers. Adapted from the work in [80].

The microcantilever systems are simple mechanical devices with the dimension in the micrometer regime. Often, for ease of handling, the devices are attached at one end to a chip or support. Compared to other conventional sensors, microcantilevers are often preferred sensors due to the numerous advantages such as high precision, high sensitivity, rapid response, large dynamic response, miniaturization, high reliability, and large-scale integration [81–83]. Most importantly, these sensors can be microfabricated and mass produced [84], greatly lowering their costs. The wide use of microcantilever sensors has been contributed to their high sensitivity. The high sensitivity to surface phenomena is contributed by their large surface area-to-mass ratios [85]. Due to their high sensitivities, the response characteristics of the microcantilevers such as phase, amplitude of deflection, frequency changes, and quality factors can be easily detected using either electrical or optical detection means. Thus, it is possible to measure small forces in the pico-newton regime with relative ease. Some of the areas of applications are outlined in Table 1.

Table 1. Areas of applications for the microcantilevers.

S/No	Areas of Application	Examples
1	Biomedical applications [86,87]	Biosensors (DNA, antibodies proteins, viruses, and microorganisms) Diagnostics pH sensors
2	High frequency resonators [88–90]	Chemical sensors
3	Food production and safety [91]	Detection of heavy metals in water To detect concentrations of herbicides Changes in pH
4	RF switching [92–94]	Broadband switches Switches for wireless communication
5	Atomic force microscopy [3,95,96]	Live cells Reaction processes of DNA Biomolecules
6	Environmental monitoring [97]	Temperature detection Humidity detection Heat changes
7	Read and write storage devices [98]	Storage devices
8	Home land security [99]	Detection of terrorism weapons Explosives detection Monitor missile storage and maintenance needs
9	Energy [100,101]	Energy harvesters

3. Modes of Operations of Cantilever-Based AFM

The AFM typically consists of several components including: scanner, a cantilever with a sharp probe, a light source, an electronic feedback controller used to maintain a given set-point and a position sensitive photodetector (PSPD). Figure 3 shows a classic example of an AFM setup. A flexible cantilever typically fabricated from silicon or silicon nitride with a sharp tip at the free end is brought into close proximity with the sample (several angstroms) where it interacts with the sample surface due to the existence of the Van der Waal's forces. This interaction force, usually in the nano-newton (nN) range, causes a deflection of the flexible cantilever. The PSPD monitors and measures the amount of deflection of the flexible cantilever in proportion to the strength of the interactions. A transducer, usually piezoelectric stack actuator enables positioning of sample in the lateral direction and the cantilever probe in the out-of-plane direction with very precise motions. The AFM controller through an electronic feedback loop is then used to regulate the tip-sample interaction and to maintain a constant separation between the tip and the sample. The output of this feedback loop can be used to obtain topographical information. This simple instrument has turned out to be one of the most powerful tool that allows visualization of objects at nanoscale.

Different microscopes that are used to extract information about a sample surface have different "modes" of operations. For example, the backscattered or secondary electrons in scanning electron microscopes are utilized in order to image and provide information about topography and chemical compositions, respectively. On the other hand, the optical microscopes can operate in a polarized, dark-field, bright-field, or phase contrast mode, depending on the optical elements used in its operation. Similarly, the atomic force microscope can operate in a number of different modes. A few of the modes are highlighted in this section. The various operational modes are derived from different methods of exciting a cantilever or whether the microcantilever is in contact with the sample surface or not while in operation.

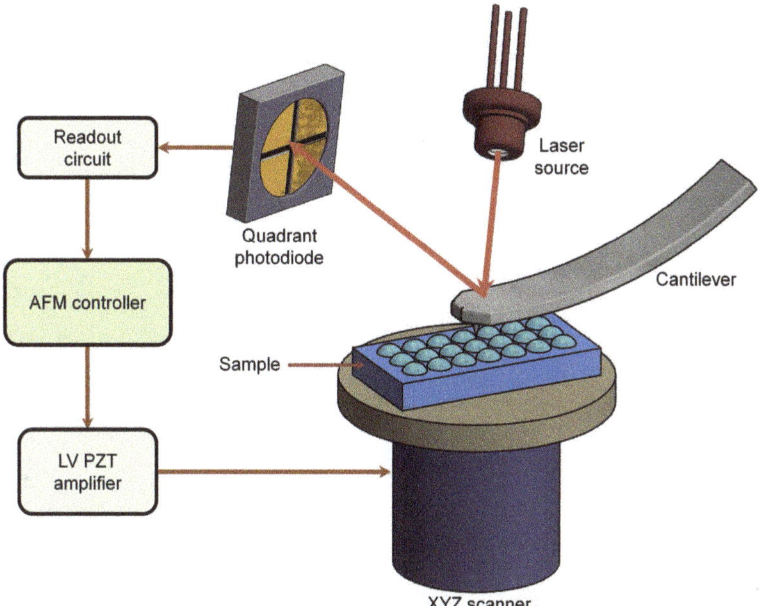

Figure 3. General schematic of a cantilever-based atomic force microscope (AFM) with the laser reflecting onto the photodetector.

Contact mode (CM) is the mother of all the imaging modes and still the most popular mode frequently used in many commercial atomic force microscopes [102]. Here, a sharp probe on the lower side of the micro-machined cantilever is constantly in contact with the sample surface as shown in Figure 4a. Therefore, the interaction between the cantilever and the sample is repulsive. Any variations in the sample topography are detected using an integrated optical sensor [103–106] that senses the deflection in the micro-machined cantilever owing to the variation in the interaction forces between the sharp tip and the sample. Significant frictional forces are generated when the cantilever raster scans the sample surface due to the applied force in the vertical direction. In contact mode, the operator is able to track stiffer and rougher surfaces better at higher scan speeds [107].

Non-contact mode (NCM) AFM has transformed the field of atomic force microscopy, thanks to Martin [108] and his research group who pioneered the technique just a year after the AFM was discovered. NCM operates in the attractive regime of the force–distance curve. Concisely, the cantilever together with a attached sharp probe oscillates above the sample surface at a preset scan speed [109,110] as shown in Figure 4b. During the scanning operation, the tip and sample distance should be maintained constant for the entire scanning period. This is made possible by tracking the changes in the phase, amplitude, or frequency of the cantilever induced as a result of the attractive forces (pico-Newton range). This interaction force is used in the feedback loop [108,111]. The small interaction forces offer the ability to image soft samples without damaging them. Furthermore, unless the tip crashes onto the sample surface, the probe remains undamaged and sharp during the whole scanning operation thereby increasing the operational lifetime of the probe. However, NCM usually has a limitation of slower scan speeds than contact mode in order to remedy the adsorbed fluid layer which is sometimes excessively thick to guarantee effective measurements [112]. This mode is not frequently used for biological sample characterization.

Friction force microscopy (FFM) is a form of static mode (contact mode). Here, the microcantilever tip and the sample surface is brought into repulsive contact. The FFM mode is often used for measuring the friction of a surface as the cantilever twists side to side by a torque, measured as the probe raster scans along the sample surface. The torsional changes are simultaneously recorded using the photodetector as a twisting of the cantilever together with the topographic features measured as a normal bending as shown in Figure 4c. FFM is commonly used to obtain a qualitative frictional contrast of the surface. However, the surface frictional coefficient can be calculated with an appropriate calibration of the lateral cantilever spring constant.

During the tapping mode operation of the AFM, simultaneous phase and topographical images can be acquired. In phase imaging (PI), the system monitors the phase lag between the signal that drives the cantilever oscillation and its output signal (see Figure 4d). Phase images can be used for assessing the information about the composition, adhesion, and viscoelastic properties of a sample surface.

Kelvin probe force microscopy (KPFM) was first recorded about three decades ago and enables imaging of surface potential at nanoscale for a variety of materials. In this technique, a contact potential difference is measured between a conductive AFM probe and the sample of interest. In KPFM, an external bias is applied to negate the contact potential difference while monitoring the cantilever amplitude at the resonant frequency (see Figure 4e).

When a microcantilever is coated with a magnetic material, the AFM can be used to study magnetic domain structures of a surface with high resolution up to 10 nm depending on the sharpness of the probe. The magnetized oscillating sharp probe first scans the surface to get topographical information, followed by an elevation of the probe off the surface by a set distance and recording the long range magnetic interaction force resulting in the magnetic force microscopy (MFM). A set-up showing the basic principle of MFM is shown in Figure 4f. In addition to magnetic forces, Van der Waals forces also act on the sample however they are weaker in magnitude. Thus, in MFM, the Van der Waals forces can be tapped to obtain the topography of the samples.

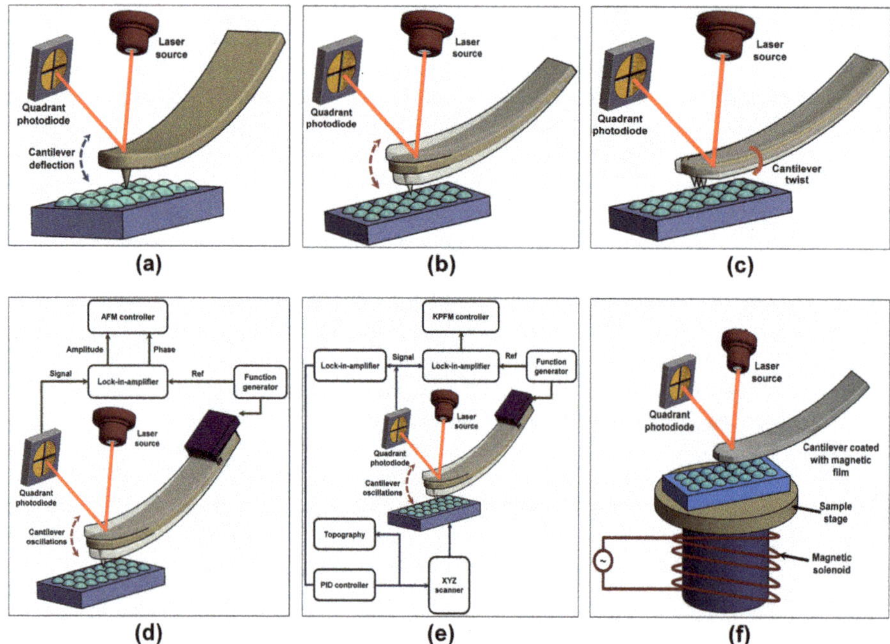

Figure 4. The AFM working modes: (**a**) contact mode, (**b**) tapping/dynamic mode, (**c**) frictional force microscopy, (**d**) phase mode, (**e**) Kelvin probe force microscopy, and (**f**) magnetic force microscopy.

In the mechanical mapping mode (MMM), the AFM measures the sample stiffness, in terms of Young's modulus values, through a nanoindentation technique. In AFM nanoindentation, the AFM collects indentation-force curves on the sample of interest. The obtained indentation–force curves can be fitted using linear elastic contact mechanical models, such as the Hertz model, in order to estimate Young's modulus. However, the technique can be very slow with one indentation per second taking over 60 min for single modulus mapping. Significant improvement in this aspect has been achieved by some AFM manufacturers now able to acquire high resolution force maps within a few minutes. At such a high acquisition speed, however, indentations of a viscoelastic time-dependent materials may lack accuracy. In addition, the tip radius, which is vitally important for the modulus calculation, will be increasingly inaccurate during the course of imaging.

Another imaging mode that has been developed for materials with high elastic modulus is the contact resonance imaging (CRI). In CRI mode, the sample is oscillated at the resonance frequency while the microcantilever tip is in contact with the sample. Mostly, the technique has been applied in the study of biological materials. CRI can provide information about the nano-mechanical properties from very small volumes. Moreover, the fact that CRI can measure viscous as well as elastic properties of materials makes it a suitable tool for studying composite materials.

In the multifrequency force microscopy (MFFM), multiple cantilever frequencies (higher harmonics and/or higher flexural eigenmodes) are excited to provide information about the tip-sample nonlinearities are recorded [113,114]. With MFFM, there is a potential of overcoming many hurdles including high throughput, material properties, and spatial resolution. Multiharmonic mode uses changes in the amplitude, the phase of the oscillator, and other appropriate harmonics in order to offer quantitative local property maps [115]. The mode enables the concurrent mapping of Young's modulus and the deformation and the topography of the sample [116]. Therefore, it can be used for investigating complex cellular and biomolecular structures to offer an in-depth quantitative multiparametric characterizations [117].

Viscoelastic mapping microscopy (VMM) is another mode that has its roots in research from multifrequency and bimodal AFM. This technique is a dynamic force-based mode that provides both imaging of the topography and maps of nanomechanical properties of soft-matter surfaces. In VMM, the cantilever is oscillated at two eigenmode frequencies. The first mode enables the recording of the surface features and loss tangent data whereas the second mode enables the recording of the frequency variations which can be used to obtain the stiffness of the sample. Thus, it is possible to obtain both the topography and map of the nanomechanical properties of soft-matter surfaces such as the contact stiffness and the modulus of elasticity [118]. Some advantages of this imaging mode are high spatial resolution, fast scanning, and low forces applied to the specimen.

Peak force tapping (PFT) was first introduced by Su and his co-workers [119] and is believed to possess advantages including the use of sinusoidal waves and subtraction of the background algorithm that allows the elimination of the parasitic deflection signal. With the use of PFT, it is possible to obtain the nanomechanical properties of samples at a faster scan speed with a very low minimum peak force in addition to high resolution mapping. The smaller force control is helpful in preventing any possible damage to the soft biological samples. Another benefit offered by the PFT is brought about by the use of tailor-made peak force microcantilevers with a longer tip length which allows a substantial distance between the cantilever and the sample, thus helping in the minimization of the hydrodynamic forces and background signal during operations [120]. Additionally, the PFT can also be employed in the study of biophysical properties by recording the single force–distance curve when the microcantilever probe is made to approach and retract from the sample surface. Thus, it is possible to characterize various mechanical properties, not necessarily limited to the adhesion force and dissipation energy.

The exhaustive list of AFM imaging modes is very long. Many of the recently developed modes are used for studying a number of biological samples, including proteins; small biological fibrils, like lipid membranes; amyloid fibrils; and viruses with the microcantilever as the sensing element [121,122].

Moreover, it has been demonstrated that malign and benign cell lines present significant differences in their viscoelastic response using MMM [123].

4. Methods of Cantilever Detection

Also of importance in the microcantilever systems are the readout methods that enable the determination of cantilever's mechanical state at any specific time. This can be done with a good accuracy using either optical or electrical techniques. The optical methods that have been adopted in atomic force microscopy are typically laser-based and include optical lever techniques and laser interferometry [124]. These techniques can be used to detect a deflection of the microcantilever in the sub-nanometer regime. Other than optical readouts, electronic readouts comprising capacitance [125], piezoresistivity [126,127], piezoelectricity [128–130], and metal-oxide semiconductor field effect [131] have also been used for cantilever array detections; they show a good progress but are limited in performance by microfabrication complexity and lack biocompatibility.

4.1. Electron Tunneling Method

In its infancy, the atomic force microscope microcantilever deflection was measured using the electron tunneling phenomenon. Here, the exponential dependence of the tunneling current between the scanning tunneling microscope (STM) tip and the cantilever on their separation distance is monitored. Typically, when a sharp, conducting tip is brought close to a conductive or a semi-conductive sample, electrons begin to tunnel from the sample to the tip or vice versa depending on the polarity of the bias voltage. The tunneling current varies with the tip–sample distance, and this variation in the tunneling current is the detector signal used to obtain the AFM images. This method offers a very high sensitivity but its main disadvantages are the reliance of electron tunneling on the surface conditions, difficulty in alignment especially in non-ambient conditions. Other limitations include the undesirable dependence of the tunneling detection on the effective spring constant, and changes in thermal drifts which greatly affect the force measurements. A schematic of the electron tunneling for measuring a cantilever deflection is shown in Figure 5a.

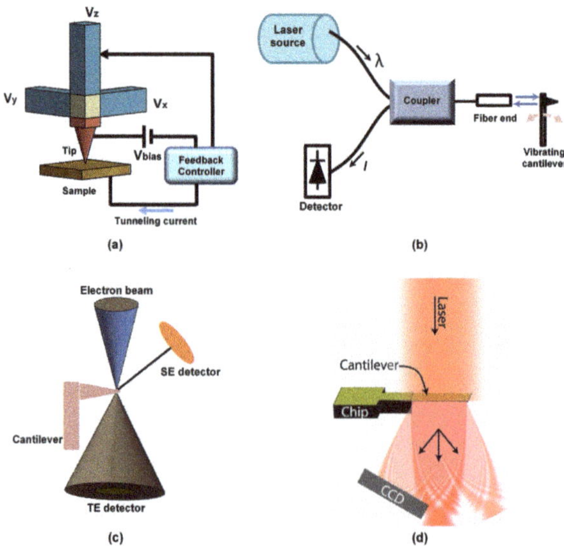

Figure 5. Schematics of (**a**) electron tunneling, (**b**) interferometric, (**c**) electron beam [132], and (**d**) optical diffraction microcantilever deflection detection systems [133].

4.2. Interferometry Method

Interferometric displacement detection (see Figure 5b) is another method for measuring the displacement of a micro-cantilever [134–136] with sub-nanometer accuracy and with high resolution, but it is bulky and expensive. In fiber-optic interferometry, the optical interference in the micro-sized cavity between the cantilever and the properly cleaved edge of a single-mode optical fiber is used to detect small cantilever deflections. It is based on guiding the light entirely through an optical fiber and using a beam splitter to route light beams while the cleaved end and the reflective surface of the cantilever act as mirrors to produce the interfering patterns. The rationale behind this concept is that light is always delivered and collected through the same aperture that is several micrometers in diameter [137]. The high sensitivity and precision of a correctly calibrated displacement measurement makes optical interferometry a suitable method for measuring the small displacements of a cantilever. However, using optical fibers may induce additional imaging errors due to thermal drifts when the imaging duration is long in addition to special handling of the equipment to prevent stress during the positioning procedure [138].

4.3. Electron Beam Detection Method

The development of small cantilevers for high-speed AFM requires that the spot size of the laser beam directed to the back of the cantilever to be small (~1 µm). If not, the laser will spill-over to the sample surface causing problems with the detection of small cantilever deflections. Wagner et al. [132] proposed an electron beam, instead of laser beam, for detection of the small deflections of the cantilever. The electron beam is focused into a smaller spot size of few nanometers nearly 100 times smaller than the spot size of the laser in the optical lever scheme permitting the detection of the deflection of smaller AFM cantilevers with ease (see Figure 5c).

4.4. Optical Diffraction Grating

Optical diffraction grating has been implemented for detecting the deflection of the microcantilever as shown in Figure 5d. Here, the reflected laser light forms a diffraction pattern in which the intensity is proportional to the cantilever deflection in atomic force microscopy [133,139,140].

4.5. Piezoelectric Method

In the piezoelectricity method, the electrical potential causes a mechanical stress on the microcantilever. The piezoelectric detectors have the advantages of consuming less power, easy to scale, possibility to be used in liquid environments, portability, and ability to withstand environmental damping. In addition, they can perform the dual function of actuation and sensing. Efficient actuation and elimination of optical interference from stray reflected light by the sample common in optical beam deflection method is a definite advantage. The on-chip actuation has the benefits of allowing multiple arrays of cantilevers on the same chip and permit feedback control at high frequencies [141]. However, thicker piezoelectric films are required for a significant output signals and also an electrical connection has to be made to the microcantilever. Recently, Moore and his co-workers [142] attempted to optimize the geometry of the piezoelectric microcantilever sensors to allow further miniaturization of such devices. They were able to achieve increased sensitivity and resonant frequency using optimized cantilever geometries compared to the conventional rectangular geometries. They formulated a means for utilizing the higher modes for the piezoelectric cantilevers by maximizing the microcantilever deflection and the measured piezoelectric charge response through strain partial distribution. Thus, they were able to increase the sensitivities of both the actuator and the sensor with a reduced sensor noise. Ruppert et al. [143] also demonstrated a method for optimizing the piezoelectric cantilever for multimode operations by altering the layout of the transducer depending on the strain mode shape without feed-through cancellation. A schematic typical of piezoelectric microcantilever deflection detection is shown in Figure 6a.

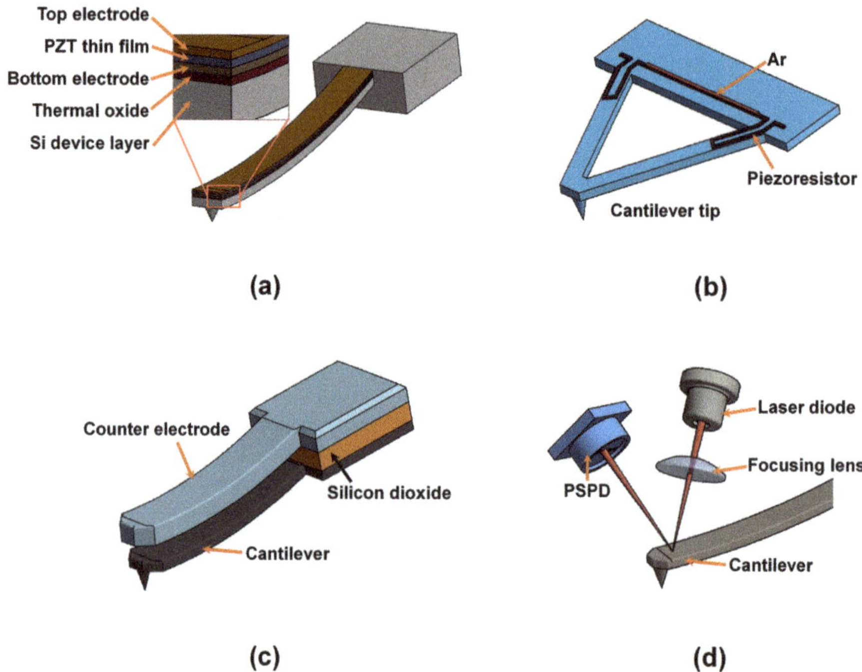

Figure 6. Schematics of (**a**) piezoelectric (adapted from [144]), (**b**) piezoresistive, (**c**) capacitive (adapted from [145]), and (**d**) optical lever microcantilever deflection detection systems [133].

4.6. Piezoresistive Method

The pioneering work in the use of piezoresistors to sense the microcantilever deflection was proven by Tortonese et al. [146] from Stanford University in 1991. Numerous piezoresistive cantilevers have been developed since then by different researchers [147,148]. The idea is to position the p-doped thin resistors at high stress locations along the length of the beam [149]. Due to the piezoresistive effect, mechanical stress, induced within the resistors, leads to changes in their specific resistance. By biasing, via a fixed current, this change is converted into an electrical voltage signal (see Figure 6b). The stress sensitivity of the p-doped resistors linearly depends on the operating current. A typical example of a material that exhibits such characteristics is the doped single crystal silicon [150,151]. A deflection of the microcantilever induces stresses and therefore strains in the piezoresistor resulting into a change in resistance. Usually, these types of detectors are appropriate for an array of microcantilevers sensors and lab-on-chip devices. However, such sensors require sophisticated electronics to minimize parasitic effects and temperature drift as well as to maximize the signal-to-noise ratio. Other limitations making this method unpopular are poor sensitivity, thermal drifts and conductance, and thermal and electronic fluctuation noises [152].

A significant improvement in performance of such cantilevers with respect to piezoresistive deflection sensitivity and temperature stability has been achieved by using an integrated Wheatstone bridge configuration [149,153]. For example, Yu and his co-workers [154] used the 192 Wheatstone bridges to improve the sensitivity and noise levels of the piezoresistive microcantilevers made from single-crystal, microcrystalline, and amorphous silicon by varying the geometry, doping levels, and the annealing temperatures to achieve improved noise levels by up to 65%. Rasmussen et al. [155] also used a mathematical model to improve the sensitivity of a piezoresistive read-out system and was able to achieve a minimum detectable surface stress range [156].

The advantage of piezoresistive detection scheme compared to standard optical techniques is that neither additional optical components nor laser alignment are needed. Moreover, the read-out electronics can be integrated on the same chip using CMOS fabrication process [157]. The piezoresistive detection is unaffected by optical artifacts arising from the surrounding medium. The piezoresistive read-out can also be accomplished by an integrated gold resistor [158]. Xia et al. [159] have developed coated active scanning probes with piezoresistive deflection detection capable of imaging in opaque liquids devoid of the need of an optical system. The "Positive 20" polymer used for coating can withstand harsh chemical environments with high acidity (e.g., 35% sulfuric acid).

4.7. Capacitive Detection Scheme

The capacitive detection method involves the measurement of the capacitance between two electrodes. Usually the separation distance between the targeted electrodes influences the sensitivity owing to the inverse proportionality of the measured capacitance and the physical distance between the electrodes. The capacitive detectors mostly find wide application in gaseous media due to the sensitivity of the device to changes in the effective dielectric constant of the media between the two electrodes. However, this detection mechanism is not commonly used because of its many limitations [160]. Accurate measurement of the microcantilever deflection requires that the dielectric material between the electrodes remains constant throughout the experiment, although this is not always possible. Moreover, miniaturization of the capacitive cantilever has the limitation of lowering the overall sensitivity because of the direct proportionality of the capacitance and the electrode areas [161,162]. Some of the outstanding advantages of the capacitive detection system are high sensitivity, absolute displacement measurements and simple electronic design configurations [161]. A schematic representation of a capacitive detection scheme is shown in Figure 6c.

4.8. Optical Lever Method

Meyer and Amer [105] pioneered the optical beam deflection (OBD) technique in 1988, and it has proven to be a very reliable and simple method for detecting cantilever deflections [103]. Generally, the cantilever deflection is measured from the displacement of the reflected laser beam from the back of the cantilever with a quadrant photodiode. The reflected beam forms an optical lever system which amplifies small cantilever displacements. The movements of the beam are detected by using a position sensing photodiode, typically a quadrant photodiode. Sub-nanometer deflection sensitivity is routinely achievable using the OBD sensor. Compared to other displacement measurement methods, ease of implementation, ability to use a variety of cantilevers, ease of alignment, and low sensor noise levels make the OBD sensor the most adopted deflection sensor in commercially available AFMs. The schematic of a typical OBD sensor consisting of a laser source, a reflective cantilever, and a position sensing photodiode (PSPD) is shown in Figure 6d. In principle, as the free end of the cantilever bends, the position of the laser spot on the position sensing photodiode changes. Due to the fact that the distance between the cantilever and the detector is large, a small movement of the cantilever causes a significantly larger change in the laser spot position on the photodetector.

One of the problems that OBD presents to the user is the need for metal-coating the cantilever backside after fabrication to improve laser reflectivity. This procedure may induce unwanted deformation due to the bimetallic effect [132,138] but is helpful in some aspects as will be discussed later in the photothermal excitation section. Laser alignment of the three elements involved (cantilever, photosensitive photodiode, and the laser source) is also a tedious exercise for any new cantilever loaded in the AFM head [163].

5. Microcantilever Excitation Methods

Operating the microcantilever especially in the dynamic mode requires a clean resonance of the cantilever. Spurious resonances from the mechanical elements in the microscope are common problem especially when the piezo-acoustic method is used. The results are undesirable artifacts in the acquired images. Therefore, in order to improve the quality of the images for various AFM applications using microcantilever as the sensing element, the choice of cantilever excitation is important. In this section, the recent advances for exciting the cantilevers and strategies are discussed.

5.1. Magnetic Excitation

Magnetic excitation is one of the mechanisms used to drive AFM cantilevers and different approaches have been developed [164–166]. Basically, the object of this mechanism is to create a magnetic cantilever or probe that is driven outwardly by coil or solenoid. The early atomic force microscope cantilever probes were made from magnetic materials such as iron wires before silicon cantilevers became widespread, and this allowed a simple means of magnetic excitation. Attaching a magnet onto the surface of a cantilever using glue is one of the traditional methods of providing magnetic properties to a cantilever [166]. However, the additional mass caused by the magnet and the epoxy for mounting has the disadvantage of reducing the resonant frequency of the free microcantilever. Moreover, the difficulty of crushing the magnets to the desired size and mounting on the cantilever surface using epoxy was a problem. In order to overcome these limitations, the backside of the cantilever is usually coated with a very thin layer (between 0.03 and 0.04 μm thick) of a magnetic material such as cobalt using cathodic sputtering [167]. Despite the appealing nature of magnetic excitation resulting in clearer resonant peaks in liquid environments, it has several drawbacks [168]. Problems of reproducibility because of varying geometries of the magnet and its magnetic properties. The mechanical properties of the cantilever are altered as a result of the integrated magnet and the uniform repetitive magnetic cantilever production is not easy. Bending angle and stiffness are also altered by the coating. The sample may be contaminated by the magnetic metal ions. The process requires additional expensive equipment for deposition of the metal coating. Additionally, the electromagnet might cause local heating to the liquid cell. Lately, magnetostrictive actuation has been proposed where a change in the magnetic state results in a dimensional change of the magnetic material. For low frequency cantilevers typically less than 1 MHz, it has proven to be the most efficient method for multi-mode actuation especially in a liquid environment.

5.2. Brownian Motion

The collisions of liquid particles with the cantilever from Brownian motion can also excite the cantilever, thermally yielding a smooth cantilever response. However, the Brownian motion signal is hardly greater than the AFM sensor noise and therefore wrong measurements may be obtained [169,170].

5.3. Sample Excitation

Some researchers have also attempted to excite the sample rather than cantilever [171]. The existence of complicated dynamics and sub-harmonics makes this technique very difficult to achieve [172].

5.4. Electrostatic Actuation

Electrostatic actuation is very versatile with the capabilities of actuating in both in-plane and out-of-plane directions. Here, the interaction forces between a conducting sample as well as a conducting cantilever probe are regulated by a bias voltage between the two [173]. One major limitation of electrostatic excitation method is the fact that both the cantilever and the sample need to be conductive. This condition greatly limits the sample that can be imaged as well as the materials that can be used for manufacturing the probe. Moreover, weak interaction forces require that a flexible cantilever must be used. Desbiolles et al. [174] demonstrated a method of exciting an encased cantilever using electrostatic technique (see Figure 7a) with a built-in electrode yielding smooth frequency resonance peaks both in air and liquid. The advantages of the built-in electrode drive and casing eliminates the need for any alignment, the use of only ac signal helps in reduction of the electrolytic production of gas bubbles, low noise, small cantilever amplitudes, thus reducing the tip–sample interaction forces and a reliable means to interpret the tip–sample interaction.

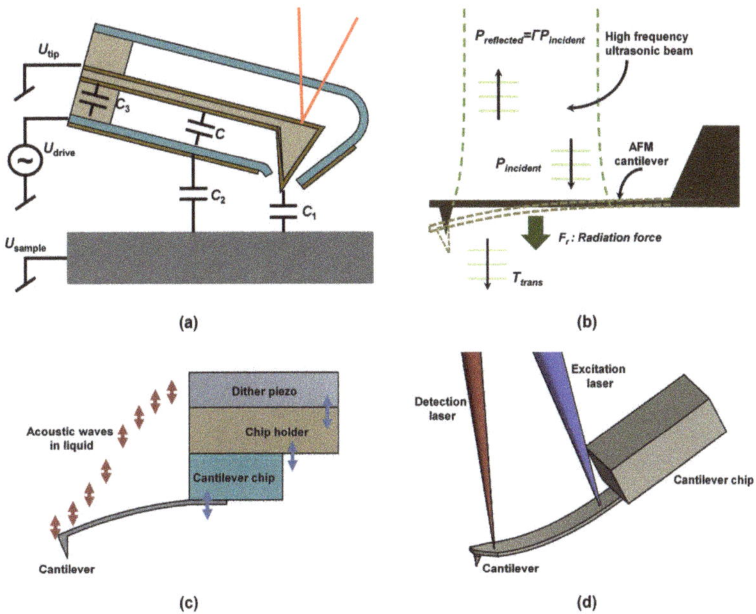

Figure 7. (**a**) A cross section showing the encased cantilever for electrostatic excitation. The capacitance C_1, C_2, and C_3 are parasitic capacitances whereas C is used for actuation [174], (**b**) acoustic radiation pressure excitation, (**c**) piezo-acoustic, and (**d**) photothermal excitation methods.

5.5. Acoustic Radiation Pressure Method

Some researchers have tried the acoustic radiation pressure method to excite the microcantilevers. One important merit of this technique is the ability to excite cantilevers of different materials and arbitrary shapes [175]. Basically, excitation is achieved by Langevin acoustic radiation pressure [176] which is created when a target cantilever is placed in the path of an acoustic wave beam at frequencies 100–300 MHz as shown in Figure 7b. When this pressure is focused at the focal plane of the lens, localized forces are generated to excite as well as evaluate the dynamic and static characteristics of the cantilever.

5.6. Piezo-Acoustic Excitation

In piezo-acoustic excitation, a small piezoelectric actuator is positioned close to the cantilever to indirectly excite the cantilevers. This method is by far the most common in AFMs. This is partly because of ease of implementation, ease of operation, and cost effectiveness. Usually, several parts are involved in the excitation process because the piezo-actuator cannot be mounted directly on to the cantilever. Therefore, the excitation begins from the cantilever holder to the cantilever via the chip on which the cantilever is mounted. Although this technique works relatively good in both air and vacuum environment, this in-direct excitation of the cantilever results in mixed resonances due to mechanical impedances of the piezo, cantilever holder, and cantilever base. This leads to the so called "forest of peaks" in liquid environment where the quality factor is low [177]. The unwanted mechanical resonances may not only affect the detection laser in the optical beam bounced technique but also makes it extremely difficult to choose the correct resonant frequency of the cantilever [178] because of the complex mechanical coupling. Moreover, it has been noted that piezo acoustic excitation may cause sample transience and movement of the molecular sample due to sonication [165]. Attempts to minimize the forest of peaks in low Q-factor environments by designing special cantilever holders have been fair at best [179,180]. Another method that have been proposed to help minimize the forest of peaks is by integrating a piezoelectric material such as zinc oxide (ZnO) on the cantilever [181]. The piezoelectric material provides a means of exciting the cantilevers at fast speed (greater 10 kHz) in tapping mode. The ZnO actuator can have dual function of exciting the cantilever and providing motion in the Z-direction for the tip–sample distance regulation. It is always desired that the measured quantity should be the variation in tip movement alone. However, this is not the absolute case in piezo-acoustic excitation. This is due to cantilever bending which is not exactly equivalent of the tip motion [168]. A schematic of a typical piezo-acoustic cantilever excitation method is shown in Figure 7c.

5.7. Photothermal Excitation

The photothermal excitation method is based on the fact that the microcantilevers can be easily modified by coating the upper surface with a thin layer of a different material. Owing to their difference in the coefficient of thermal expansion, when the composite material is subjected to a temperature change, the microcantilever deflects. In atomic force microscopy, power modulation of a focused laser beam at the back of the microcantilever at a designated drive frequency [182] or joule heating [183] are the two major methods used to achieve the desired heating. Photothermal excitation method has a few drawbacks including low displacement and low efficiency [184] and the difficulty of exciting the higher modes. The advantages of photothermal excitation are however enormous when compared to other conventional; high bandwidth [185,186], sharp resonant peak in liquid and ease of implementation as shown in Figure 7d and the ability to use as fabricated microcantilevers without coating. One problem with the bimorph microcantilevers is the fact that longitudinal thermal diffusion inhibits the lateral bending in diffusion direction. However, the implementation of the photothermal excitation on single crystal by Nishida and his co-workers [186] a decade ago was a major breakthrough in solving this problem. Thus it is possible to precisely excite microcantilever modes of higher frequencies. Another common method of exciting different modes in single crystal microcantilevers is through varying the position at which the focused laser spot hits the back of the cantilever.

5.8. Optical Excitation

Miyahara and his co-workers [187] have proposed a new method for exciting a microcantilever sensor by combining two laser in a single-mode optical fiber using a filter wavelength division multiplexer (FWDM) to achieve both excitation and detection. With the set-up it was possible to eliminate the spurious mechanical resonances associated with the piezo-acoustic excitation method (see Figure 8a). The interference of the returning light from the back of the cantilever and the fiber end goes back to the FWDM that helps to block the reflected excitation laser signal and only allows the detection laser to pass to the photodetector through an optical circulator. Modulation was achieved by modulating the drive current with a power combiner. This method allows an easy modification to the existing AFMs that use the fiber-optic interferometers for detecting the microcantilevers.

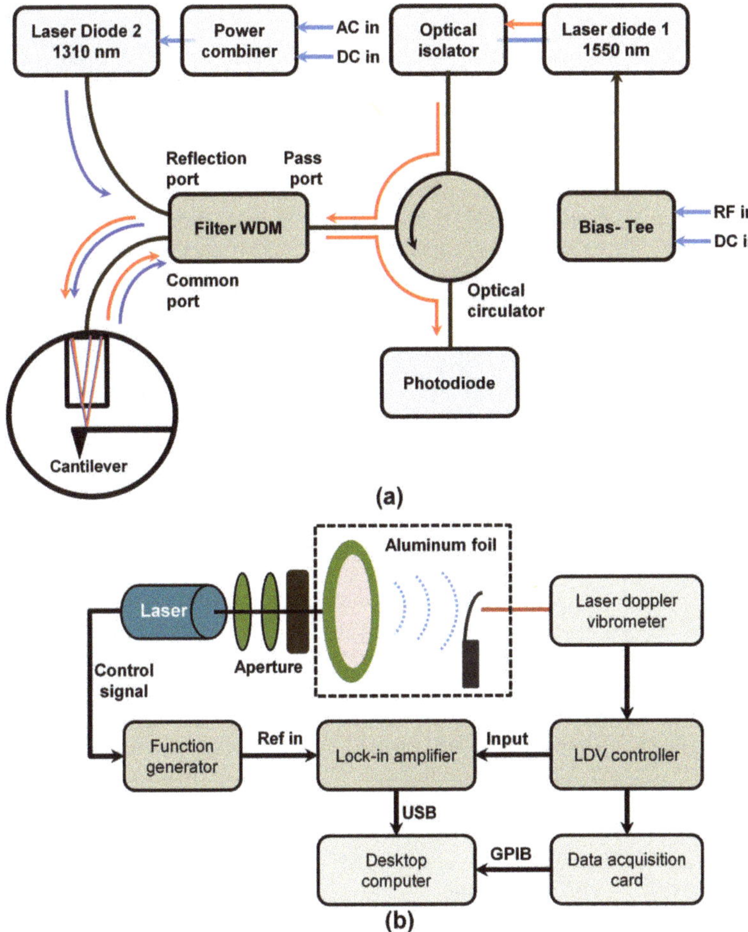

Figure 8. (**a**) A schematic of the optical excitation method for the microcantilever proposed by Miyahara et al. [187]. (**b**) Schematic of the experimental set-up for the laser induced photoacoustic excitation method for the microcantilever [188].

5.9. Laser Induced Photoacoustic Excitation

Remote excitations of microcantilever based sensors by laser-induced photoacoustic (PA) waves have recently been reported by Gao et al. [188]. This excitation technique shown in Figure 8b typically relies on the generation of PA waves from an optical absorber, followed by effective delivery of these propagating PA waves on the lever surface through a medium. It may enable microcantilevers to be used as photoacoustic sensors and presents itself as a substitute method for detecting small signals by eliminating the heating effect common in other optical excitation methods. However, these potential applications call for a comprehensive understanding of the microcantilever response to the laser-induced PA waves.

6. Fabrication, Modification, and Functionalization of AFM Microcantilevers

6.1. Fabrication

Cheap, miniature, and reproducible fabrication of microcantilevers has been possible from silicon, silicon nitride, silicon oxide, or silicon-on-insulator (SOI) by taking an advantage of the batch silicon micromachining techniques developed for integrated circuits (IC) and CMOS process technologies [189–192]. In fact, wide applications of microcantilevers in industries and most research facilities have been made possible by the fact that they can be mass-produced and they are easy to be miniaturized. The microcantilevers are available in various dimensions, shapes, and sensitivity. Often, the geometry of the microcantilever is dictated by the mode of detection. The dimensions of the microcantilevers range from 100 to 500 microns in length and below 5 micrometers in thickness. Typical shapes are the "T" (rectangular) or the "V" (triangular) with a sharp tip mounted on the free end. In the recent past, the need for small high bandwidth cantilevers has risen for high-speed atomic force microscopy, and similar technologies have been used for their production [193–195]. Many investigators with a full access to well-established micromachining facilities have delved in the fabrication of microcantilevers.

Other researchers have attempted the fabrication of microcantilevers using organic-based materials such as SU-8 and Polydimethylsiloxane (PDMS) [53,196] because of their low modulus of elasticity and versatile and simple processing procedures. By using the bottom-up approaches, the microfabrication process of the SU-8 microcantilevers has a high output. Using polymer cantilevers has been shown to outperform the silicon or silicon nitride microcantilevers particularly concerning the imaging speed of the atomic force microscopy (AFM) in tapping mode by up to one order of magnitude. However, the polymer microcantilever tips do not often have the required sharpness and durability for imaging in contact or contact resonance mode [197]. Therefore, a way to combine the high imaging bandwidth of polymer cantilevers with the sharp and wear-resistant tips is necessary for a future adoption of polymer cantilevers in routine AFM uses [198]. An attempt by Martin-Olmos and co-workers [199] to coat SU-8 microcantilever and the tips with wear-resistant graphene was unsuccessful in creating sharp tips.

The SU-8 polymer microcantilevers have been applied in the study of different biological phenomenon. High-resolution AFM images of DNA plasmid molecules have been presented by Genotel and co-workers [200]. Additionally, the polymeric SU-8 microcantilevers have been applied in high speed amplitude modulation AFM and shown improved performance due to their high mechanical bandwidth and low mechanical quality factor (Q-factor) [201]. In a recent article, Kramer et al. [202] proposed a simple method of fabricating ready-to-use micro-fluidic microcantilevers by using a combination of two-photon polymerization and stereolithography 3D additive manufacturing processes. The method offers an inexpensive, fast and more flexible way of fabricating the microcantilevers. A microcantilever of dimensions 564 μm long, 30 μm wide, 30 μm thick was fabricated with a spring constant of about 0.0037 N/mm. The reported micro-fluidic microcantilevers were used to puncture the cell membrane and aspiration of a single cell.

6.2. Microcantilever Tip Fabrication

A majority of the imaging and surface characterization done using an atomic force microscope are carried out with microcantilever probes as the sensing elements. They form part of the consumable items required for the running of the AFM especially when high spatial resolution imaging is needed. Several methods have been used to create the sharp probes on the microcantilever suitable for high resolution imaging. Zenhausern et al. [203] used scanning electron microscope (SEM) to fabricate sharp carbon tips at the end of commercial silicon nitride cantilevers through electron beam induced deposition (EBID) technique. Akiyama and his co-worker reported a successful fabrication of a sharp tip with a radius of curvature of less than 5 nm in a microcantilever using the focused ion-beam (FIB) method [204]. Tay and Thong used a simple field emission induced growth (FEIG) of a tungsten nanowire enabling the production of sharp and robust high-aspect ratio microcantilever probes for AFM applications. They were able to achieve probe lengths up to 1500 nm with a tip radius of less than 2 nm [205]. Dremov and his co-workers demonstrated the fabrication of robust, conductive microcantilever tips suitable for scanning contrast or Kelvin probe force microscopy using a single multiwalled carbon nanotube (MWCNT) by employing dielectrophoresis technique from the MWCNT suspension [206].

Lee and his co-workers [207] demonstrated a process for fabrication of photopolymerizable hydrogel nanoprobes with tunable mechanical properties, allowing an easy encapsulation of nanomaterial with differing sizes and different possibilities of functionalization [208–210]. Additionally, the hydrogel material on account of its softness could provide a good microcantilever for biological and soft matter AFM applications [211]. The hydrogel-based cantilevers are found to have widely tunable and low mechanical stiffness suitable for sensitive nanomechanical measurements of soft matter. The multifunctional and programmable capabilities of the hydrogel nanoprobes were also demonstrated including temperature sensing, material delivery, and local heating. The process involves using ultraviolet light-induced curing of a pre-polymer solution introduced into a mold in order to fabricate the tipless hydrogel cantilever. The tipless microcantilever is then brought into contact with a tip mold filled with a pre-polymer solution. Curing is achieved by exposure of the hydrogel in the tip mold using a secondary ultraviolet resulting in a strong connection between the tip and the cantilever before coating to increase reflectivity [207]. The hydrogel filled tip mold can be optionally deformed to apply a compressive strain to enable tunable tip sharpness and high aspect ratio. A summary of the fabrication method for the hydrogel AFM micro-cantilever is shown in Figure 9.

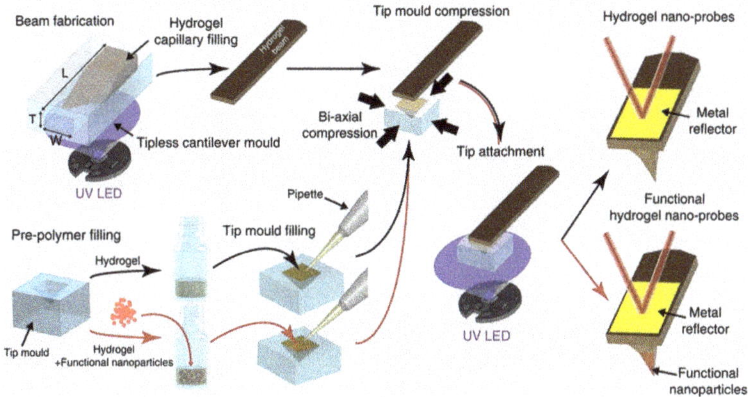

Figure 9. Fabrication method for hydrogel AFM probes. A tipless hydrogel cantilever is first prepared by ultraviolet curing of the pre-polymer solution introduced into the cantilever beam mold. The tipless hydrogel cantilever then makes contact with a tip mold filled with pre-polymer solution with or without encapsulated functional elements, followed by a second round of ultraviolet exposure to cure the hydrogel in the tip mold. This results in the firm attachment between the cantilever and tip. Before the second ultraviolet exposure, the hydrogel-filled tip mold can be optionally deformed by applying bi-axial compressive strains to facilitate tunable tip sharpness and aspect ratio. Reprinted from the work in [207].

6.3. AFM Microcantilever Modification

As-fabricated microcantilevers work effectively in many AFM applications. In fact, it has been demonstrated that when they are operated in the dynamic flexural mode, they exhibit relatively good sensitivities [212–214]. The principle of operation of such microcantilevers is based on a shift in the resonant frequencies owing to the fluid moved by the microcantilevers during vibration. For density sensor applications, a decrease in mass density of the fluid surrounding the microcantilever causes the equivalent effective mass of the microcantilever to decrease, thereby causing the resonant frequency to decrease or vice versa [215]. The advantages of the uncoated microcantilevers are numerous such as reductions in aging effects, thermal drift and longtime response [216]. The thermal drift is a result of increased heat from the surface due to a temperature gradient.

The uncoated microcantilevers offer low sensitivity and are non-selective when used in special sensor applications for gas detection or density measurement. However, the uncoated microcantilevers suffer from a low level of reflected laser power from the back of the cantilever. In the application of the microcantilevers for photothermal excitation, the most common principle used is the bimorph that requires the surface of the microcantilever to be coated with a secondary material having a different coefficient of linear expansion. The reflective metal coating with a thickness of a few tens of nanometers offers a benefit of amplifying the reflected laser beam off the microcantilever surface thereby enhancing the signal-to-noise ratio. In addition, the coated microcantilever can prevent the interference between the reflected beam from a very reflective sample. The main coating materials used in microcantilevers are gold, platinum, and aluminum. Although the use of aluminum for surface coating is cheap and provides good reflectivity, it is not suitable for use in most biological buffers or solvents because it is highly unstable or even dissolvable in a liquid environment. The use of gold on the other hand offers stability because it is biologically and chemically inert. Platinum is often used for electrical or magnetic measurements. Recently, Xia et al. [159] tried different polymer materials such as M-Bond 610, 2K-Epoxy, M-Bond 43B, and Positiv 20 for active AFM microcantilevers using dip coating process. Positiv 20 polymer gave superior outcome in terms of coating layer thickness, good bond capability and less corrosiveness to chemical attack. The developed polymer coated active cantilevers allowed imaging in opaque liquid environments such as crude oil, vinegar, and immersion test in blood sample.

6.4. AFM Probe Functionalization

The atomic force microscope microcantilever tips have the flexibility of being functionalized for chemical and biological applications to allow the attachment of the sensing molecules. Prior to functionalization, sometimes the cantilevers are gold coated to provide a convenient platform for chemical or biological functionalization by taking advantage of thiol-gold chemistry [217]. The customization possibilities for tips are endless. With the functionalized tips, the AFM is capable of providing sensitive tool for measuring and mapping surface chemistry and quantifying repulsive and adhesion forces related to the biological samples and inorganic materials. This is made possible by controlling the chemical interactions between the AFM tip and the sample. Functionalization typically involves chemical modifications of the tips using particular functional groups in order to carry out a specific function in the system. Before functionalization, the tips should be carefully inspected for quality in terms of the material, tip radius, shape and size, resonant frequency, and spring constant. When a low quality tips are used, it can lead to imaging artifacts. The silicon nitride cantilevers are preferred for studies involving molecular recognition [218]. Their biggest advantage is the commercial availability of several different silica precursors highly suitable for decorating the AFM tips with the desired functional groups.

A number of techniques have been proposed to functionalize the AFM microcantilevers for use as chemical or biological sensors. For example, Daza et al. [219] attempted the functionalization of a reliable and robust AFM microcantilever tips by using the activated vapor silanization (AVS) process. The functionalized tips were able to withstand repetitive interactions with a model graphite substrate under relatively harsh conditions with no damages to the tip. The process involved pre-heating the tip to create a high density of hydroxyl (-OH) groups on the surface. The hydroxyl (-OH) groups may then react with an organosilicon compound such as aminopropyltrietoxisilane (APTES) terminated in a reactive group such as amine. More sophisticated functionalization methods have also been proposed and explored, such as plasma enhanced chemical deposition (PECVD). The use of PECVD-functionalization, however, requires an activation step of the substrate, which can be performed by creating oxygen-containing plasma before starting the functionalization process [220]. This activation step is supposed to create a high density of hydroxyl groups on the surface to which the APTES molecules may bind covalently. Even though PECVD-functionalization allows deposition of thick layers, the thickness of the functionalized thin film tends to be restricted to a few nanometers (5–10 nm) [219].

Other processes that have been used to functionalize AFM microcantilever tips include self-assembled monolayer (SAM), vacuum thermal evaporation, and sputtering. The self-assembled monolayer functionalization method involves dropping specific reagents on the tip of the microcantilevers and rinsing after a time duration with a different reagent. This is followed by immersion into a solution or ultra-pure water. During the process, functional groups such as $-CH_3$, $-COOH$, or organosiloxane monolayers are formed on the tip under the controlled conditions on a gold coated surface. By using the self-assembled monolayers to functionalize the AFM tips, a window of opportunities have been opened that enables understanding diverse interfacial phenomena, self-organization, and structure–property relationship [221–223]. Sputtering on the other hand is used to functionalize the microcantilever AFM tips to induce specific properties such as ferroelectricity, thermal, and electrical conduction and optical reflectivity [224].

Operation of the atomic force microscope in the colloidal probe mode has also proven to be effective in quantitatively measuring the nanoscale interactions at biopolymer interfaces, drainage of thin films, lubrication theory, mechanical properties of cells and deformation of colloidal droplets [225–228]. It involves attaching a colloid sphere below the microcantilever thus allowing the measurement of the surface phenomena with sub-nanometer and pico-newton resolution.

The functionalized AFM tips using various binding groups have been used widely in the past to study interfacial interactions. For example, Ma and his co-workers [229] investigated the generated adhesive force between a hydrophobic microcantilever tip and immobilized oligopeptides surface. It is possible to quantify and identify the receptor–ligand interactions usually in the range below 100 pN [230]. Different force spectroscopy techniques such as optical tweezers, atomic force microscopes, and biomembrane force probe have been used to obtain quantitative information about the adhesion force below nN range [231]. Often, the optical tweezers and the biomembrane force probes methods are less preferred because they are limited in the detachable adhesion force range [232]. For atomic force microscope techniques, the single-cell force spectroscopy mode is employed to study the cell to cell interactions mostly carried out in their physiological buffer solutions and conditions. Additionally, it has a significantly large range of detachable forces up to 1 µN in addition to the precise temporal and spatial control over the experiments. For example, Zhang and co-workers [233] used a soft microcantilever functionalized with the cancer cell using biotin-conA brought in contact with the endothelial cell monolayer grown on a surface allowing the detection global adhesion strength and breakup of receptor–ligand bonds.

7. High-Speed Imaging

In spite of many positive aspects, one of the most limiting disadvantages of typical atomic force microscopes is the slow scanning speed. For most commercial atomic force microscopes, image acquisition takes several to tens of minutes [234] since the line scan speed is typically around 1 Hz. Fortunately, there have been many improvements in the imaging speeds of AFM, especially in the past decade. Several technological hurdles should be overcome to improve imaging speeds, and these include the slow data acquisition systems [235,236], low resonant frequency of the nano-positioners and scanners [234,237,238], low bandwidth of the feedback controller [234,237], and low resonant frequency of the microcantilevers [235,236,239,240]. An effective means to excite the microcantilevers in the MHz regime is also needed.

Extending the speed capabilities of AFM has inspired many researchers to do an extensive work in this area. Significant efforts have been put on developing high bandwidth scanners and cantilevers [235,241], high bandwidth cantilever deflection detection systems [242–245], and fast and robust feedback system with Z-scanner [109,246]. The heart of atomic force microscope is the microcantilever sensor that interacts with the sample to measure the desired surface features. Interestingly, the microcantilever was the biggest obstacle for raising the speed of the AFM due to the limited bandwidth of typically available cantilevers. High bandwidth of the microcantilevers was achieved by the advent of the robust, commercially available ultra-small cantilevers that enabled the reduction of the overall microcantilever mass [181,237,247]. Typical dimensions of the ultra-short microcantilevers are a few microns in length, about 10 times smaller than conventional cantilevers, a resonance frequency above 1 MHz and a low force constant typically in a few nN/m [248] compared to the conventional tapping mode cantilevers. A summary of the properties of the commonly available ultra-short cantilevers (AC10 and AC7) and regular cantilevers (MLCT-E and AC40) are compared in Table 2. The miniature cantilevers also have low spring constant (k), reduced coefficient of viscous drag (β), and low quality factor (Q). The low quality factors and high resonance frequencies are required for the ultra-short cantilevers to have a small response time. The total thermal noise $\sqrt{\frac{k_B T}{k_c}}$, where k_B is Boltzmann's constant, k_c is the spring constant, and T is the temperature in Kelvin, is distributed over frequencies up to slightly above the resonance frequency, f_c. Thus, a cantilever with a higher f_c has a lower noise density.

Table 2. The comparison of the properties of commonly available regular (MLCT-E and AC40) and ultra-short (AC10 and AC7) microcantilevers.

Property	MLCT-E	AC40	AC10	AC7
Shape	V-shaped	Rectangle	Rectangle	Rectangle
Length (µm)	140	38	8	6
Width (µm)	18	16	2	2
Thickness (nm)	600	180	130	130
K (pN/nm)	112	102	87	592
f_0 in liquid (kHz)	7	31	431	1231
Q-factor in liquid	1.7	1.6	0.8	0.7
β (pNs/µm)	4.59	0.82	0.03	0.05

AC10 and AC7 are the commonly used probes because they have silicon nitride tips that allow easy functionalization compared to the carbon AFM tips deposited using the electron beam method [3]. The ultra-short cantilevers require a small laser spot size for the detection of the cantilever deflection. The small laser spot required for the ultra-short cantilevers are provided for either by using the power micro-lenses or using a microscope objective [235,237].

The development of high-speed atomic force microscopy (HSAFM) has enabled the generation of AFM images at video rate and recording of force–distance curves at high speeds [237,249–251]. The introduction of AFM to capture the live actions of biomolecules at high spatial and temporal resolutions has been demonstrated by HSAFM [252,253]. AFM-based recognition imaging and force spectroscopy allow unbinding force mapping of receptor–ligand interaction sites on a lipid membrane at the single molecule level [254].

HSAFM is also a force spectroscopy tool. In force spectroscopy, the force–distance curves are obtained. Typically, there are different force spectroscopy approaches based on the experimental setup such as the functionalization of the tip or the type of distance modulation used. Single-cell force spectroscopy and single-molecule force spectroscopy are used in the study of biomolecular or cell adhesion processes at the single-biomolecule level [117,255,256]. Peak force tapping and force volume methods are the other two force spectroscopy methods applied in the study of the nanomechanical response of polymers, cells, inorganic, and organic interfaces [257].

High speed capabilities have been useful in the study of time-dependent dynamic and kinetic processes that involve melting, crystallization, growth, and annealing of several surfaces including polymers, crystals, and biological molecules [3]. The high-speed AFMs provide a way for understanding the mechanical properties of biological systems and processes at the nanoscale [252,258]. In fact, many biological processes present in many organisms occur over a short time scale. It is possible to visualize cellular dynamics and various proteins at video rates [249,259]. Kodera and his co-workers demonstrated a real-time observation of walking myosin V on an actin filament [250]. Yu et al. [260] applied the high-speed atomic force microscope with ultra-short cantilevers to unfold the individual bacteriorhodopsin molecules in a native lipid bilayer. Matusovsky and co-workers [261] studied the 3-state model of activation of cardiac thin filaments isolated as a complex and deposited on a mica-supported lipid bilayer. They realized that the successful imaging of the regulatory proteins tropomyosin and troponin complexes is dependent on the force applied by the cantilever tip because of their low affinity to F-actin. Thus, a small force should be applied neither to break the electrostatic bonds within the regulatory units of the cardiac thin filaments nor reconstituted F-actin–tropomyosin–troponin complex.

8. Microcantilever Sensors in AFM Applications

The applications of AFM microcantilevers are enormous, ranging from solving problems in different areas such as energy, health care, and agriculture, to handling environmental and process industrial issues. For example, nano-biosensors have been used to monitor the treatment procedures and detection of contaminants and heavy metals in industrial processes [91,262]. The microcantilever nano-biosensors are easy to use, sensitive, small, fast, and versatile in terms of detection and monitoring [263]. Some of the limitations of nano-biosensors, however, include the possibility of multi-agent detections by the conversion of bimolecular activity into a measurable quantity and disturbance from the fluid medium during the measuring and temperature control. Typically, the microcantilever deflection or the frequency shift due to the mass change is used to determine the concentration of the target parameters [264].

Rigo et al. [265] developed an efficient, highly sensitive nano-biosensor by functionalizing a microcantilever with urease enzyme, and they were able to detect heavy metals such as cobalt, zinc, nickel, and lead in water. The nano-biosensor was able to achieve a detection limit of parts per billion for the 30 days of storage showing a relatively good stability. The functionalization process was performed on the upper surface of a gold coated silicon cantilever using self-assembled monolayers (SAM) process, by cross-linking agents 1-ethyl-3-(3-dimethylaminopropyl) carbodiimide (EDC), and N-hydroxysuccinimide (NHS). The heavy metals present in the water solution bind to the active site groups of the urease enzyme by reacting with the sulfhydryl groups. The reaction causes a stress tension on the cantilever surface, resulting in a deflection measured by the voltage change of the cantilever nano-biosensor.

Muenchen et al. [266] functionalized a microcantilever for use as a biosensor using peroxidase from vegetables for the detection of glyphosopahe herbicide with a wide spectral range. The deposition of the peroxidase enzyme on the cantilever was done using the self-assembled monolayers (SAM). The adsorption of the glyphosate resulted in a change in the surface tension causing a conformal change in the structure of the peroxidase enzyme.

Rezaee et al. [267] presented a numerical model of an electrically actuated biosensor for identification and characterization of different bio-particles. The process involved coating the microcantilever with receptor chemicals followed by biasing before analyzing the pull-in instability characteristics.

Sutter et al. [268] combined high-speed atomic force microscopy and X-ray crystallography to study the structure and dynamics of the bacteria micro-compartments shell facet assembly at the molecular resolution. Diverse insights into the structure revealed the formation of single layer sheets of a uniform orientation from pre-assembled shell hexamers. The hexamers could also dissociate and combine into an assembled sheet showing the flexibility in the intermolecular interaction. Having a better concept of the bacteria micro-compartments help researchers understand their control and potential use in nanoreactors and molecular scaffolds.

Possas-Abreu et al. [269] recently detected the binding of 2-isobutyl-3-methoxypyrazine to the immobilized odorant binding proteins (OBP) using a grafted OBP on a diamond micro-cantilever by applying MEMS technology. From their work, an approximated 108 molecules of 2-isobutyl-3-methoxypyrazine was bound to the immobilized OBPs showing the possibility of using them as reliable vapor biosensors.

Improvement in the binding efficiency of microcantilever array biosensor has been demonstrated by Liu et al. [270] using the Yersinia detection method. They introduced an antibody to increase the capture efficiency by enhancing the binding sites and reaction efficiency.

Bertke et al. [271] developed a sensitive micro-cantilever based particulate matter detector with a combined electrostatic on-chip ultra-fine particle collection and separation. The microcantilevers had collection electrodes in order to attract the charged particles naturally and an integrated microchannnel to enhance the efficacy of the particle collection. The detection limit for the miniature sensors is about 1 µg/m^{-3}.

Guillaume-Gentil et al. [272] presented a simple method for extracting the endogenous soluble elements from single cells using fluidic force microscopy for further analysis. The process involves the insertion of the microcantilever tip with a triangular aperture of about 400 nm on the front side of the pyramidal tip inside the single cells. After which the extracted fluid fills the probe with the help of a negative pressure. Quantification of the extracted endogenous elements was made possible by using an integrated optical microscopy. Because of the gentle and controlled force offered by the flexible microcantilever during the operation of fluid force microscopy, it was observed that even after the extraction of large volumes of cytoplasm molecules, it was possible for the cells to undergo cell divisions and stay alive. The method demonstrates that there is a potential of extracting smidgen elements for molecular analyses. In addition, it is possible to use undiluted samples for third generation sequencing technologies, building and analysis of the artificial cells and determination of epigenetic changes.

Microcantilever-based biosensors arrays have become to be reliable and very precise instruments for the detection of cancer diseases. Wang et al. [273] reported antibody functionalized microcantilever arrays for the detection of liver cancer. They reduced the adsorption-induced variation of the cantilever stiffness by making a micro-cavity at the end of the microcantilever for local immobilization of the antibody. In addition to the analytical model, they were able to increase the detection sensitivity of the mass of the detected antigen and the overall accuracy of the liver cancer biomarker detection.

In another article, Kamble et al. [274] reported the detection biomarker for early diagnosis of diabetes using piezoresistive microcantilevers and inter-digitated electrodes. The principle is based on the high sensitivity and selectivity of tungsten trioxide towards acetone in an environment filled with selected volatile organic compounds. Screen printing was used to deposit the tungsten trioxide on the inter-digitated electrode fingers, and the resistance measurement was done by using an electrometer. This piezoresistive-based microcantilever work showed the high sensitivity of 2.1 towards 10 ppm acetone at 250 °C.

Recently, Kim et al. [275] reported a universal means of measuring the binding affinities of nivolumab antibody drug towards the target. The method involved coating the surface of a tipless microcantilever with nanocapsules followed by the immobilization of the nivolumab molecules through binding between the antibody and the target protein. The nivolumab-coated AFM cantilever and the T lymphocytes on which programmed cell death 1 molecules expressed are used for investigations. In the experiment, the rupture forces between the programmed cell death 1 molecules and the nivolumab molecules on the microcantilever were monitored. It was demonstrated that this method could allow a comparison of the affinities of different antibody drugs towards a single cell because it does not involve a chemical treatment.

Korayem et al. [276] recently used the microcantilever-based atomic force microscope to obtain various mechanical and physical properties of the head and neck cancer cells. These properties include the modulus of elasticity, cell topography, and viscoelastic properties. From the measurements, the average adhesion force recorded in contact mode for a cantilevers operated in air was 2.47 nN. The research is a step ahead towards characterizing head and neck cancer cells in a heterogeneous population.

9. Conclusions and Prospects

This review article presents the recent developments in the microcantilevers and their applications in various fields. It has been shown that microcantilevers play a pivotal role in the detection of various phenomena using atomic force microscope. A variety of methods for detecting the deflection of the microcantilever have been discussed and the improvements in the recent years have been done to accommodate the ultra-short microcantilevers. It is possible to fabricate microcantilevers both from silicon and selected polymers sensitive to bending moments owing to their lower spring constants. The high sensitivities of microcantilevers have made it possible to investigate complex and advanced chemical and biological problems. Different methods for coating and functionalization of the microcantilever surface for chemical and biological purposes have been assessed.

There is a constant progress in the microcantilever applications with novel detection strategies being developed for higher sensitivities in the atto-newton regime and easier operations. Latest applications of microcantilever-based chemical and biological sensors have been presented. The sensors are reproducible, cost effective for fabrication, robust, easy to handle, power efficient, and small.

The developments achieved in the last few years in both hardware and software for the atomic force microscope has enabled imaging at unprecedented speeds. Additionally, the measurements of the mechanical properties and other surface phenomena in air, aqueous media and at cryogenic conditions have also been conducted with relative ease. Biomaterials and soft matters that seemed impossible to image in the past is now possible by the development of the ultra-fast, flexible microcantilevers.

It is evident that the microcantilever-based sensor is still a work in progress allowing researchers to explore more areas of applications. Further research is required for the development and realization of more robust microcantilever systems for the future applications.

Author Contributions: B.O.A.; wrote and edited the manuscript, Y.J.L. reviewed, edited and corrected the manuscript. The project administration and funding acquisition relevant for the manuscript preparation was performed by Y.J.L. All authors have read and agreed to the published version of the manuscript.

Funding: This research was supported by the Basic Science Research Program through the National Research Foundation of Korea (NRF) funded by the Ministry of Science and ICT (NRF-2018R1A2B6008264).

Conflicts of Interest: The authors declare no conflict of interest.

Abbreviations

The following abbreviations are used in this manuscript.

ACM	Aperture Correction Microscopy
AFM	Atomic Force Microscope
APTES	Aminopropyltrietoxisilane
CM	Contact Mode
CMI	Contact Resonance Imaging
CMOS	Complementary Metal Oxide Semiconductor
EBID	Electron Beam Induced Deposition
EDC	1-ethyl-3-(3-dimethylaminopropyl) carbodiimide
FEIG	Field Emission Induced Growth
FIB	Focused Ion-Beam
FLIM	Fluorescence Lifetime Imaging Microscopy
FMM	Friction Mode Microscopy

FRET	Förster Resonance Energy Transfer
FWDM	Filter Wavelength Division Multiplexer
HSAFM	High-Speed Atomic Force Microscopy
IC	Integrated Circuits
KPFM	Kelvin Probe Force Microscopy
LSCM	Laser Scanning Confocal Microscopy
MEMS	Microelectromechanical Systems
MFFM	Multifrequency Force Microscopy
MFM	Magnetic Force Microscopy
MMM	Mechanical Mapping Mode
MWCNT	Multiwalled Carbon Nanotube
NCM	Non-Contact Mode
NHS	N-Hydroxysuccinimide
OBD	Optical Beam Deflection
OBP	Odorant Binding Proteins
OCLSM	Optical Confocal Laser Scanning Microscopy
OFM	Optical Fluorescent Microscopy
PDMS	Polydimethylsiloxane
PECVD	Plasma Enhanced Chemical Vapor Deposition
PI	Phase Imaging
TERS	Tip-Enhanced Raman Spectroscopy
SAM	Self-Assembly Monolayer
SEM	Scanning Electron Microscopy
SOI	Silicon-on-Insulator
SNOM	Scanning Near-Field Optical Microscopy
SRFM	Super-Resolution Fluorescence Microscopy
STEDM	Correlative Stimulated Emission Depletion Microscopy
STORM	Stochastic Optical Reconstruction Microscopy
TEM	Transmission Electron Microscopy
TIRFM	Total Internal Reflection Fluorescence Microscopy
VMM	Viscoelastic Mapping Microscopy

References

1. Kwon, T.; Gunasekaran, S.; Eom, K. Atomic force microscopy-based cancer diagnosis by detecting cancer-specific biomolecules and cells. *Biochim. Biophys. Acta-Rev. Cancer* **2019**, *1871*, 367–378. [CrossRef] [PubMed]
2. Ruggeri, F.S.; Šneideris, T.; Vendruscolo, M.; Knowles, T.P.J. Atomic force microscopy for single molecule characterisation of protein aggregation. *Arch. Biochem. Biophys.* **2019**, *664*, 134–148. [CrossRef] [PubMed]
3. Valotteau, C.; Sumbul, F.; Rico, F. High-speed force spectroscopy: microsecond force measurements using ultrashort cantilevers. *Biophy. Rev.* **2019**, *11*, 689–699. [CrossRef] [PubMed]
4. Kasas, S.; Ruggeri, F.S.; Benadiba, C.; Maillard, C.; Stupar, P.; Tournu, H.; Dietler, G.; Longo, G. Detecting nanoscale vibrations as signature of life. *Proc. Natl. Acad. Sci. USA* **2015**, *112*, 378–381. [CrossRef]
5. Guillaume-Gentil, O.; Potthoff, E.; Ossola, D.; Franz, C.M.; Zambelli, T.; Vorholt, J.A. Force-controlled manipulation of single cells: from AFM to FluidFM. *Trends Biotechnol.* **2014**, *32*, 381–388. [CrossRef]
6. Beaussart, A.; El-Kirat-Chatel, S. Microbial adhesion and ultrastructure from the single-molecule to the single-cell levels by atomic force microscopy. *Cell Surf.* **2019**, *5*, 100031. [CrossRef]
7. Kojima, T.; Husari, A.; Dieterle, M.P.; Fontaine, S.; Prucker, O.; Tomakidi, P.; Rühe, J. PnBA/PDMAA-based iron-loaded micropillars allow for discrete cell adhesion and analysis of actuation-related molecular responses. *Adv. Mater. Interfaces* **2020**, *7*, 1901806. [CrossRef]
8. Dickinson, L.E.; Rand, D.R.; Tsao, J.; Eberle, W.; Gerecht, S. Endothelial cell responses to micropillar substrates of varying dimensions and stiffness. *J. Biomed. Mater. Res. A* **2012**, *100*, 1457–1466. [CrossRef]
9. Doll, J.C.; Harjee, N.; Klejwa, N.; Kwon, R.; Coulthard, S.M.; Petzold, B.; Goodman, M.B.; Pruitt, B.L. SU-8 force sensing pillar arrays for biological measurements. *Lab Chip* **2009**, *9*, 1449–1454. [CrossRef]

10. Luka, G.; Ahmadi, A.; Najjaran, H.; Alocilja, E.; DeRosa, M.; Wolthers, K.; Malki, A.; Aziz, H.; Althani, A.; Hoorfar, M. Microfluidics integrated biosensors: A leading technology towards lab-on-a-chip and sensing applications. *Sensors* **2015**, *15*, 30011–30031. [CrossRef]

11. Li, W.; Ouyang, R.; Zhang, W.; Zhou, S.; Yang, Y.; Ji, Y.; Feng, K.; Liang, X.; Xiao, M.; Miao, Y. Single walled carbon nanotube sandwiched Ni-Ag hybrid nanoparticle layers for the extraordinary electrocatalysis toward glucose oxidation. *Electrochim. Acta* **2016**, *188*, 197–209. [CrossRef]

12. Mi, S.; Xia, J.; Xu, Y.; Du, Z.; Sun, W. An integrated microchannel biosensor platform to analyse low density lactate metabolism in HepG2 cells in vitro. *RSC Adv.* **2019**, *9*, 9006–9013. [CrossRef]

13. Drake, B.; Prater, C.B.; Weisenhorn, A.L.; Gould, S.A.C.; Albrecht, T.R.; Quate, C.F.; Cannell, D.S.; Hansma, H.G.; Hansma, P.K. Imaging crystals, polymers, and processes in water with the atomic force microscope. *Science* **1989**, *243*, 1586–1589. [CrossRef] [PubMed]

14. Muller, D.J.; Dufrêne, Y.F. Atomic force microscopy: A nanoscopic window on the cell surface. *Trends Cell Biol.* **2011**, *21*, 461–469. [CrossRef] [PubMed]

15. Pillet, F.; Chopinet, L.; Formosa, C.; Dague, É. Atomic force microscopy and pharmacology: From microbiology to cancerology. *Biochim. Biophys. Acta* **2014**, *1840*, 1028–1050. [CrossRef] [PubMed]

16. Moropoulou, A.; Zendri, E.; Ortiz, P.; Delegou, E.T.; Ntoutsi, I.; Balliana, E.; Becerra, J.; Ortiz, R. Scanning microscopy techniques as an assessment tool of materials and interventions for the protection of built cultural heritage. *Scanning* **2019**. [CrossRef]

17. Wang, Y.; Chen, X.; Cao, H.; Deng, C.; Cao, X.; Wang, P. A structural study of Escherichia coli cells using an in situ liquid chamber TEM technology. *J. Anal. Methods Chem.* **2015**.

18. Medalia, O.; Weber, I.; Frangakis, A.S.; Nicastro, D.; Gerisch, G.; Baumeister, W. Macromolecular architecture in eukaryotic cells visualized by cryoelectron tomography. *Science* **2002**, *298*, 1209–1213. [CrossRef]

19. Leis, A.; Rockel, B.; Andrees, L.; Baumeister, W. Visualizing cells at the nanoscale. *Trends Biochem. Sci.* **2009**, *34*, 60–70. [CrossRef]

20. Liv, N.; van Oosten Slingeland, D.S.; Baudoin, J.P.; Kruit, P.; Piston, D.W.; Hoogenboom, J.P. Electron microscopy of living cells during in situ fluorescence microscopy. *ACS Nano* **2016**, *10*, 265–273. [CrossRef]

21. De Jonge, N.; Peckys, D.B. Live cell electron microscopy is probably impossible. *ACS Nano* **2016**, *10*, 9061–9063. [CrossRef] [PubMed]

22. Kennedy, E.; Nelson, E.M.; Tanaka, T.; Damiano, J.; Timp, G. Live bacterial physiology visualized with 5 nm resolution using scanning transmission electron microscopy. *ACS Nano* **2016**, *10*, 2669–2677. [CrossRef] [PubMed]

23. Staunton, J.R.; Doss, B.L.; Lindsay, S.; Ros, R. Correlating confocal microscopy and atomic force indentation reveals metastatic cancer cells stiffen during invasion into collagen I matrices. *Sci. Rep.* **2016**, *6*, 1–15. [CrossRef] [PubMed]

24. Newton, R.; Delguste, M.; Koehler, M.; Dumitru, A.C.; Laskowski, P.R.; Müller, D.J.; Alsteens, D. Combining confocal and atomic force microscopy to quantify single-virus binding to mammalian cell surfaces. *Nat. Protoc.* **2017**, *12*, 2275–2292. [CrossRef]

25. Ramachandran, S.; Arce, F.T.; Patel, N.R.; Quist, A.P.; Cohen, D.A.; Lal, R. Structure and permeability of ion-channels by integrated AFM and waveguide TIRF microscopy. *Sci. Rep.* **2014**, *4*, 1–5. [CrossRef]

26. Christenson, W.; Yermolenko, I.; Plochberger, B.; Camacho-Alanis, F.; Ros, A.; Ugarova, T.P.; Ros, R. Combined single cell AFM manipulation and TIRFM for probing the molecular stability of multilayer fibrinogen matrices. *Ultramicroscopy* **2014**, *136*, 211–215. [CrossRef]

27. Maki, K.; Han, S.W.; Hirano, Y.; Yonemura, S.; Hakoshima, T.; Adachi, T. Real-time TIRF observation of vinculin recruitment to stretched α-catenin by AFM. *Sci. Rep.* **2018**, *8*, 1–8. [CrossRef]

28. Miranda, A.; Martins, M.; De Beule, P.A. Simultaneous differential spinning disk fluorescence optical sectioning microscopy and nanomechanical mapping atomic force microscopy. *Rev. Sci. Instrum.* **2015**, *86*, 093705. [CrossRef]

29. Harke, B.; Chacko, J.V.; Haschke, H.; Canale, C.; Diaspro, A. A novel nanoscopic tool by combining AFM with STED microscopy. *Opt. Nanoscopy* **2012**, *1*, 1–6. [CrossRef]

30. Curry, N.; Ghézali, G.; Kaminski Schierle, G.S.; Rouach, N.; Kaminski, C.F. Correlative STED and atomic force microscopy on live astrocytes reveals plasticity of cytoskeletal structure and membrane physical properties during polarized migration. *Front. Cell. Neurosci.* **2017**, *11*, 104. [CrossRef]

31. Cosentino, M.; Canale, C.; Bianchini, P.; Diaspro, A. AFM-STED correlative nanoscopy reveals a dark side in fluorescence microscopy imaging. *Sci. Adv.* **2019**, *5*, eaav8062. [CrossRef] [PubMed]

32. Hu, D.; Micic, M.; Klymyshyn, N.; Suh, Y.D.; Lu, H.P. Correlated topographic and spectroscopic imaging beyond diffraction limit by atomic force microscopy metallic tip-enhanced near-field fluorescence lifetime microscopy. *Rev. Sci. Instrum.* **2003**, *74*, 3347–3355. [CrossRef]

33. Micic, M.; Hu, D.; Suh, Y.D.; Newton, G.; Romine, M.; Lu, H.P. Correlated atomic force microscopy and fluorescence lifetime imaging of live bacterial cells. *Colloids Surface B.* **2004**, *34*, 205–212. [CrossRef]

34. Odermatt, P.D.; Shivanandan, A.; Deschout, H.; Jankele, R.; Nievergelt, A.P.; Feletti, L.; Davidson, M.W.; Radenovic, A.; Fantner, G.E. High-resolution correlative microscopy: Bridging the gap between single molecule localization microscopy and atomic force microscopy. *Nano Lett.* **2015**, *15*, 4896–4904. [CrossRef] [PubMed]

35. Hirvonen, L.M.; Cox, S. STORM without enzymatic oxygen scavenging for correlative atomic force and fluorescence superresolution microscopy. *Methods Appl. Fluoresc.* **2018**, *6*, 045002. [CrossRef]

36. Gómez-Varela, A.I.; Stamov, D.R.; Miranda, A.; Alves, R.; Barata-Antunes, C.; Dambournet, D.; Drubin, D.G.; Paiva, S.; De Beule, P.A. Simultaneous co-localized super-resolution fluorescence microscopy and atomic force microscopy: Combined SIM and AFM platform for the life sciences. *Sci. Rep.* **2020**, *10*, 1–10. [CrossRef]

37. Chan, K.A.; Kazarian, S.G. Tip-enhanced Raman mapping with top-illumination AFM. *Nanotechnology* **2011**, *22*, 175701. [CrossRef]

38. Kumar, N.; Su, W.; Veselý, M.; Weckhuysen, B.M.; Pollard, A.J.; Wain, A.J. Nanoscale chemical imaging of solid–liquid interfaces using tip-enhanced Raman spectroscopy. *Nanoscale* **2018**, *10*, 1815–1824. [CrossRef]

39. Zhang, M.; Wang, R.; Zhu, Z.; Wang, J.; Tian, Q. Experimental research on the spectral response of tips for tip-enhanced raman spectroscopy. *J. Opt.* **2013**, *15*, 055006. [CrossRef]

40. Ingham, J.; Craig, T.; Smith, C.I.; Varro, A.; Pritchard, D.M.; Barrett, S.D.; Martin, D.S.; Harrison, P.; Unsworth, P.; Kumar, J.D.; et al. Submicron infrared imaging of an oesophageal cancer cell with chemical specificity using an IR-FEL. *Biomed. Phys. Eng. Express* **2018**, *5*, 015009. [CrossRef]

41. Halliwell, D.E.; Morais, C.L.; Lima, K.M.; Trevisan, J.; Siggel-King, M.R.; Craig, T.; Ingham, J.; Martin, D.S.; Heys, K.A.; Kyrgiou, M.; et al. Imaging cervical cytology with scanning near-field optical microscopy (SNOM) coupled with an IR-FEL. *Sci. Rep.* **2016**, *6*, 29494. [CrossRef]

42. Troian, B.; Boscolo, R.; Ricci, G.; Lazzarino, M.; Zito, G.; Prato, S.; Andolfi, L. Ultra-structural analysis of human spermatozoa by aperture scanning near-field optical microscopy. *J. Biophotonics* **2020**, *13*, e2418. [CrossRef] [PubMed]

43. He, Y.; Lu, M.; Cao, J.; Lu, H.P. Manipulating protein conformations by single-molecule AFM-FRET nanoscopy. *ACS Nano* **2012**, *6*, 1221–1229. [CrossRef] [PubMed]

44. Binnig, G.; Quate, C.F.; Gerber, C. Atomic force microscope. *Phys. Rev. Lett.* **1986**, *56*, 930. [CrossRef]

45. Fritz, J.; Baller, M.K.; Lang, H.P.; Rothuizen, H.; Vettiger, P.; Meyer, E.; Güntherodt, H.J.; Gerber, C.; Gimzewski, J.K. Translating biomolecular recognition into nanomechanics. *Science* **2000**, *288*, 316–318. [CrossRef] [PubMed]

46. Wan, Z.; Lee, S.; Koo, K.; Kim, K. Nanowire-based sensors for biological and medical applications. *IEEE Trans. Nanobiosci.* **2016**, *15*, 186–199. [CrossRef]

47. Pramanik, C.; Saha, H. Low pressure piezoresistive sensors for medical electronics applications. *Mater. Manuf. Process.* **2006**, *21*, 233–238.

48. Dong, Y.; Gao, W.; Zhou, Q.; Zheng, Y.; You, Z. Characterization of the gas sensors based on polymer-coated resonant microcantilevers for the detection of volatile organic compounds. *Anal. Chim. Acta* **2010**, *671*, 85–91. [CrossRef]

49. Lavrik, N.V.; Sepaniak, M.J.; Datskos, P.G. Cantilever transducers as a platform for chemical and biological sensors. *Rev. Sci. Instrum.* **2004**, *75*, 2229–2253. [CrossRef]

50. Bashir, R. BioMEMS: State-of-the-art in detection, opportunities and prospects. *Adv. Drug Deliv. Rev.* **2004**, *56*, 1565–1586. [CrossRef]

51. Kenkel, S.; Mittal, S.; Bhargava, R. Closed-loop atomic force microscopy-infrared spectroscopic imaging for nanoscale molecular characterization. *Nat. Commun.* **2020**, *11*, 1–10. [CrossRef]

52. Ziegler, C. Cantilever-based biosensors. *Anal. Bioanal. Chem.* **2004**, *379*, 946–959. [CrossRef] [PubMed]

53. Tomizawa, Y.; Dixit, K.; Daggett, D.; Hoshino, K. Biocompatible cantilevers for mechanical characterization of zebrafish embryos using image analysis. *Sensors* **2019**, *19*, 1506. [CrossRef]

54. Brand, U.; Xu, M.; Doering, L.; Langfahl-Klabes, J.; Behle, H.; Bütefisch, S.; Ahbe, T.; Peiner, E.; Völlmeke, S.; Frank, T.; et al. Long slender piezo-resistive silicon microprobes for fast measurements of roughness and mechanical properties inside micro-holes with diameters below 100 μm. *Sensors* **2019**, *19*, 1410. [CrossRef] [PubMed]

55. Markidou, A.; Shih, W.Y.; Shih, W.H. Soft-materials elastic and shear moduli measurement using piezoelectric cantilevers. *Rev. Sci. Instrum.* **2005**, *76*, 064302. [CrossRef] [PubMed]

56. Oden, P.I.; Chen, G.Y.; Steele, R.A.; Warmack, R.J.; Thundat, T. Viscous drag measurements utilizing microfabricated cantilevers. *Appl. Phys. Lett.* **1996**, *68*, 3814–3816. [CrossRef]

57. Berger, R.; Gerber, C.; Gimzewski, J.K.; Meyer, E.; Güntherodt, H.J. Thermal analysis using a micromechanical calorimeter. *Appl. Phys. Lett.* **1996**, *69*, 40–42. [CrossRef]

58. Arakawa, E.T.; Lavrik, N.V.; Rajic, S.; Datskos, P.G. Detection and differentiation of biological species using microcalorimetric spectroscopy. *Ultramicroscopy* **2003**, *97*, 459–465. [CrossRef]

59. Cherian, S.; Thundat, T. Determination of adsorption-induced variation in the spring constant of a microcantilever. *Appl. Phys. Lett.* **2002**, *80*, 2219–2221. [CrossRef]

60. Pei, J.; Tian, F.; Thundat, T. Novel glucose biosensor based on the microcantilever. *Mater. Res. Soc. Symp. Proc.* **2003**, *776*, 243–247. [CrossRef]

61. Subramanian, A.; Oden, P.I.; Kennel, S.J.; Jacobson, K.B.; Warmack, R.J.; Thundat, T.; Doktycz, M.J. Glucose biosensing using an enzyme-coated microcantilever. *Appl. Phys. Lett.* **2002**, *81*, 385–387. [CrossRef]

62. Hansen, K.M.; Thundat, T. Microcantilever biosensors. *Methods* **2005**, *37*, 57–64. [CrossRef]

63. Ma, R.H.; Lee, C.Y.; Wang, Y.H.; Chen, H.J. Microcantilever-based weather station for temperature, humidity and flow rate measurement. Microsyst Technol. **2008**, *14*, 971–977. [CrossRef] [PubMed]

64. Wang, J.; Wu, W.G.; Huang, Y.; Hao, Y.L. Microcantilever humidity sensor based on embedded nMOSFET with < 100 >-crystal-orientation channel. *Proc. IEEE Sens.* **2009**, *50730009*, 727–730. [CrossRef]

65. Priyadarisshini, B.; Sindhanaiselvi, D. Microcantilever as humidity sensor. In Proceedings of the Conference on Emerging Devices and Smart Systems, Tamilnadu, India, 2–3 March 2018.

66. Strembicke, D.; Robinson, A.M.; Vermeulen, F.E.; Seto, M.; Brown, K.B. Humidity measurement using resonating CMOS microcantilever structures. *Can. Conf. Electr. Comput. Eng.* **1999**, *3*, 1658–1661.

67. Steffens, C.; Manzoli, A.; Leite, F.L.; Fatibello, O.; Herrmann, P.S.P. Atomic force microscope microcantilevers used as sensors for monitoring humidity. *Microelectron. Eng.* **2014**, *113*, 80–85.

68. Wu, L.; Cheng, T.; Zhang, Q.C. A bi-material microcantilever temperature sensor based on optical readout. *Meas. J. Int. Meas. Confed.* **2012**, *45*, 1801–1806. [CrossRef]

69. Egger, M.; Ohnesorge, F.; Weisenhorn, A.L.; Heyn, S.P.; Drake, B.; Prater, C.B.; Gould, S.A.C.; Hansma, P.K.; Gaub, H.E. Wet lipid-protein membranes imaged at submolecular resolution by atomic force microscopy. *J. Struct. Biol.* **1990**, *103*, 89–94. [CrossRef]

70. Bianco, S.; Cocuzza, M.; Ferrero, S.; Giuri, E.; Piacenza, G.; Pirri, C.F.; Ricci, A.; Scaltrito, L.; Bich, D.; Merialdo, A.; et al. Silicon resonant microcantilevers for absolute pressure measurement. *J. Vac. Sci. Technol. B Microelectron. Nanometer Struct.* **2006**, *24*, 1803–1809. [CrossRef]

71. Keskar, G.; Elliott, B.; Gaillard, J.; Skove, M.J.; Rao, A.M. Using electric actuation and detection of oscillations in microcantilevers for pressure measurements. *Sens. Actuators Phys.* **2008**, *147*, 203–209. [CrossRef]

72. Albrecht, T.R.; Akamine, S.; Carver, T.E.; Quate, C.F. Microfabrication of cantilever styli for the atomic force microscope. *J. Vac. Sci. Technol. A Vac. Surf. Film* **1990**, *8*, 3386–3396. [CrossRef]

73. Barnes, J.R.; Stephenson, R.J.; Welland, M.E.; Gerber, C.; Gimzewski, J.K. Photothermal spectroscopy with femtojoule sensitivity using a micromechanical device. *Nature* **1994**, *372*, 79–81. [CrossRef]

74. Boisen, A.; Thaysen, J.; Jensenius, H.; Hansen, O. Environmental sensors based on micromachined cantilevers with integrated read-out. *Ultramicroscopy* **2000**, *82*, 11–16. [CrossRef]

75. Singamaneni, S.; McConney, M.E.; LeMieux, M.C.; Jiang, H.; Enlow, J.O.; Bunning, T.J.; Naik, R.R.; Tsukruk, V.V. Polymer-silicon flexible structures for fast chemical vapor detection. *Adv. Mater.* **2007**, *19*, 4248–4255. [CrossRef]

76. Huber, C.; Pina, M.P.; Morales, J.J.; Mehdaoui, A.A. A multiparameter gas-monitoring system combining functionalized and non-functionalized microcantilevers. *Micromachines* **2020**, *11*, 283. [CrossRef]

77. Grogan, C.; Amarandei, G.; Lawless, S.; Pedreschi, F.; Lyng, F.; Benito-Lopez, F.; Raiteri, R.; Florea, L. Silicon microcantilever sensors to detect the reversible conformational change of a molecular switch, spiropyan. *Sensors* **2020**, *20*, 854. [CrossRef]

78. Boisen, A.; Dohn, S.; Keller, S.S.; Schmid, S.; Tenje, M. Cantilever-like micromechanical sensors. *Rep. Prog. Phys.* **2011**, *74*, 036101. [CrossRef]
79. Diksha, S.; Tripathi, N. Microcantilever: An efficient tool for biosensing applications. *Int. J. Intell. Syst. Appl.* **2017**, *9*, 63–74. [CrossRef]
80. Sepaniak, M.; Datskos, P.; Lavrik, N.; Tipple, C. Peer reviewed: Microcantilever transducers: A new approach in sensor technology. *Anal. Chem.* **2002**, *74*, 568A–575A.
81. McKendry, R.; Zhang, J.; Arntz, Y.; Strunz, T.; Hegner, M.; Lang, H.P.; Baller, M.K.; Certa, U.; Meyer, E.; Güntherodt, H.J.; et al. Multiple label-free biodetection and quantitative DNA-binding assays on a nanomechanical cantilever array. *Proc. Natl. Acad. Sci. USA* **2002**, *99*, 9783–9788. [CrossRef]
82. Lange, D.; Hagleitner, C.; Hierlemann, A.; Brand, O.; Baltes, H. Cantilever arrays on a single chip: Mass-sensitive detection of volatile organic compounds. *Anal. Chem.* **2002**, *74*, 3084–3095. [CrossRef]
83. Chow, E.M.; Yaralioglu, G.G.; Quate, C.F.; Kenny, T.W. Characterization of a two-dimensional cantilever array with through-wafer electrical interconnects. *Appl. Phys. Lett.* **2002**, *80*, 664–666. [CrossRef] [PubMed]
84. Fagan, B.C.; Tipple, C.A.; Xue, Z.; Sepaniak, M.J.; Datskos, P.G. Modification of micro-cantilever sensors with sol-gels to enhance performance and immobilize chemically selective phases. *Talanta* **2000**, *53*, 599–608. [CrossRef]
85. Long, Z.; Kou, L.; Sepaniak, M.J.; Hou, X. Recent advances in gas phase microcantileverbased sensing. *Rev. Anal. Chem.* **2013**, *32*, 135–158. [CrossRef]
86. Patkar, R.; Vinchurkar, M.; Ashwin, M.; Adami, A.; Giacomozzi, F.; Lorenzelli, L.; Baghini, M.S.; Rao, V.R. Microcantilever based dual mode biosensor for agricultural applications. *IEEE Sens. J.* **2019**. [CrossRef]
87. Gopinath, P.G.; Anitha, V.R.; Mastani, S.A. Microcantilever based biosensor for disease detection applications. *J. Med. Bioeng.* **2015**, *4*, 307–311. [CrossRef]
88. Cai, T. Theoretical Analysis of Torsionally Vibrating Microcantilevers for Chemical Sensor Applications in Viscous Liquids. Ph.D. Thesis, Marquette University, Milwaukee, WI, USA, 2013. [CrossRef]
89. Canavese, G.; Ricci, A.; Gazzadi, G.C.; Ferrante, I.; Mura, A.; Marasso, S.L.; Ricciardi, C. Resonating behaviour of nanomachined holed microcantilevers. *Sci. Rep.* **2015**, *5*, 1–6.
90. Sohi, A.N.; Nieva, P.M. Frequency response of curved bilayer microcantilevers with applications to surface stress measurement. *J. Appl. Phys.* **2016**, *119*, 044503. [CrossRef]
91. Ejeian, F.; Etedali, P.; Mansouri-Tehrani, H.A.; Soozanipour, A.; Low, Z.X.; Asadnia, M.; Taheri-Kafrani, A.; Razmjou, A. Biosensors for wastewater monitoring: A review. *Biosens. Bioelectron.* **2018**, *118*, 66–79. [CrossRef]
92. Cheulkar, L.N.; Sawant, V.B.; Mohite, S.S. Evaluating performance of thermally curled microcantilever RF MEMS switches. *Mater. Today Proc.* **2019**, *27*, 12–18. [CrossRef]
93. Basha, S.J.; Krishna, M.H.S.; Praharsha, C.A.; Babu, P.H.; Karthikeya, V.; Srinivas, Y.; Lakshmi, D.R.; Rao, S. Microcantilever based RF MEMS switch for wireless communication. *Microelectron. Solid State Electron.* **2016**, *5*, 1–6. [CrossRef]
94. Choi, J.Y. RF MEMS switch using silicon cantilevers. In Proceedings of the EU-Korea Conference on Science and Technology, Berlin, Germany, 19 October 2008.
95. Leitner, M.; Fantner, G.E.; Fantner, E.J.; Ivanova, K.; Ivanov, T.; Rangelow, I.; Ebner, A.; Rangl, M.; Tang, J.; Hinterdorfer, P. Single-molecule AFM characterization of individual chemically tagged DNA tetrahedra. *ACS Nano* **2011**, *5*, 7048–7054.
96. Shibata, M.; Uchihashi, T.; Ando, T.; Yasuda, R. Long-tip high-speed atomic force microscopy for nanometer-scale imaging in live cells. *Sci. Rep.* **2015**, *5*, 1–7. [CrossRef] [PubMed]
97. Vashist, S.K.; Holthofer, H. Microcantilevers for sensing applications. *Meas. Control* **2010**, *43*, 84–88. [CrossRef] [PubMed]
98. Severi, S.; Heck, J.; Chou, T.K.; Belov, N.; Park, J.S.; Harrar, D.; Jain, A.; Van Hoof, R.; Du Bois, B.; De Coster, J.; et al. CMOS-integrated poly-sige cantilevers with read/write system for probe storage device. In Proceedings of the International Conference on Solid-State Sensors, Actuators and Microsystems, Denver, CO, USA, 21–25 June 2009. [CrossRef]
99. Pinnaduwage, L.A.; Ji, H.F.; Thundat, T. Moore's law in homeland defense: An integrated sensor platform based on silicon microcantilevers. *IEEE Sens. J.* **2005**, *5*, 774–785.

100. Praveenkumar, S.; Nath, S.S.; Dinesh, R.G.; Ramya, S.; Priya, M. Design optimization and simulation of micro-electro-mechanical system based solar energy harvester for low voltage applications. *J. Renew. Sustain. Energy* **2018**, *10*, 053503. [CrossRef]

101. Gupta, R.; Rana, L.; Sharma, A.; Gupta, V.; Tomar, M. Fabrication of micro-cantilever and its theoretical validation for energy harvesting applications. *Microsyst. Technol.* **2019**, *25*, 4249–4256. [CrossRef]

102. Park Systems, XE-70 High accuracy small sample SPM. In *User's Manual*; Park Systems: Suwon, Korea, 2013. [CrossRef]

103. Meyer, G.; Amer, N.M. Novel optical approach to atomic force microscopy. *Appl. Phys. Lett.* **1988**, *53*, 1045–1047.

104. Alexander, S.; Hellemans, L.; Marti, O.; Schneir, J.; Elings, V.; Hansma, P.K.; Longmire, M.; Gurley, J. An atomic-resolution atomic-force microscope implemented using an optical lever. *J. Appl. Phys.* **1989**, *65*, 164–167. [CrossRef]

105. Meyer, G.; Amer, N.M. Optical-beam-deflection atomic force microscopy: The NaCl (001) surface. *Appl. Phys. Lett.* **1990**, *56*, 2100–2101. [CrossRef]

106. Putman, C.A.J.; De Grooth, B.G.; Van Hulst, N.F.; Greve, J. A detailed analysis of the optical beam deflection technique for use in atomic force microscopy. *J. Appl. Phys.* **1992**, *72*, 6–12. [CrossRef]

107. Marrese, M.; Guarino, V.; Ambrosio, L. Atomic force microscopy: a powerful tool to address scaffold design in tissue engineering. *J. Funct. Biomater.* **2017**, *8*, 7. [CrossRef]

108. Martin, Y.; Williams, C.C.; Wickramasinghe, H.K. Atomic force microscope–force mapping and profiling on a sub 100-Å scale. *J. Appl. Phys.* **1987**, *61*, 4723–4729. [CrossRef] [PubMed]

109. Kodera, N.; Sakashita, M.; Ando, T. Dynamic proportional-integral-differential controller for high-speed atomic force microscopy. *Rev. Sci. Instrum.* **2006**, *77*, 083704. [CrossRef]

110. Kodera, N.; Yamashita, H.; Ando, T. Active damping of the scanner for high-speed atomic force microscopy. *Rev. Sci. Instrum.* **2005**, *76*, 053708. [CrossRef]

111. Wickramasinghe, H.K. Scanning probe microscopy: Current status and future trends. *J. Vac. Sci. Technol. Vac. Surf. Film* **1990**, *8*, 363–368. [CrossRef]

112. Maver, U.; Velnar, T.; Gaberšček, M.; Planinšek, O.; Finšgar, M. Recent progressive use of atomic force microscopy in biomedical applications. *TrAC Trends Anal. Chem.* **2016**, *80*, 96–111. [CrossRef]

113. Harcombe, D.M.; Ruppert, M.G.; Ragazzon, M.R.P.; Fleming, A.J. Higher-harmonic AFM Imaging with a high-bandwidth multifrequency lyapunov filter. In Proceedings of the International Conference on Advanced Intelligent Mechatronics (AIM), Munich, Germany, 3–7 July 2017. [CrossRef]

114. Shamsudhin, N.; Rothuizen, H.; Nelson, B.J.; Sebastian, A. Multi-frequency atomic force microscopy: A system-theoretic approach. *IFAC Proc. Vol.* **2014**, *47*, 7499–7504.

115. Krisenko, M.O.; Cartagena, A.; Raman, A.; Geahlen, R.L. Nanomechanical property maps of breast cancer cells as determined by multiharmonic atomic force microscopy reveal SYK-dependent changes in microtubule stability mediated by MAP1B. *Biochemistry* **2015**, *54*, 60–68. [CrossRef]

116. Benaglia, S.; Gisbert, V.G.; Perrino, A.P.; Amo, C.A.; Garcia, R. Fast and high-resolution mapping of elastic properties of biomolecules and polymers with bimodal AFM. *Nat. Protoc.* **2018**, *13*, 2890–2907. [CrossRef]

117. Dufrêne, Y.F.; Martínez-Martín, D.; Medalsy, I.; Alsteens, D.; Müller, D.J. Multiparametric imaging of biological systems by force-distance curve-based AFM. *Nat. Methods* **2013**, *10*, 847–854. [CrossRef]

118. Chawla, G.; Solares, S.D. Mapping of conservative and dissipative interactions in bimodal atomic force microscopy using open-loop and phase-locked-loop control of the higher eigenmode. *Appl. Phys. Lett.* **2011**, *99*, 2011–2014. [CrossRef] [PubMed]

119. Chanmin Su, J.S.; Hu, Y.; Hu, S.; Ma, J. Method and Apparatus of Using Peak Force Tapping Mode to Measure Physical Properties of a Sample. U.S. Patent No. 8,650,660, 13 December 2008. [CrossRef]

120. Schillers, H.; Medalsy, I.; Hu, S.; Slade, A.L.; Shaw, J.E. PeakForce Tapping resolves individual microvilli on living cells. *J. Mol. Recognit.* **2016**, *29*, 95–101.

121. Al-Rekabi, Z.; Contera, S. Multifrequency AFM reveals lipid membrane mechanical properties and the effect of cholesterol in modulating viscoelasticity. *Proc. Natl. Acad. Sci. USA* **2018**, *115*, 2658–2663. [CrossRef]

122. Zheng, Z.; Xu, R.; Ye, S.; Hussain, S.; Ji, W.; Cheng, P.; Li, Y.; Sugawara, Y.; Cheng, Z. High harmonic exploring on different materials in dynamic atomic force microscopy. *Sci. China Technol. Sci.* **2018**, *61*, 446–452. [CrossRef] [PubMed]

123. Rother, J.; Nöding, H.; Mey, I.; Janshoff, A. Atomic force microscopy-based microrheology reveals significant differences in the viscoelastic response between malign and benign cell lines. *Open Biol.* **2014**, *4*, 140046. [CrossRef]

124. Yang, Q.; Ma, Q.; Herum, K.M.; Wang, C.; Patel, N.; Lee, J.; Wang, S.; Yen, T.M.; Wang, J.; Tang, H.; et al. Array atomic force microscopy for real-time multiparametric analysis. *Proc. Natl. Acad. Sci. USA* **2019**, *116*, 5872–5877. [CrossRef]

125. Kim, S.J.; Ono, T.; Esashi, M. Capacitive resonant mass sensor with frequency demodulation detection based on resonant circuit. *Appl. Phys. Lett.* **2006**, *88*, 1–3. [CrossRef]

126. Rogers, B.; Manning, L.; Sulchek, T.; Adams, J. Improving tapping mode atomic force microscopy with piezoelectric cantilevers. *Ultramicroscopy* **2004**, *100*, 267–276. [CrossRef]

127. Dukic, M.; Winhold, M.; Schwalb, C.H.; Adams, J.D.; Stavrov, V.; Huth, M.; Fantner, G.E. Direct-write nanoscale printing of nanogranular tunnelling strain sensors for sub-micrometre cantilevers. *Nat. Commun.* **2016**, *7*, 1–7. [CrossRef]

128. Itoh, T.; Suga, T. Piezoelectric sensor for detecting force gradients in atomic force microscopy. *Jpn. J. Appl. Phys.* **1994**, *33*, 334. [CrossRef]

129. Ruiz-Díez, V.; Toledo, J.; Hernando-García, J.; Ababneh, A.; Seidel, H.; Sánchez-Rojas, J.L. A geometrical study on the roof tile-shaped modes in AlN-based piezoelectric microcantilevers as viscosity–density sensors. *Sensors* **2019**, *19*, 658. [CrossRef]

130. Fischeneder, M.; Oposich, M.; Schneider, M.; Schmid, U. Tuneable Q-factor of MEMS cantilevers with integrated piezoelectric thin films. *Sensors* **2018**, *18*, 3842. [CrossRef] [PubMed]

131. Shekhawat, G.; Tark, S.H.; Dravid, V.P. MOSFET-embedded microcantilevers for measuring deflection in biomolecular sensors. *Science* **2006**, *311*, 1592–1595. [CrossRef] [PubMed]

132. Wagner, R.; Woehl, T.J.; Keller, R.R.; Killgore, J.P. Detection of atomic force microscopy cantilever displacement with a transmitted electron beam. *Appl. Phys. Lett.* **2016**, *109*, 043111. [CrossRef]

133. Hermans, R.I.; Dueck, B.; Ndieyira, J.W.; McKendry, R.A.; Aeppli, G. Optical diffraction for measurements of nano-mechanical bending. *Sci. Rep.* **2016**, *6*, 1–12. [CrossRef]

134. Andreeva, N.V. Atomic force microscopy with interferometric method for detection of the cantilever displacement and its application for low-temperature studies. *Ferroelectrics* **2018**, *525*, 178–186. [CrossRef]

135. Putman, C.A.J.; de Grooth, B.G.; van Hulst, N.F.; Greve, J. A theoretical comparison between interferometric and optical beam deflection technique for the measurement of cantilever displacement in AFM. *Ultramicroscopy* **1992**, *42–44 Pt 2*, 1509–1513. [CrossRef]

136. Routley, B.; Fleming, A. High sensitivity interferometer for on-axis detection of AFM cantilever deflection. In Proceedings of the International Conference on Manipulation, Automation and Robotics at Small Scales, Paris, France, 18–22 July 2016. [CrossRef]

137. Von Schmidsfeld, A.; Nörenberg, T.; Temmen, M.; Reichling, M. Understanding interferometry for micro-cantilever displacement detection. *Beilstein J. Nanotechnol.* **2016**, *7*, 841–851.

138. Rugar, D.; Mamin, H.J.; Guethner, P. Improved fiber-optic interferometer for atomic force microscopy. *Appl. Phys. Lett.* **1989**, *55*, 2588–2590. [CrossRef]

139. Yaralioglu, G.G.; Atalar, A.; Manalis, S.R.; Quate, C.F. Analysis and design of an interdigital cantilever as a displacement sensor. *J. Appl. Phys.* **1998**, *83*, 7405–7415. [CrossRef]

140. Manalis, S.R.; Minne, S.C.; Atalar, A.; Quate, C.F. Interdigital cantilevers for atomic force microscopy. *Appl. Phys. Lett.* **1996**, *69*, 3944–3946. [CrossRef]

141. Doll, J.C.; Pruitt, B.L. Design of piezoresistive versus piezoelectric contact mode scanning probes. *J. Micromech. Microeng.* **2010**, *20*, 095023. [CrossRef]

142. Moore, S.I.; Ruppert, M.G.; Yong, Y.K. An optimization framework for the design of piezoelectric AFM cantilevers. *Precis. Eng.* **2019**, *60*, 130–142. [CrossRef]

143. Ruppert, M.G.; Moore, S.I.; Zawierta, M.; Fleming, A.J.; Putrino, G; Yong, Y.K. Multimodal atomic force microscopy with optimized higher eigenmode sensitivity using on-chip piezoelectric actuation and sensing. *Nanotechnology* **2019**, *30*, 085503. [CrossRef]

144. Liu, H.; Zhang, S.; Kathiresan, R.; Kobayashi, T.; Lee, C. Development of piezoelectric microcantilever flow sensor with wind-driven energy harvesting capability. *Appl. Phys. Lett.* **2012**, *100*, 2012–2015. [CrossRef]

145. Brugger, J.; Buser, R.A.; de Rooij, N.F. Micromachined atomic force microprobe with integrated capacitive read-out. *J. Micromech. Microeng.* **1992**, *2*, 218–220. [CrossRef]

146. Tortonese, M.; Yamada, H.; Barrett, R.C.; Quate, C.F. Atomic force microscopy using a piezoresistive cantilever. In Proceedings of the International Conference on Solid-State Sensors and Actuators, San Francisco, CA, USA, 24–27 June 1991. [CrossRef]

147. Gotszalk, T.; Linnemann, R.; Rangelow, I.W.; Dumania, P.; Grabiec, P.B. AFM with piezoresistive Wheatstone bridge cantilever: Noise performance and applications in contact and noncontact mode. *Met. Microsys. Phys. Technol. Appl.* **1996**, *2780*, 376–379.

148. Harley, J.A.; Kenny, T.W. High-sensitivity piezoresistive cantilevers under 1000 Å thick. *Appl. Phys. Lett.* **1999**, *75*, 289–291.

149. Rangelow, I.W.; Skocki, S.; Dumania, P. Plasma etching for micromechanical sensor applications. *Microelectron. Eng.* **1994**, *23*, 365–368. [CrossRef]

150. Tortonese, M.; Barrett, R.C.; Quate, C.F. Atomic resolution with an atomic force microscope using piezoresistive detection. *Appl. Phys. Lett.* **1993**, *62*, 834–836. [CrossRef]

151. Oden, P.I.; Datskos, P.G.; Thundat, T.; Warmack, R.J. Uncooled thermal imaging using a piezoresistive microcantilever. *Appl. Phys. Lett.* **1996**, *69*, 3277–3279. [CrossRef]

152. Thaysen, J.; Boisen, A.; Hansen, O.; Bouwstra, S. Atomic force microscopy probe with piezoresistive read-out and a highly symmetrical Wheatstone bridge arrangement. *Sens. Actuators Phys.* **2000**, *83*, 47–53. [CrossRef]

153. Linnemann, R.; Gotszalk, T.; Hadjiiski, L.; Rangelow, I.W. Characterization of a cantilever with an integrated deflection sensor. *Thin Solid Films* **1995**, *264*, 159–164. [CrossRef]

154. Yu, X.; Thaysen, J.; Hansen, O.; Boisen, A. Optimization of sensitivity and noise in piezoresistive cantilevers. *J. Appl. Phys.* **2002**, *92*, 6296–6301. [CrossRef]

155. Rasmussen, P.A.; Thaysen, J.; Hansen, O.; Eriksen, S.C.; Boisen, A. Optimised cantilever biosensor with piezoresistive read-out. *Ultramicroscopy* **2003**, *97*, 371–376. [CrossRef]

156. Thaysen, J. Cantilever for Bio-Chemical Sensing Integrated In a Microliquid Handling System. Ph.D. Thesis, Technical University of Denmark (DTU), Lyngby, Denmark, 1999. [CrossRef]

157. Lange, D.; Akiyama, T.; Hagleitner, C.; Tonin, A.; Hidber, H.R.; Niedermann, P.; Staufer, U.; De Rooij, N.F.; Brand, O.; Baltes, H. Parallel scanning AFM with on-chip circuitry in CMOS technology. In Proceedings of the International MEMS Conference on Micro Electro Mechanical Systems, Orlando, FL, USA, 21 January 1999.

158. Thaysen, J.; Yalçinkaya, A.D.; Vettiger, P.; Menon, A. Polymer-based stress sensor with integrated readout. *J. Phys. D Appl. Phys.* **2002**, *35*, 2698–2703.

159. Xia, F.; Yang, C.; Wang, Y.; Youcef-Toumi, K.; Reuter, C.; Ivanov, T.; Holz, M.; Rangelow, I.W. Lights out! nano-scale topography imaging of sample surface in opaque liquid environments with coated active cantilever probes. *Nanomaterials* **2019**, *9*, 1013. [CrossRef]

160. Goeders, K.M.; Colton, J.S.; Bottomley, L.A. Microcantilevers: Sensing chemical interactions via mechanical motion. *Chem. Rev.* **2008**, *108*, 522–542. [CrossRef]

161. Britton, C.L., Jr.; Jones, R.L.; Oden, P.I.; Hu, Z.; Warmack, R.J.; Smith, S.F.; Bryan, W.L.; Rochelle, J.M. Multiple-input microcantilever sensors. *Ultramicroscopy* **2000**, *82*, 17–21. [CrossRef]

162. Forsen, E.; Abadal, G.; Ghatnekar-Nilsson, S.; Teva, J.; Verd, J.; Sandberg, R.; Svendsen, W.; Perez-Murano, F.; Esteve, J.; Figueras, E.; et al. Ultrasensitive mass sensor fully integrated with complementary metal-oxide-semiconductor circuitry. *Appl. Phys. Lett.* **2005**, *87*, 1–4. [CrossRef]

163. Azcona, F.J.; Jha, A.; Yáñez, C.; Atashkhooei, R.; Royo, S. Microcantilever displacement measurement using a mechanically modulated optical feedback interferometer. *Sensors* **2016**, *16*, 997. [CrossRef]

164. Lee, B.; Prater, C.B.; King, W.P. Lorentz force actuation of a heated atomic force microscope cantilever. *Nanotechnology* **2012**, *23*, 055709. [CrossRef] [PubMed]

165. Revenko, I.; Proksch, R. Magnetic and acoustic tapping mode microscopy of liquid phase phospholipid bilayers and DNA molecules. *J. Appl. Phys.* **1999**, *87*, 526. [CrossRef] [PubMed]

166. Florin, E.; Radmacher, M.; Fleck, B.; Gaub, H.E. Atomic force microscope with magnetic force modulation. *Rev. Sci. Instrum.* **1994**, *65*, 639–643. [CrossRef]

167. Mazeran, P.E.; Loubet, J.L. Force modulation with a scanning force microscope: an analysis. *Tribol. Lett.* **1997**, *3*, 125–132. [CrossRef]

168. Xu, X.; Raman, A. Comparative dynamics of magnetically, acoustically, and Brownian motion driven microcantilevers in liquids. *J. Appl. Phys.* **2007**, *102*, 034303. [CrossRef]

169. Motamedi, R.; Wood-Adams, P.M. Influence of fluid cell design on the frequency response of afm microcantilevers in liquid media. *Sensors* **2008**, *8*, 5927–5941. [CrossRef]

170. Carbone, G.; Pierro, E.; Soria, L. Microcantilever dynamics: Effect of Brownian excitation in liquids. In Proceedings of the SEM Annual Conference Exposition on Experimental and Applied Mechanics, Hyatt Regency Albuquerque, Albuquerque, New Mexico, 1–4 June 2009. [CrossRef]

171. Berg, J.; Briggs, G.A.D. Nonlinear dynamics of intermittent-contact mode atomic force microscopy. *Phys. Rev. B* **1997**, *55*, 14899–14908.

172. Salapaka, S.; Dahleh, M.; Mezić, I. On the dynamics of a harmonic oscillator undergoing impacts with a vibrating platform. *Nonlinear Dyn.* **2001**, *24*, 333–358. [CrossRef]

173. Umeda, K.; Oyabu, N.; Kobayashi, K.; Hirata, Y.; Matsushige, K.; Yamada, H. High-resolution frequency-modulation atomic force microscopy in liquids using electrostatic excitation method. *Appl. Phys. Express* **2010**, *3*, 065205. [CrossRef]

174. Desbiolle, B.X.E.; Furlan, G.; Schwartzberg, A.M.; Ashby, P.D.; Ziegler, D. Electrostatically actuated encased cantilevers. *Beilstein J. Nanotechnol.* **2018**, *9*, 1381–1389. [CrossRef]

175. Degertekin, F.L.; Hadimioglu, B.; Sulchek, T.; Quate, C.F. Actuation and characterization of atomic force microscope cantilevers in fluids by acoustic radiation pressure. *Appl. Phys. Lett.* **2001**, *78*, 1628–1630. [CrossRef] [PubMed]

176. Chu, B.; Apfel, R.E. Acoustic radiation pressure produced by a beam of sound. *J. Acoust. Soc. Am.* **1982**, *72*, 1673–1687. [CrossRef]

177. Kiracofe, D.; Raman, A. Quantitative force and dissipation measurements in liquids using piezo-excited atomic force microscopy: a unifying theory. *Nanotechnology* **2011**, *22*, 485502. [CrossRef]

178. Rogers, B.; York, D.; Whisman, N.; Jones, M.; Murray, K.; Adams, J.D.; Sulchek, T.; Minne, S.C. Tapping mode atomic force microscopy in liquid with an insulated piezoelectric microactuator. *Rev. Sci. Instrum.* **2002**, *73*, 3242–3244. [CrossRef]

179. Maali, A.; Hurth, C.; Cohen-Bouhacina, T.; Couturier, G.; Aimé, J.P. Improved acoustic excitation of atomic force microscope cantilevers in liquids. *Appl. Phys. Lett.* **2006**, *88*, 163504. [CrossRef]

180. Carrasco, C.; Ares, P.; de Pablo, P.J.; Gómez-Herrero, J. Cutting down the forest of peaks in acoustic dynamic atomic force microscopy in liquid. *Rev. Sci. Instrum.* **2008**, *79*, 126106. [CrossRef]

181. Rogers, B.; Sulchek, T.; Murray, K.; York, D.; Jones, M.; Manning, L.; Malekos, S.; Beneschott, B.; Adams, J.D.; Cavazos, H.; et al. High speed tapping mode atomic force microscopy in liquid using an insulated piezoelectric cantilever. *Rev. Sci. Instrum.* **2003**, *74*, 4683–4686. [CrossRef]

182. Ilic, B.; Krylov, S.; Craighead, H.G. Theoretical and experimental investigation of optically driven nanoelectromechanical oscillators. *J. Appl. Phys.* **2010**, *107*, 034311. [CrossRef]

183. Pedrak, R.; Ivanov, T.; Ivanova, K.; Gotszalk, T.; Abedinov, N.; Rangelow, I.W. Micromachined atomic force microscopy sensor with integrated piezoresistive sensor and thermal bimorph actuator for high-speed tapping-mode atomic force microscopy phase-imaging in higher eigenmodes. *J. Vac. Sci. Technol. Microelectron. Nanometer Struct. Process. Meas. Phenom.* **2003**, *21*, 3102–3107. [CrossRef]

184. Angelov, T.; Roeser, D.; Ivanov, T.; Gutschmidt, S.; Sattel, T.; Rangelow, I.W. Thermo-mechanical transduction suitable for high-speed scanning probe imaging and lithography. *Microelectron. Eng.* **2016**, *154*, 1–7. [CrossRef]

185. Michels, T.; Guliyev, E.; Klukowski, M.; Rangelow, I. Micromachined self-actuated piezoresistive cantilever for high speed SPM. *Microelectron. Eng.* **2012**, *97*, 265–268. [CrossRef]

186. Nishida, S.; Kobayashi, D.; Kawakatsu, H.; Nishimori, Y. Photothermal excitation of a single-crystalline silicon cantilever for higher vibration modes in liquid. *J. Vac. Sci. Technol. B Microelectron. Nanometer Struct.* **2009** *27*, 964–968. [CrossRef]

187. Miyahara, Y.; Griffin, H.; Roy-Gobeil, A.; Belyansky, R.; Bergeron, H.; Bustamante, J.; Grutter, P. Optical excitation of atomic force microscopy cantilever for accurate spectroscopic measurements. *EPJ Tech. Instrum.* **2020**, *7*, 1–11. [CrossRef]

188. Gao, N.; Zhao, D.; Jia, R.; Liu, D. Microcantilever actuation by laser induced photoacoustic waves. *Sci. Rep.* **2016**, *6*, 1–7. [CrossRef]

189. Ono, M.; Lange, D.; Brand, O.; Hagleitner, C.; Baltes, H. A complementary metal oxide semiconductor field effect transistor compatible atomic force microscopy tip fabrication process and integrated atomic force microscopy cantilevers fabricated with this process. *Ultramicroscopy* **2002**, *91*, 9–20. [CrossRef]

190. Franks, W.; Lange, D.; Lee, S.; Hierlemann, A.; Spencer, N.; Baltes, H. Nanochemical surface analyzer in CMOS technology. *Ultramicroscopy* **2002**, *91*, 21–27. [CrossRef]

191. Rangelow, I.W.; Grabiec, P.; Gotszalk, T.; Edinger, K. Piezoresistive SXM sensors. *Surf. Interface Anal.* **2002**, *33*, 59–64. [CrossRef]

192. Miyahara, Y.; Deschler, M.; Fujii, T.; Watanabe, S.; Bleuler, H. Non-contact atomic force microscope with a PZT cantilever used for deflection sensing, direct oscillation and feedback actuation. *Appl. Surf. Sci.* **2002**, *188*, 450–455. [CrossRef]

193. Bachels, T.; Schäfer, R. Microfabricated cantilever-based detector for molecular beam experiments. *Rev. Sci. Instrum.* **1998**, *69*, 3794–3797. [CrossRef]

194. Chand, A.; Viani, M.B.; Schäffer, T.E.; Hansma, P.K. Microfabricated small metal cantilevers with silicon tip for atomic force microscopy. *J. Microelectromech. Syst.* **2000**, *9*, 112–116. [CrossRef]

195. Viani, M.B.; Schäffer, T.E.; Chand, A.; Rief, M.; Gaub, H.E.; Hansma, P.K. Small cantilevers for force spectroscopy of single molecules. *J. Appl. Phys.* **1999**, *86*, 2258–2262. [CrossRef]

196. Du, B.; Xu, X.; He, J.; Guo, K.; Huang, W.; Zhang, F.; Zhang, M.; Wang, Y. In-fiber collimator-based fabry-perot interferometer with enhanced vibration sensitivity. *Sensors* **2019**, *19*, 435. [CrossRef]

197. Sidler, K. Fabrication and Characterization of SU-8 Cantilevers with Integrated Tips Designed for Dip-Pen Nanolithography. Master's Thesis, ETH, Eidgenössische Technische Hochschule Zürich, Mechanical Engineering, Zürich, Switzerland, 2006. [CrossRef]

198. Hosseini, N.; Neuenschwander, M.; Peric, O.; Andany, S.H.; Adams, J.D.; Fantner, G.E. Integration of sharp silicon nitride tips into high-speed SU8 cantilevers in a batch fabrication process. *Beilstein J. Nanotechnol.* **2019**, *10*, 2357–2363.

199. Martin-Olmos, C.; Rasool, H.I.; Weiller, B.H.; Gimzewski, J.K. Graphene MEMS: AFM probe performance improvement. *ACS Nano* **2013**, *7*, 4164–4170. [CrossRef]

200. Genolet, G.; Despont, M.; Vettiger, P.; Anselmetti, D.; De Rooij, N.F. All-photoplastic, soft cantilever cassette probe for scanning force microscopy. *J. Vac. Sci. Technol. B Microelectron. Nanometer Struct.* **2000**, *18*, 617–620. [CrossRef]

201. Adams, J.D.; Erickson, B.W.; Grossenbacher, J.; Brugger, J.; Nievergelt, A.; Fantner, G.E. Harnessing the damping properties of materials for high-speed atomic force microscopy. *Nat. Nanotechnol.* **2016**, *11*, 147–151. [CrossRef]

202. Kramer, R.C.; Verlinden, E.J.; Angeloni, L.; Van Den Heuvel, A.; Fratila-Apachitei, L.E.; Van Der Maarel, S.M.; Ghatkesar, M.K. Multiscale 3D-printing of microfluidic AFM cantilevers. *Lab Chip* **2020**, *20*, 311–319. [CrossRef]

203. Zenhausern, F.; Adrian, M.; Ten Heggeler-Bordier, B.; Ardizzoni, F.; Descouts, P. Enhanced imaging of biomolecules with electron beam deposited tips for scanning force microscopy. *J. Appl. Phys.* **1993**, *73*, 7232–7237. [CrossRef]

204. Akiyama, K.; Eguchi, T.; An, T.; Fujikawa, Y.; Yamada-Takamura, Y.; Sakurai, T.; Hasegawa, Y. Development of a metal-tip cantilever for noncontact atomic force microscopy. *Rev. Sci. Instrum.* **2005**, *76*, 033705. [CrossRef]

205. Tay, A.B.H.; Thong, J.T.L. Fabrication of super-sharp nanowire atomic force microscope probes using a field emission induced growth technique. *Rev. Sci. Instrum.* **2004**, *75*, 3248–3255. [CrossRef]

206. Dremov, V.; Fedoseev, V.; Fedorov, P.; Grebenko, A. Fast and reliable method of conductive carbon nanotube-probe fabrication for scanning probe microscopy. *Rev. Sci. Instrum.* **2015**, *86*, 053703. [CrossRef]

207. Lee, J.S.; Song, J.; Kim, S.O.; Kim, S.; Lee, W.; Jackman, J.A.; Kim, D.; Cho, N.J.; Lee, J. Multifunctional hydrogel nano-probes for atomic force microscopy. *Nat. Commun.* **2016**, *7*, 1–14. [CrossRef] [PubMed]

208. Imran, A.B.; Esaki, K.; Gotoh, H.; Seki, T.; Ito, K.; Sakai, Y.; Takeoka, Y. Extremely stretchable thermosensitive hydrogels by introducing slide-ring polyrotaxane cross-linkers and ionic groups into the polymer network. *Nat. Commun.* **2014**, *5*, 1–8. [CrossRef]

209. Gou, M.; Qu, X.; Zhu, W.; Xiang, M.; Yang, J.; Zhang, K.; Wei, Y.; Chen, S. Bio-inspired detoxification using 3d-printed hydrogel nanocomposites. *Nat. Commun.* **2014**, *5*, 1–9. [CrossRef]

210. Purcell, B.P.; Lobb, D.; Charati, M.B.; Dorsey, S.M.; Wade, R.J.; Zellars, K.N.; Doviak, H.; Pettaway, S.; Logdon, C.B.; Shuman, J.A.; et al. Injectable and bioresponsive hydrogels for on-demand matrix metalloproteinase inhibition. *Nat. Mater.* **2014**, *13*, 653–661. [CrossRef]

211. Ivanovska, I.L.; Miranda, R.; Carrascosa, J.L.; Wuite, G.J.L.; Schmidt, C.F. Discrete fracture patterns of virus shells reveal mechanical building blocks. *Proc. Natl. Acad. Sci. USA* **2011**, *108*, 12611–12616. [CrossRef]

212. Rosario, R.; Mutharasan, R. Piezoelectric excited millimeter sized cantilever sensors for measuring gas density changes. *Sens. Actuators B Chem.* **2014**, *192*, 99–104. [CrossRef]

213. Zhao, L.; Xu, L.; Zhang, G.; Jiang, Z.; Zhao, Y.; Wang, J.; Wang, X.; Liu, Z. In-situ measurement of fluid density rapidly using a vibrating piezoresistive microcantilever sensor without resonance occurring. *IEEE Sens. J.* **2014**, *14*, 645–650. [CrossRef]

214. Bircher, B.A.; Duempelmann, L.; Renggli, K.; Lang, H.P.; Gerber, C.; Bruns, N.; Braun, T. Real-time viscosity and mass density sensors requiring microliter sample volume based on nanomechanical resonators. *Anal. Chem.* **2013**, *85*, 8676–8683. [CrossRef]

215. Boudjiet, M.T.; Cuisset, V.; Pellet, C.; Bertrand, J.; Dufour, I. Preliminary results of the feasibility of hydrogen detection by the use of uncoated silicon microcantilever-based sensors. *Int. J. Hydrogen Energy* **2014**, *39*, 20497–20502. [CrossRef] [PubMed]

216. Boudjiet, M.T.; Bertrand, J.; Mathieu, F.; Nicu, L.; Mazenq, L.; Leichle, T.; Heinrich, S.M.; Pellet, C.; Dufour, I. Geometry optimization of uncoated silicon microcantilever-based gas density sensors. *Sens. Actuators Chem.* **2015**, *208*, 600–607. [CrossRef]

217. Chen, G.; Ning, X.; Park, B.; Boons, G.J.; Xu, B. Simple, clickable protocol for atomic force microscopy tip modification and its application for trace ricin detection by recognition imaging. *Langmuir* **2009**, *25*, 2860–2864. [CrossRef]

218. Arafat, A.; Schroën, K.; de Smet, L.C.; Sudhölter, E.J.; Zuilhof, H. Tailor-made functionalization of silicon nitride surfaces. *J. Am. Chem. Soc.* **2004**, *126*, 8600–8601. [CrossRef]

219. Daza, R.; Colchero, L.; Corregidor, D.; Elices, M.; Guinea, G.V.; Rojo, F.J.; Pérez-Rigueiro, J. Functionalization of atomic force microscopy cantilevers and tips by activated vapour silanization. *Appl. Surf. Sci.* **2019**, *484*, 1141–1148. [CrossRef]

220. Volcke, C.; Gandhiraman, R.P.; Gubala, V.; Doyle, C.; Fonder, G.; Thiry, P.A.; Cafolla, A.A.; James, B.; Williams, D.E. Plasma functionalization of AFM tips for measurement of chemical interactions. *J. Colloid Interface Sci.* **2010**, *348*, 322–328. [CrossRef]

221. Carrascosa, L.G.; Moreno, M.; Álvare, M.; Lechuga, L.M. Nanomechanical biosensors: A new sensing tool. *TrAC Trends Anal. Chem.* **2006**, *25*, 196–206. [CrossRef]

222. Ito, T.; Ibrahim, S.; Grabowska, I. Chemical-force microscopy for materials characterization. *TrAC-Trends Anal. Chem.* **2010**, *29*, 225–233. [CrossRef]

223. Smith, D.A.; Connell, S.D.; Robinson, C.; Kirkham, J. Chemical force microscopy: Applications in surface characterisation of natural hydroxyapatite. *Anal. Chim. Acta* **2003**, *479*, 39–57. [CrossRef]

224. Shchukin, D.G.; Möhwald, H. Smart nanocontainers as depot media for feedback active coatings. *Chem. Commun.* **2011**, *47*, 8730. [CrossRef]

225. Vinogradova, O.I.; Yakubov, G.E. Dynamic effects on force measurements Lubrication and the atomic force microscope. *Langmuir* **2003**, *19*, 1227–1234. [CrossRef] [PubMed]

226. Alonso, J.L.; Goldmann, W.H. Feeling the forces: Atomic force microscopy in cell biology. *Life Sci.* **2003**, *72*, 2553–2560. [CrossRef]

227. Gillies, G.; Prestidge, C.A.; Attard, P. An AFM study of the deformation and nanorheology of cross-linked PDMS droplets. *Langmuir* **2002**, *18*, 1674–1679. [CrossRef]

228. Gillies, G.; Prestidge, C.A. Interaction forces, deformation and nano-rheology of emulsion droplets as determined by colloid probe AFM. *Adv. Colloid Interface Sci.* **2004**, *108–109*, 197–205. [CrossRef]

229. Ma, C.D.; Acevedo-Velez, C.; Wang, C.; Gellman, S.H.; Abbott, N.L. Interaction of the hydrophobic tip of an atomic force microscope with oligopeptides immobilized using short and long tethers. *Langmuir* **2016**, *32*, 2985–2995. [CrossRef]

230. Pfreundschuh, M.; Alsteens, D.; Wieneke, R.; Zhang, C.; Coughlin, S.R.; Tampe, R.; Kobilka, B.K.; Müller, D.J. Identifying and quantifying two ligand-binding sites while imaging native human membrane receptors by AFM. *Nat. Commun.* **2015**, *6*, 1–7. [CrossRef]

231. Litvinov, R.I.; Shuman, H.; Bennett, J.S.; Weisel, J.W. Binding strength and activation state of single fibrinogen-integrin pairs on living cells. *Proc. Natl. Acad. Sci. USA* **2002**, *99*, 7426–7431. [CrossRef]

232. Friedrichs, J.; Legate, K.R.; Schubert, R.; Bharadwaj, M.; Werner, C.; Müller, D.J.; Benoit, M. A practical guide to quantify cell adhesion using single-cell force spectroscopy. *Methods* **2013**, *60*, 169–178. [CrossRef]

233. Zhang, X.; Wojcikiewicz, E.; Moy, V. Force spectroscopy of the leukocyte function-associated antigen-1/intercellular adhesion molecule-1 interaction. *Biophys. J.* **2002**, *83*, 2270–2279. [CrossRef]

234. Fleming, A.J. Dual-stage vertical feedback for high-speed scanning probe microscopy. *IEEE Trans. Control Syst. Technol.* **2011**, *19*, 156–165. [CrossRef]
235. Fantner, G.E.; Schitter, G.; Kindt, J.H.; Ivanov, T.; Ivanova, K.; Patel, R.; Holten-Andersen, N.; Adams, J.; Thurner, P.J.; Rangelow, I.W.; et al. Components for high speed atomic force microscopy. *Ultramicroscopy* **2006**, *106*, 881–887. [CrossRef]
236. Fantner, G.E.; Hegarty, P.; Kindt, J.H.; Schitter, G.; Cidade, G.A.G.; Hansma, P.K. Data acquisition system for high speed atomic force microscopy. *Rev. Sci. Instrum.* **2005**, *76*, 026118. [CrossRef] [PubMed]
237. Ando, T.; Uchihashi, T.; Fukuma, T. High-speed atomic force microscopy for nano-visualization of dynamic biomolecular processes. *Prog. Surf. Sci.* **2008**, *83*, 337–437. [CrossRef]
238. Alunda, B.O.; Lee, Y.J.; Park, S. A novel two-axis parallel-kinematic high-speed piezoelectric scanner for atomic force microscopy. *J. Korean Phys. Soc.* **2016**, *69*, 691–696. [CrossRef]
239. Yang, C.; Yan, J.; Dukic, M.; Hosseini, N.; Zhao, J.; Fantner, G.E. Design of a high-bandwidth tripod scanner for high speed atomic force microscopy. *Scanning* **2016**, *38*, 889–900. [CrossRef]
240. Adams, J.D.; Nievergelt, A.; Erickson, B.W.; Yang, C.; Dukic, M.; Fantner, G.E. High-speed imaging upgrade for a standard sample scanning atomic force microscope using small cantilevers. *Rev. Sci. Instrum.* **2014**, *85*, 093702. [CrossRef]
241. Leitner, M.; Fantner, G.E.; Fantner, E.J.; Ivanova, K.; Ivanov, T.; Rangelow, I.; Ebner, A.; Rangl, M.; Tang, J.; Hinterdorfer, P. Increased imaging speed and force sensitivity for bio-applications with small cantilevers using a conventional AFM setup. *Micron.* **2012**, *43*, 1399–1407. [CrossRef]
242. Alunda, B.O.; Otieno, L.O.; Chepkoech, M.; Byeon, C.C.; Lee, Y.J. Comparative study of trans-linear and trans-impedance readout circuits for optical beam deflection sensors in atomic force microscopy. *J. Korean Phys. Soc.* **2019**, *74*, 88–93. [CrossRef]
243. Fukuma, T.; Kimura, M.; Kobayashi, K.; Matsushige, K.; Yamada, H. Development of low noise cantilever deflection sensor for multienvironment frequency-modulation atomic force microscopy. *Rev. Sci. Instrum.* **2005**, *76*, 053704. [CrossRef]
244. Enning, R.; Ziegler, D.; Nievergelt, A.; Friedlos, R.; Venkataramani, K.; Stemmer, A. A high frequency sensor for optical beam deflection atomic force microscopy. *Rev. Sci. Instrum.* **2011**, *82*, 043705. [CrossRef]
245. Fukuma, T. Wideband low-noise optical beam deflection sensor with photothermal excitation for liquid-environment atomic force microscopy. *Rev. Sci. Instrum.* **2009**, *80*, 023707. [CrossRef] [PubMed]
246. Ando, T.; Kodera, N.; Uchihashi, T.; Miyagi, A.; Nakakita, R.; Yamashita, H.; Matada, K. High-speed atomic force microscopy for capturing dynamic behavior of protein molecules at work. *e-J. Surf. Sci. Nanotechnol.* **2005**, *3*, 384–392. [CrossRef] [PubMed]
247. Tabak, F.C.; Disseldorp, E.C.M.; Wortel, G.H.; Katan, A.J.; Hesselberth, M.B.S.; Oosterkamp, T.H.; Frenken, J.W.M.; van Spengen, W.M. MEMS-based fast scanning probe microscopes. *Ultramicroscopy* **2010**, *110*, 599–604. [CrossRef]
248. Richter, C.; Weinzierl, P.; Krause, O.; Engl, W.; Penzkofer, C.; Irmer, B.; Sulzbach, T. Microfabricated ultrashort cantilever probes for high speed AFM. In Proceedings of the International Society for Optics and Photonics, Prague, Czech Republic, 5 May 2011. [CrossRef]
249. Ando, T.; Kodera, N.; Takai, E.; Maruyama, D.; Saito, K.; Toda, A. A high-speed atomic force microscope for studying biological macromolecules. *Proc. Natl. Acad. Sci. USA* **2001**, *98*, 12468–12472.
250. Kodera, N.; Yamamoto, D.; Ishikawa, R.; Ando, T. Video imaging of walking myosin v by high-speed atomic force microscopy. *Nature* **2010**, *468*, 72–76. [CrossRef]
251. Rico, F.; Gonzalez, L.; Casuso, I.; Puig-Vidal, M.; Scheuring, S. High-speed force spectroscopy unfolds titin at the velocity of molecular dynamics simulations. *Science* **2013**, *342*, 741–743. [CrossRef]
252. Ando, T. High-speed atomic force microscopy and its future prospects. *Biophys. Rev.* **2018**, *10*, 285–292. [CrossRef]
253. Heath, G.R.; Scheuring, S. Advances in high-speed atomic force microscopy (HS-AFM) reveal dynamics of transmembrane channels and transporters. *Curr. Opin. Struct. Biol.* **2019**, *57*, 93–102. [CrossRef]
254. Koehler, M.; Fis, A.; Gruber, H.J.; Hinterdorfer, P. AFM-based force spectroscopy guided by recognition imaging: A new mode for mapping and studying interaction sites at low lateral density. *Methods Protoc.* **2019**, *2*, 6. [CrossRef]

255. Noy, A.; Friddle, R.W. Practical single molecule force spectroscopy: How to determine fundamental thermodynamic parameters of intermolecular bonds with an atomic force microscope. *Methods* **2013**, *60*, 142–150. [CrossRef]

256. Helenius, J.; Heisenberg, C.P.; Gaub, H.E.; Muller, D.J. Single-cell force spectroscopy. *J. Cell Sci.* **2008**, *121*, 1785–1791. [CrossRef] [PubMed]

257. Butt, H.J.; Cappella, B.; Kappl, M. Force measurements with the atomic force microscope: Technique, interpretation and applications. *Surf. Sci. Rep.* **2005**, *59*, 1–152. [CrossRef] [PubMed]

258. Ando, T.; Uchihashi, T.; Scheuring, S. Filming biomolecular processes by high-speed atomic force microscopy. *Chem. Rev.* **2014**, *114*, 3120–3188. [CrossRef]

259. Humphris, A.D.L.; Miles, M.J.; Hobbs, J.K. A mechanical microscope: High-speed atomic force microscopy. *Appl. Phys. Lett.* **2005**, *86*, 1–3. [CrossRef] [PubMed]

260. Yu, H.; Siewny, M.G.W.; Edwards, D.T.; Sanders, A.W.; Perkins, T.T. Hidden dynamics in the unfolding of individual bacteriorhodopsin proteins. *Science (80)* **2017**, *355*, 945–950. [CrossRef]

261. Matusovsky, O.S.; Mansson, A.; Persson, M.; Cheng, Y.S.; Rassier, D.E. High-speed AFM reveals subsecond dynamics of cardiac thin filaments upon Ca2+ activation and heavy meromyosin binding. *Proc. Natl. Acad. Sci. USA* **2019**, *116*, 16384–16393. [CrossRef]

262. Frutos-Puerto, S.; Miró, C.; Pinilla-Gil, E. Nafion-protected sputtered-bismuth screen-printed electrode for on-site voltammetric measurements of Cd (II) and Pb (II) in natural water samples. *Sensors* **2019**, *19*, 279. [CrossRef]

263. Muenchen, D.K.; Martinazzo, J.; de Cezaro, A.M.; Rigo, A.A.; Brezolin, A.N.; Manzoli, A.; de Lima Leite, F.; Steffens, C.; Steffens, J. Pesticide detection in soil using biosensors and nanobiosensors. *Biointerface Res. Appl.* **2016**, *6*, 6. [CrossRef]

264. Rotake, D.; Darji, A.D. Heavy metal ion detection in water using MEMs based sensor. *Mater. Today Proc.* **2018**, *5*, 1530–1536.

265. Rigo, A.A.; Cezaro, A.M.D.; Muenchen, D.K.; Martinazzo, J.; Brezolin, A.N.; Hoehne, L.; Steffens, C. Cantilever nanobiosensor based on the enzyme urease for detection of heavy metals. *Braz. J. Chem. Eng.* **2019**, *36*, 1429–1437. [CrossRef]

266. Muenchen, D.K.; Martinazzo, J.; Brezolin, A.N.; de Cezaro, A.M.; Rigo, A.A.; Mezarroba, M.N.; Manzoli, A.; De Lima Leite, F.; Steffens, J.; Steffens, C. Cantilever Functionalization using peroxidase extract of low cost for glyphosate detection. *Appl. Biochem. Biotechnol.* **2018**, *186*, 1061–1073.

267. SoltanRezaee, M.; Bodaghi, M. Simulation of an electrically actuated cantilever as a novel biosensor. *Sci. Rep.* **2020**, *10*, 1–14. [CrossRef]

268. Sutter, M.; Faulkner, M.; Aussignargues, C.; Paasch, B.C.; Barrett, S.; Kerfeld, C.A.; Liu, L.N. Visualization of bacterial microcompartment facet assembly using high-speed atomic force microscopy. *Nano Lett.* **2016**, *16*, 1590–1595. [CrossRef] [PubMed]

269. Possas-Abreu, M.; Rousseau, L.; Ghassemi, F.; Lissorgues, G.; Habchi, M.; Scorsone, E.; Cal, K.; Persaud, K. Biomimetic diamond MEMS sensors based on odorant-binding proteins: Sensors validation through an autonomous electronic system. In Proceedings of the International Symposium on Olfaction and Electronic Nose (ISOEN), Montreal, QC, Canada, 28–31 May 2017. [CrossRef] [PubMed]

270. Liu, X.; Wang, L.; Zhao, J.; Zhu, Y.; Yang, J.; Yang, F. Enhanced binding efficiency of microcantilever biosensor for the detection of Yersinia. *Sensors* **2019**, *19*, 3326.

271. Bertke, M.; Xu, J.; Setiono, A.; Kirsch, I.; Uhde, E.; Peiner, E. Fabrication of a microcantilever-based aerosol detector with integrated electrostatic on-chip ultrafine particle separation and collection. *Micromech. Microeng.* **2020**, *30*, 014001. [CrossRef]

272. Guillaume-Gentil, O.; Grindberg, R.V.; Kooger, R.; Dorwling-Carter, L.; Martinez, V.; Ossola, D.; Pilhofer, M.; Zambelli, T.; Vorholt, J.A. Tunable single-cell Extraction for molecular analyses. *Cell* **2016**, *166*, 506–516. [CrossRef]

273. Wang, J.; Zhu, Y.; Zhang, J.; Yang, J. Development of microcantilever sensors for liver cancer detection. *Adv. Cancer Prev.* **2016**, *1*, 1000103. [CrossRef]

274. Kamble, C.; Panse, M.S. Microdevices for low-level acetone gas sensing using tungsten trioxides. *IETE J. Res.* **2019**, 1–8. [CrossRef]

Sensors **2020**, *20*, 4784

275. Kim, H.; Hoshi, M.; Iijima, M.; Kuroda, S.; Nakamura, C. Development of a universal method for the measurement of binding affinities of antibody drugs towards a living cell based on AFM force spectroscopy. *Anal. Methods* **2020**, *12*, 2922–2927. [CrossRef]
276. Korayem, M.H.; Heidary, K.; Rastegar, Z. The head and neck cancer (HN-5) cell line properties extraction by AFM. *J. Biol. Eng.* **2020**, *14*, 1–15. [CrossRef]

Article

Contact Resonance Atomic Force Microscopy Using Long, Massive Tips

Tony Jaquez-Moreno, Matteo Aureli and and Ryan C. Tung *

Mechanical Engineering Department, University of Nevada, Reno, 1664 N. Virginia St, Reno, NV 89557-0312, USA; tonyj@nevada.unr.edu (T.J.-M.); maureli@unr.edu (M.A.)
* Correspondence: rtung@unr.edu

Received: 28 October 2019; Accepted: 12 November 2019; Published: 15 November 2019

Abstract: In this work, we present a new theoretical model for use in contact resonance atomic force microscopy. This model incorporates the effects of a long, massive sensing tip and is especially useful to interpret operation in the so-called trolling mode. The model is based on traditional Euler–Bernoulli beam theory, whereby the effect of the tip as well as of the sample in contact, modeled as an elastic substrate, are captured by appropriate boundary conditions. A novel interpretation of the flexural and torsional modes of vibration of the cantilever, when not in contact with the sample, is used to estimate the inertia properties of the long, massive tip. Using this information, sample elastic properties are then estimated from the in-contact resonance frequencies of the system. The predictive capability of the proposed model is verified via finite element analysis. Different combinations of cantilever geometry, tip geometry, and sample stiffness are investigated. The model's accurate predictive ranges are discussed and shown to outperform those of other popular models currently used in contact resonance atomic force microscopy.

Keywords: cantilever based sensors; atomic force microscopy; contact resonance; trolling mode

---

1. Introduction

Contact Resonance (CR) atomic force microscopy (AFM) is a relatively new, popular measurement technique used to characterize nanoscale material properties. CR AFM relies on analyzing the coupled vibrations of an AFM cantilever probe that is resonated while in permanent, net-repulsive contact with a sample of interest. CR AFM has been used to characterize the properties of thin metallic films [1] and polymer blends [2]. CR AFM has also been used to measure the viscoelastic loss tangents of polymer blends [3], study the effect of relative humidity on the viscoelastic properties of organic thin films [4], and conduct photorheological measurements to study curing kinetics of polymers [5]. Additionally, CR AFM has been used to measure buried, subsurface nanostructures [6–9] that are not visible in typical topographic AFM measurements. Finally, the principles of contact resonance have been used to enhance other popular modes of AFM, such as electrochemical strain microscopy (ESM) [10–12] and piezoresponse force microscopy [13,14] (PFM), and researchers have developed new experimental measurement procedures and techniques for CR AFM that aim to increase the accuracy of these coupled methods [15].

The underlying theoretical model of CR AFM utilizes the Euler–Bernoulli (EB) beam model. To date, researchers have included the effects of tip offset [16], tip height effects [17], normal and lateral contact springs [17], Poisson's ratio of the sample material [18], and sample viscoelasticity effects [19]. More recent modeling efforts have included the effect of using U-shaped cantilever probes [20] and using a Timoshenko beam model in the theoretical framework [21].

Recently, AFM cantilever sensor designs have included large sensing tips. For example, the qPlus sensor [22] uses a massive tip affixed to a quartz tuning fork tine and is capable of conducting extremely sensitive measurements. Long, massive tips have been affixed to AFM cantilevers and are used in operational modes such as "Trolling Mode" [23,24] to measure material properties of polymers and living cells. By using long sensing tips, researchers are able to remove the main cantilever body from liquid environments [25–27], thereby reducing unwanted hydrodynamic forces on the cantilever [28–34] and reducing extraneous noise sources prevalent in liquid AFM imaging environments [35]. Notwithstanding the practical importance that these techniques are gaining and their potential to open up new sensing modalities in CR AFM, rigorous analyses of the effect that the long, massive tip has on the system dynamics are so far lacking in the established literature. Therefore, incomplete understanding of the behavior and idiosyncrasies of cantilever-based sensors endowed with long, massive tips is limiting their applications and adoption in key sensing areas.

To bridge this knowledge gap, in this work, we analyze the behavior of AFM cantilever probes with long, massive tips to determine their effect on the surface-coupled vibrations of the system. To this aim, we modify the traditional EB model for cantilever vibration with a new set of boundary conditions that models both the presence of a long, massive tip (via the transverse force and moment that the tip, modeled as a rigid body, exerts on the cantilever) and contact with an elastic sample. Since the effect of the tip is only included in the boundary conditions, possible dynamics of the tip are not explicitly captured. However, a procedure to estimate the effective inertia and moment of inertia of the tip, as seen by the cantilever, is proposed based on a novel interpretation of flexural and torsional modes of vibration of the structure when not in contact with the sample. Contact with an elastic sample is modeled via an orthogonal set of springs, coupled to the cantilever tip, capable of linear elastic response in the transverse and in-plane directions.

An estimation procedure for the sample stiffness is then proposed based on analysis of the free and in-contact resonance frequencies of the system. The dynamics of the system are also investigated using a finite element model simulation to verify the proposed model and assess the impact of the modeling hypotheses. Particular interest is placed on the flexibility of the tip and its effect on the accuracy of the prediction. The proposed model is shown to be superior to traditional models which ignore inertia and moment of inertia of the long, massive tip for a broad range of system dimensions and stiffness parameters. Thus, the proposed model paves the way for correct interpretation of trolling mode CR AFM experiments.

The remainder of the paper is organized as follows. In Section 2, we develop the theoretical model for flexural and torsional vibrations and introduce the characteristic equations on which the estimation procedure hinges. In Section 3, we detail our numerical experiments conducted in lieu of physical experiments on fabricated cantilever sensors. Results and discussions are presented in Section 4, where we discuss the limits of applicability of the proposed model. Conclusions are reported in Section 5.

2. Theory and Model Development

In this section, we develop a simple model for a cantilever beam endowed with a long, massive tip in contact with an elastic substrate, representative of typical CR AFM configurations in trolling mode operations. To maintain a realistic model, with manageable complexity, we introduce a set of assumptions whose validity will be analyzed in the rest of the paper. Figure 1 depicts a schematic representation of the idealized system under study. Small amplitude vibrations are considered throughout.

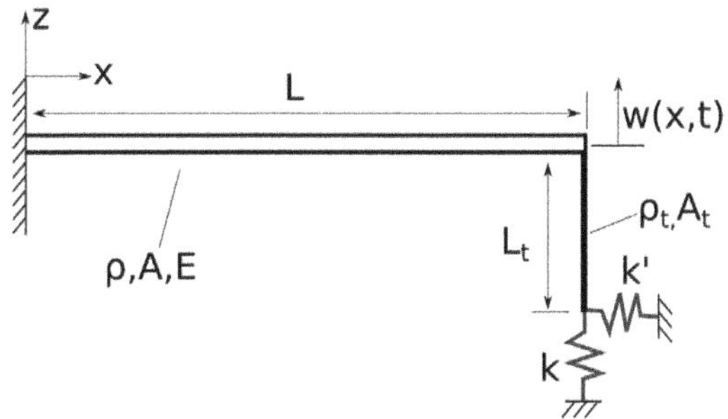

Figure 1. Euler–Bernoulli beam model of a cantilever with a long, massive tip in contact with an elastic substrate. Here, the tip is assumed to be rigid. The substrate is modeled through linear springs.

2.1. The Flexural Problem

With reference to Figure 1, we first focus on flexural vibrations of the beam in the xz-plane, exclusively. The beam is assumed to be of an isotropic and homogeneous material, with Young's modulus E and Poisson's ratio ν. Furthermore, ρ denotes the mass density (per unit volume) of the beam; A is the rectangular cross-sectional area, assumed to be constant along the axis; and L is the length of the beam. In this model, the long sensing tip is fixed at the end of the beam, with length L_t, mass density ρ_t, and circular cross-sectional area A_t. At this stage, the tip is assumed to be rigid, and connected to two one-dimensional linear springs of constants k and k' in the z- and x-directions, respectively. These springs model the normal and lateral stiffness of the sample in contact. The equations of motion for the transverse vibrations of the unforced system are given by [36]

$$\rho A \frac{\partial^2 w(x,t)}{\partial t^2} + EI \frac{\partial^4 w(x,t)}{\partial x^4} = 0, \tag{1}$$

where I is the second area moment of inertia of the cantilever beam and $w(x,t)$ represents the transverse displacement of the beam at a given location x along the axis and a specified time t.

Translational and rotational inertia effects of the massive tip, along with sample stiffness, are incorporated into the model in Equation (1) via the following boundary conditions [36]:

$$w(0,t) = 0, \tag{2a}$$

$$\frac{\partial w}{\partial x}(0,t) = 0, \tag{2b}$$

$$EI \frac{\partial^2 w(L,t)}{\partial x^2} = -I_t \frac{\partial^3 w(L,t)}{\partial x \partial t^2} - k' L_t^2 \frac{\partial w(L,t)}{\partial x}, \tag{2c}$$

$$EI \frac{\partial^3 w(L,t)}{\partial x^3} = m_t \frac{\partial^2 w(L,t)}{\partial t^2} + kw(L,t), \tag{2d}$$

where I_t is the rotational inertia and m_t is the total mass of the sensing tip. In Equations (2c) and (2d), the sensing tip has effectively been modeled as a point-mass and point-inertia. Specifically, in Equation (2c), the cantilever end is subject to a bending moment due to the rotational inertia of the tip, along with the reaction from the lateral stiffness of the sample. Similarly, in Equation (2d), the cantilever end is subject to a shear force due to the translational inertia of the massive tip and to the normal stiffness of the sample. Note that, consistent with the assumptions of small displacements and deformations, higher order contributions to tip shear force and bending moment due to changes in length of the cantilever are neglected. It is also important to observe that any effects that may be related to deformability of the tip are ignored.

For free vibrations at a given frequency ω, the boundary condition in Equation (2c) is equivalent to the effect of a torsional spring connected to the cantilever tip, with effective torsional (dynamic) stiffness given by $K_T = k'L_t^2 - \omega^2 I_t$. Similarly, the boundary condition in Equation (2d) is equivalent to the effect of a normal spring connected to the cantilever tip, with effective (dynamic) stiffness given by $K_N = k - \omega^2 m_t$. These effective dynamic stiffnesses will be used later in the discussion of the model's performance.

Through dimensional analysis of the equations of motion and associated boundary conditions, we identify the following governing nondimensional parameters: α is the nondimensional ratio of the normal spring stiffness k to the cantilever static stiffness $k_c = (3EI)/L^3$, so that $\alpha = k/k_c$; Δ is the nondimensional tip mass given by $\Delta = m_t/(\rho AL)$; $\hat{I}_t$ is the nondimensional rotational inertia of the tip given by $\hat{I}_t = I_t/(\rho AL^3)$; ϕ is the ratio of the lateral to the normal spring stiffnesses $\phi = k'/k$; and $\ell = L_t/L$ is the ratio between the tip length and cantilever length. The limit of $\alpha = 0$ corresponds to the case of an "unsprung" cantilever, and the limit of $\Delta = \hat{I}_t = 0$ corresponds to the case of an ideally massless tip.

Next, as in standard practice [36], we assume that the solution for $w(x,t)$ is separable, that is $w(x,t) = W(x)T(t)$. Substituting this ansatz into Equation (1) results in a fourth order ordinary differential equation (ODE) in the spatial dimension x and a second order ODE in the time dimension t. The general form of the spatial solution is given by $W(x) = C_1 \cos(\lambda x) + C_2 \sin(\lambda x) + C_3 \cosh(\lambda x) + C_4 \sinh(\lambda x)$, where λ is the separation constant. The general spatial solution along with the boundary conditions in Equation (2) form the eigenvalue problem (EVP) that governs the eigenmodes and eigenfrequencies of the system. Solution of the EVP generates the characteristic equation $f(\lambda L, \alpha, \Delta, \hat{I}_t, \phi, \ell) = 0$, which describes the relationship between the natural frequencies of the system and the governing nondimensional parameters. Here, λL are the countably infinite nondimensional natural frequencies of the system given by $(\lambda L)^4 = \omega^2(\rho AL^4)/(EI)$, where ω is the dimensional natural frequency. The complete characteristic equation for transverse vibrations of the system is given by

$$
\begin{aligned}
&\left[\left(-2\Delta \hat{I}_t(\lambda L)^8 + (2 + (6\Delta \ell^2 \phi + 6\hat{I}_t)\alpha)(\lambda L)^4 - 18\ell^2\alpha^2\phi\right)\cos(\lambda L)\right. \\
&\left. + 6(\lambda L)\sin(\lambda L)\left(-\hat{I}_t(\lambda L)^6/3 + \phi\ell^2\alpha(\lambda L)^2 - \Delta(\lambda L)^4/3 + \alpha\right)\right]\cosh(\lambda L) + \\
&6(\lambda L)\left(-\hat{I}_t(\lambda L)^6/3 + \phi\ell^2\alpha(\lambda L)^2 + \Delta(\lambda L)^4/3 - \alpha\right)\sinh(\lambda L)\cos(\lambda L) + 2\Delta\hat{I}_t(\lambda L)^8 + \\
&\left(2 + (-6\Delta\ell^2\phi - 6\hat{I}_t)\alpha\right)(\lambda L)^4 + 18\ell^2\alpha^2\phi = 0.
\end{aligned}
\tag{3}
$$

Equation (3) defines the relationship between the transverse natural frequencies of vibration of the system, the sample stiffness in both the normal and lateral directions, and the tip mass and rotational inertia.

2.2. The Torsional Problem

In the development of the model, we use the freely vibrating, unsprung torsional modes of vibration of the system to estimate the rotational inertia of the massive tip. Within this approach, we continue to assume that the tip is rigid. Note that, because of its circular cross section, the tip is symmetric about the axes of rotation excited in transverse and torsional bending motions. This means that the tip rotational inertia identified from torsional oscillation can reasonably be used as a proxy for the tip rotational inertia needed in Equation (2) and, thus, in Equation (3).

With reference to the schematics in Figure 1, and focusing exclusively on torsional vibrations about the x-axis of the beam, the equations of torsional motion of the free, unsprung system are given by [36]

$$
\rho J \frac{\partial^2 \theta(x,t)}{\partial t^2} = C \frac{\partial^2 \theta(x,t)}{\partial x^2},
\tag{4}
$$

where J is the polar moment of inertia of the beam cross section, $\theta(x,t)$ is the twist angle of the beam, and C is the torsional rigidity of the beam. For a rectangular cross section of thickness h and width b, C is given by $C = \kappa G h^3 b$, where κ is given by [37]

$$\kappa = \frac{1}{3}\left(1 - \frac{192}{\pi^5}\frac{h}{b}\sum_{i=1,3,5,\ldots}^{\infty}\frac{1}{i}\tanh\left(i\pi\frac{b}{h}\right)\right) \tag{5}$$

and $G = E/[2(1+\nu)]$ is the shear modulus of the beam. For thin cross sections with $h \ll b$, κ is well approximated by the value $1/3$, see for example [34]. The boundary conditions for Equation (4) are given by

$$\theta(0,t) = 0, \tag{6a}$$

$$C\frac{\partial\theta(L,t)}{\partial x} = -I_t\frac{\partial^2\theta(L,t)}{\partial t^2}, \tag{6b}$$

which show that the free end of the cantilever is subject to a twisting torque caused by the rotational inertia of the tip. Equation (6b) suggests the existence of an additional nondimensional parameter, namely, $\hat{I}_{tor} = I_t/(\rho J L)$, which represents the nondimensional rotational inertia of the tip.

By assuming a separable solution for $\theta(x,t)$, we obtain the following characteristic equation

$$(\beta L)\cot(\beta L) - \hat{I}_{tor}(\beta L)^2 = 0, \tag{7}$$

where βL are the nondimensional natural frequencies of torsional vibration given by $\beta L = \omega_{tor}\sqrt{\rho J/CL}$, and ω_{tor} are the dimensional natural frequencies of torsional vibration. Equation (7), in the symbolic form $g(\beta L, \hat{I}_{tor}) = 0$, defines the relationship between the freely vibrating torsional modes and the rotational inertia of the massive tip.

2.3. Sample Stiffness Identification Procedure

Figure 2 schematically depicts the identification procedure used in this work. Assuming the availability of certain unsprung and sprung natural frequencies from an experiment, as well as of some basic material and geometry parameters, the proposed procedure is capable of identifying the unknown sample stiffness.

In the first step, the first freely vibrating torsional frequency $T_1 = \omega_{tor,1}$ (assumed to be measured from an experiment or otherwise available) is used as an input to solve Equation (7) for the nondimensional rotational inertia of the tip $\hat{I}_{tor}$. The measured value T_1 is converted to the nondimensional eigenvalue $\beta_1 L$ via the relationship given above in the discussion of Equation (7). Then, from Equation (7), we have

$$\hat{I}_{tor} = (\beta_1 L)^{-1}\cot(\beta_1 L). \tag{8}$$

The nondimensional value $\hat{I}_{tor}$, once determined, is then converted to the nondimensional value $\hat{I}_t$ using the relation $\hat{I}_t = \hat{I}_{tor}[(b^2 + h^2)/(12L^2)]$, where we have used the definitions of these nondimensional quantities and the fact that $A = bh$ and $J = (bh^3 + hb^3)/12$ for a rectangular cross section with width b and thickness h.

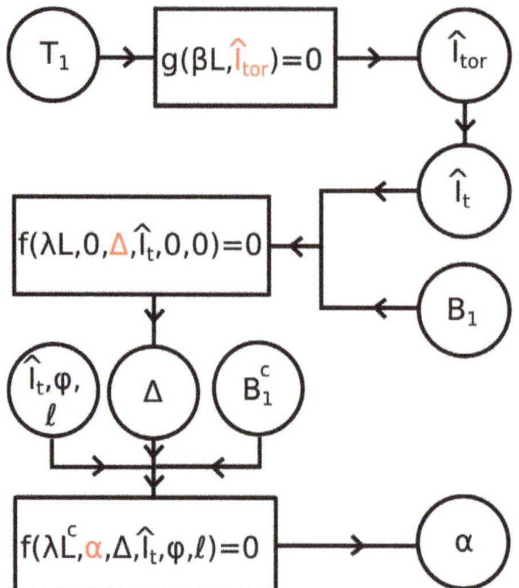

Figure 2. Schematic flowchart of the estimation procedure. At each step, the quantities highlighted in red are the unknowns to be estimated.

Next, $\hat{I}_t$ and the first freely vibrating unsprung transverse natural frequency $B_1 = \omega_1$ (assumed to be measured from an experiment or otherwise available) are used as input to solve Equation (3) with $\alpha = 0$, $\phi = 0$, and $\ell = 0$ for the nondimensional tip mass Δ. Specifically, we find from Equation (3)

$$\Delta = \frac{[1 + \cos(\lambda L)\cosh(\lambda L)] - \hat{I}_t(\lambda L)^3[\sin(\lambda L)\cosh(\lambda L) + \cos(\lambda L)\sinh(\lambda L)]}{\lambda L\{\hat{I}_t(\lambda L)^3[\cos(\lambda L)\cosh(\lambda L) - 1] + [\sin(\lambda L)\cosh(\lambda L) - \cos(\lambda L)\sinh(\lambda L)]\}}, \tag{9}$$

where λL should be evaluated at the $\lambda_1 L$ value determined from $B_1 = \omega_1$. The value B_1 is converted to the nondimensional natural frequency $\lambda_1 L$ via the relationship given above in the discussion of Equation (3). In the proposed framework, the nondimensional tip length ℓ only affects the moment generated by the lateral spring in Equation (2c) and does not influence the rotational inertia of the tip.

Finally, the estimated nondimensional mass Δ and rotational inertia $\hat{I}_t$ are used, along with the in-contact transverse natural frequency of vibration $B_1^c = \omega_1^c$, the lateral to normal stiffness ratio ϕ, and the tip length to cantilever ratio ℓ to solve Equation (3) for the nondimensional stiffness of the sample α. Equation (3) can be rearranged into the following quadratic equation in α:

$$c_2\alpha^2 + c_1\alpha + c_0 = 0, \tag{10}$$

where the coefficients of this polynomial are

$$c_2 = 9\ell^2\phi\left[\cosh(\lambda^c L)\cos(\lambda^c L) - 1\right], \tag{11a}$$

$$c_1 = -3(\lambda^c L)\left[\left((\lambda^c L)^3(\Delta\ell^2\phi + \hat{I}_t)\cos(\lambda^c L) + \sin(\lambda^c L)(\phi\ell^2(\lambda^c L)^2 + 1)\right)\cosh(\lambda^c L) + \right.$$

$$\left. \sinh(\lambda^c L)(\phi\ell^2(\lambda^c L)^2 - 1)\cos(\lambda^c L) - (\lambda^c L)^3(\Delta\ell^2\phi + \hat{I}_t)\right], \tag{11b}$$

$$c_0 = (\lambda^c L)^4\left[\left((\Delta\hat{I}_t(\lambda^c L)^4 - 1)\cos(\lambda^c L) + (\lambda^c L)\sin(\lambda^c L)(\hat{I}_t(\lambda^c L)^2 + \Delta)\right)\cosh(\lambda^c L) + \right.$$

$$\left. (\lambda^c L)\sinh(\lambda^c L)(\hat{I}_t(\lambda^c L)^2 - \Delta)\cos(\lambda^c L) - \Delta\hat{I}_t(\lambda^c L)^4 - 1\right]. \tag{11c}$$

Here, λ^c represents the in-contact eigenvalue of the problem which is related to ω_1^c, as described above in the discussion of Equation (3). It should be noted that, in this analysis, we assume that the stiffness ratio ϕ and the length ratio ℓ are known, to simplify the estimation procedure. However, Equation (3) could be solved with multiple measured in-contact natural frequencies to provide simultaneous estimates of ϕ, ℓ, and α, similar to the approaches discussed in [2,18,19,38].

When assuming the values of ϕ and ℓ and using a single measured in-contact natural frequency to estimate α, situations arise in which two real solutions of α may exist for Equation (10). This occurs when two distinct pairs of k and k' values, such that $k' = \phi k$, satisfy Equation (10) for the same in-contact natural frequency. This apparent paradox is resolved by considering the mode shape of vibration for each solution. For different pairs of k and k', different mode shapes at the same frequency can satisfy the equations of the system. Figure 3 shows such a case, where the mode shape for the larger α solution is plotted in solid black and the mode shape for the smaller α solution is plotted with a dash-dotted line. It is apparent that the solution for the lower α value is being generated by a higher order mode. Using the mode shape data from the model, along with the knowledge of which specific in-contact natural frequency is being used for property estimation, will ensure the proper α branch selection.

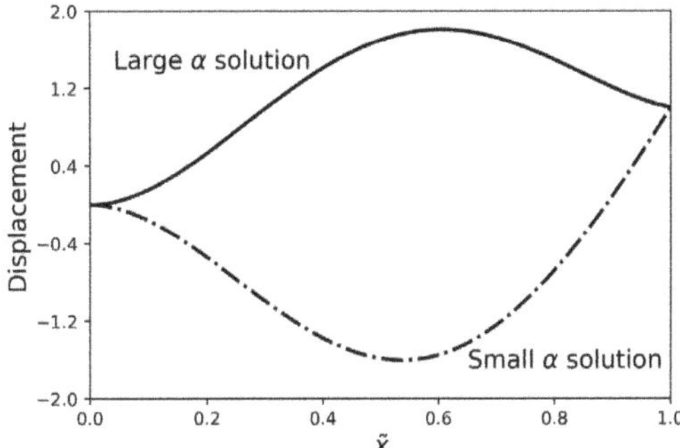

Figure 3. Mode shapes for two distinct α value solutions, at the same frequency, of Equation (10). These solutions represent distinct pairs of k and k' values. The mode shapes have been normalized such that the tip displacement equals one and the x-coordinate has been nondimensionalized by the cantilever length L such that $\tilde{x} = x/L$.

3. Numerical Experiments

To verify our identification procedure, in lieu of experimental data on the unsprung and in-contact flexural and torsional vibrations of the prototype cantilever in Figure 1, we conduct numerical experiments to simulate the system vibrational behavior via finite element analysis. A similar approach was previously employed by our group in [39]. The simulations are conducted within the ANSYS Mechanical APDL v. 17 commercial software package. Four different systems are analyzed in detail, as discussed below. The first few modes of vibration are identified for these systems, for a variety of sample stiffnesses. Finite element results on the unsprung flexural and torsional frequencies, as well as on the in-contact flexural frequencies, are then used as input in the identification procedure, as depicted in the flowchart in Figure 2.

The cantilever beam system, schematically depicted in Figure 1, is implemented in the finite element analysis via three-dimensional 2-node beam elements, with six degrees of freedom per node. The beam elements are based on Timoshenko beam theory [36] with shear deformability. Timoshenko beam theory is selected in the numerical experiments as it is expected to accurately model the vibration behavior of the real system. However, since only linear modal analyses are conducted, we do not anticipate significant discrepancies between the Timoshenko and the EB theories for the lowest modes of vibration of sufficiently slender beams.

The sample stiffness during in-contact operation is implemented via one-dimensional linear springs. Note that, as opposed to Figure 1, where we focus on a two-dimensional problem, since this implementation is completely three-dimensional we incorporate two lateral springs k' in the x- and y-directions. The origin of the Cartesian coordinate system coincides with the centroid of the cross section of the fixed end of the beam.

We assume that the cantilever material is silicon with the following properties: $E = 169$ GPa, $\rho = 2330$ kg/m^3, and $\nu = 0.25$. Similarly, we assume that platinum is used for the tip, with the following material properties: $E_t = 171$ GPa, $\rho_t = 21,450$ kg/m^3, and $\nu_t = 0.39$. Note that, different from our work in Section 2, in our numerical experiment, we assume that the tip is deformable, as it would be in a real AFM scenario. We will comment on the effect of these modeling assumptions in the next section. Throughout the numerical campaign, we set $b = 30\,\mu$m and $h = 2\,\mu$m for the cross-sectional dimensions of the beam and $d = 3\,\mu$m for the diameter of the circular cross section of the tip. We further select $\phi = k'/k = 0.8$ for the lateral to normal stiffness of the sample. This value is within the theoretically allowed bounds [40,41] and is uniquely determined given the so-called reduced Young's modulus E^* and reduced shear modulus G^* of the system. For example, assuming the sample under test is silicon, with the aforementioned properties, and that both tip and sample are comprised of linearly elastic, homogeneous, and isotropic materials, the theoretical value for $\phi = 4G^*/E^* \approx 0.8$.

Four combinations of beam length L and tip length L_t are explored as reported in Table 1. These combinations are a long cantilever with a long tip (LCLT), a long cantilever with a short tip (LCST), a short cantilever with a long tip (SCLT), and a short cantilever with a short tip (SCST). Table 1 also reports the nomenclature adopted in the rest of the paper as well as the pertinent values of the nondimensional parameter ℓ. The table further reports the so-called static and dynamic stiffness ratios, denoted as R_s and R_d, respectively, between the tip stiffness and the cantilever stiffness. Specifically, $R_s = k_c/k_t$, where the tip stiffness is calculated as $k_t = 3E_tI_t/L_t^3$ and represents the ratio of the cantilever to the tip stiffnesses in static conditions. Similarly, $R_d = \sqrt{(EI\rho_t A_t L_t^4)/(E_t I_t \rho A L^4)}$ gives a measure of the overlap between the spectrum of the cantilever and of the tip as if they were independent uncoupled systems. A small or large value of R_d indicates essentially decoupled dynamics between the cantilever and the tip. Conversely, $R_d \approx 1$ indicates large coupling between the two. In the ideal case of a rigid tip, we have $R_s = 0$ and $R_d = 0$. Thus, the larger the corresponding numbers in Table 1, the further the departure from the initial hypotheses of a rigid tip. However, significant numerical departure from these values do not necessarily indicate poor predictions, as explained later.

Table 1. Geometries explored in the numerical experiments and associated nomenclature, along with resulting values of the nondimensional parameter ℓ and of static and dynamic stiffness ratios R_s and R_d, respectively.

Tip Length Beam Length	$L_t = 10\,\mu$m	$L_t = 50\,\mu$m
$L = 150\,\mu$m	SCST ($\ell = 0.067$) $R_s = 7.36 \times 10^{-4}$ $R_d = 7.30 \times 10^{-3}$	SCLT ($\ell = 0.333$) $R_s = 9.21 \times 10^{-2}$ $R_d = 1.82 \times 10^{-1}$
$L = 300\,\mu$m	LCST ($\ell = 0.033$) $R_s = 9.21 \times 10^{-5}$ $R_d = 1.82 \times 10^{-3}$	LCLT ($\ell = 0.167$) $R_s = 1.15 \times 10^{-2}$ $R_d = 4.56 \times 10^{-2}$

The cantilever and the tip are meshed with beam elements with a uniform length of 0.1 μm. For the shortest tip length considered in this study $L_t = 10$ μm, this choice still leaves 100 elements along the axis of the tip (and significantly more along the axis of the beam), which is deemed satisfactory to capture the first few structural modes of the system. Since we are interested in the first flexural and the first torsional frequency, for each simulation case, we extract the lowest 10 structural modes. Although the exact ordering in the spectrum of flexural, torsional, and other out-of-plane vibrations depends on the particular geometric configuration, as well as on the values of the sample stiffness, the frequencies B_1, T_1, and B_1^c that are used as input in our model are always within the first ten modes and are thus available from the simulations.

Simulations are conducted for the four geometries for $\alpha = 0$, representative of the unsprung case as well as for values of α spanning the values $[10^{-3}, 10^3]$, thus capturing a broad range of sample stiffnesses, from very soft to very stiff, when compared to the static cantilever stiffness k_c.

4. Results and Discussion

4.1. Parameter Identification

Table 2 shows the predictions of the nondimensional added mass Δ and rotational inertia $\hat{I}_t$ for the four cantilever cases tested using the estimation method depicted in Figure 2. In this table, we also report values for the added mass parameter calculated as if the tip were a point mass $m_t' = \rho_t A_t L_t$, and the rotational inertia parameter calculated as if the tip were a rigid rod pinned at one of its ends, so that $I_t' = m_t L_t^2 / 3$. Note that m_t' and I_t' are, in general, different from the values used in the boundary conditions in Equation (2) and lead to nondimensional parameters, respectively indicated in Table 2 as Δ' and rotational inertia $\hat{I}_t'$. These parameters are defined as

$$\Delta' = \frac{m_t'}{m_c} = \frac{\rho_t A_t L_t}{\rho A L}, \tag{12a}$$

$$\hat{I}_t' = \frac{I_t'}{\rho A L^3} = \frac{m_t L_t^2 / 3}{\rho A L^3}. \tag{12b}$$

It is important to observe that the estimated nondimensional parameters are point-mass and point-inertia representations of the physical tip. The physical tip has spatial dimensions and inherent flexibility. Poor agreement of the estimated values with the prediction from Equation (12) does not necessarily indicate poor model performance. Specifically, for the Δ determinations, the discrepancy between the values estimated and the values determined with Equation (12a) are within 10% of each other for the long tip cases, but are very different for the short tip cases. Similarly, while discrepancies on the $\hat{I}_t$ determinations are within approximately 30% for the long tip cases, negative values are, surprisingly, observed for $\hat{I}_t$ for the short tip cases. This behavior is likely due to the model trying to capture unmodeled effects caused by the dynamics of the tip that, in simulations, can lead to negative values for $\hat{I}_{tor}$ in Equation (8).

Table 2. Estimated values of nondimensional added mass and rotational inertia $(\Delta, \hat{I}_t)$ using the method in Figure 2 for the four cantilever cases tested. Nondimensional values $(\Delta', \hat{I}_t')$ are calculated directly from assigned geometric and material properties from the previous section using Equation (12).

Cantilever	Δ	Δ'	$\hat{I}_t$	$\hat{I}_t'$
SCST	5.55×10^{-2}	7.23×10^{-2}	-1.30×10^{-5}	1.07×10^{-4}
SCLT	3.98×10^{-1}	3.62×10^{-1}	2.01×10^{-2}	1.34×10^{-2}
LCST	2.10×10^{-2}	3.62×10^{-2}	-2.39×10^{-5}	1.34×10^{-5}
LCLT	1.66×10^{-1}	1.81×10^{-1}	2.28×10^{-3}	1.67×10^{-3}

Figure 4 shows the model results of the estimation of the nondimensional stiffness α versus the assigned values of α used in the FEA simulations discussed in Section 3. Blue circles represent results from the current method (CM), as described in Figure 2, which includes the effect of tip length, mass, and rotational inertia. Red triangles represent results of the current method in which added mass and rotational inertia effects are neglected, henceforth referred to as the "massless tip model". The massless tip model can be found in works such as [18]. Finally, green squares represent the results of the current method in which tip length, mass, and rotational effects are neglected, henceforth referred to as the "no-tip model". No-tip models can be found in works such as [2,19]. Figure 5 shows the corresponding percentage error of each model relative to the prescribed α values (verification data) for the four explored cantilever geometries in Table 1.

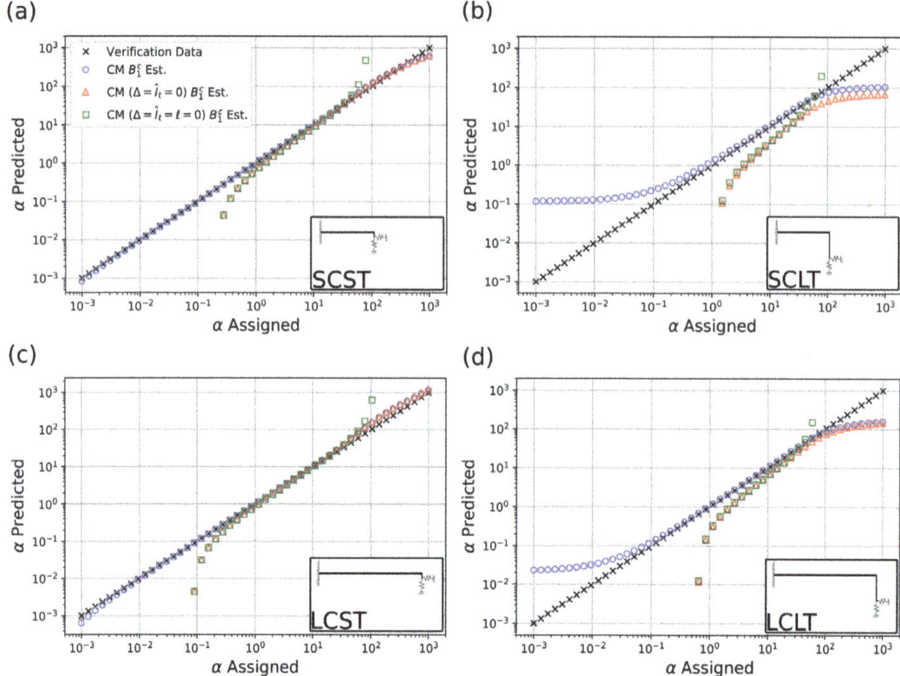

Figure 4. Model predictions versus FEA assigned values (verification data) for the four cantilever geometries in Table 1. Panels (**a–d**) correspond to different cantilever/tip geometries indicated by the inset figures. Black *x*'s represent the prescribed verification data, blue circles represent the current model proposed in this work, red triangles represent the current model with no added mass and no rotational inertia, and green squares represent the current model with no added mass, no rotational inertia, and zero tip length.

In the discussion of these results, we will first focus on the performance of the current method which incorporates tip length, added mass, and rotational inertia effects. We will then review the effects of neglecting tip length, added mass, and rotational effects.

4.2. Detection Range

The characteristic behavior of the current method result curves (blue circles) in Figure 4 can be summarized by observing that, independent of the particular geometry or cantilever case studied, the current method offers accurate detection of the assigned α value in the neighborhood of $\alpha = 1$. The range of accurate estimation varies for every case studied. In particular, we see the emergence of saturation tails at the low- and high-α ranges. The high-α saturation phenomenon was previously observed in a variety of studies, including for example [39], and is similar to the effect of replacing the free end of the cantilever with a simply supported end, as the sample stiffness increases with respect to the cantilever stiffness. That is, after a sufficiently high sample stiffness, the cantilever can no longer detect subsequent increases in stiffness and all higher stiffness can be described by the same fixity condition. Similar to the high-α range, in which the saturation effect is due to the low stiffness of the cantilever system, we posit that in the low-α range, the saturation effect is due to the low stiffness of the sample with respect to the cantilever system.

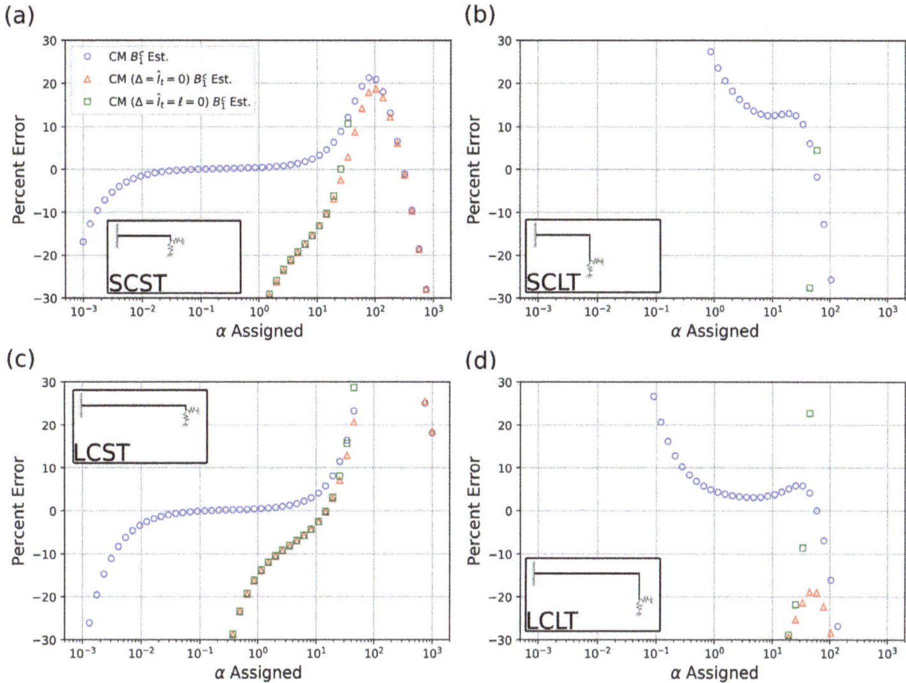

Figure 5. Percent error of the model predictions. Panels (**a–d**) correspond to different cantilever/tip geometries indicated by the inset figures.

Thus, the detection range with its two characteristic asymptotic tails, which can be identified as the range of values of α for which the current method yields essentially an "exact" prediction, see also Figure 5, can be interpreted and estimated as follows. In a fundamental sense, and neglecting several second order effects, the current method stipulates that the overall stiffness of the system, as described in Equations (1) and (2), can be described by the stiffness of the cantilever k_c in parallel with the normal and tangential (dynamic) stiffnesses K_N and K_T, introduced in the discussion of Equation (2). Neglecting, for simplicity, the contribution of the torsional spring, the overall stiffness of the system can thus be written as $k_c + K_N$. On the other hand, the finite element model, which for the purpose of this study is a proxy for a real experiment, introduces a slightly more complicated arrangement, whereby the tip stiffness k_t is to be considered in series with the (dynamic) stiffness K_N. The situation is schematically depicted in Figure 6.

Since the proposed model is required to interpret the simulation results within its assumptions, the estimation performance can be understood by equating the analytical model and the simulation stiffnesses, so that

$$k_c + \frac{k_t K_N^{(a)}}{k_t + K_N^{(a)}} = k_c + K_N^{(e)}, \tag{13}$$

where the superscripts (a) and (e) stand for assigned and estimated, respectively. After some manipulation, and using the definition of K_N, we have

$$k^{(e)} = \frac{k_t(k^{(a)} - \omega^2 m_t^{(a)})}{k_t + k^{(a)} - \omega^2 m_t^{(a)}} + \omega^2 m_t^{(e)}. \tag{14}$$

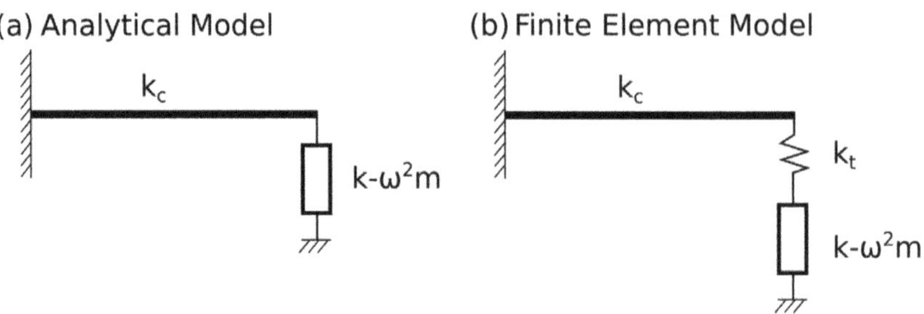

Figure 6. (a) Lumped parameter schematic of the current method's analytical model and (b) of the finite element model used for verification.

This formula allows us to explain the behavior of the estimation performance displayed in Figure 4. First, it is intuitive to assume that the model will yield better predictions as $k_t \to \infty$ or, in other words, as $R_s \to 0$. Indeed, if k_t dominates the denominator of Equation (14), we obtain the ideal case

$$\alpha^{(e)} \approx \alpha^{(a)} + \frac{\omega^2 (m_t^{(e)} - m_t^{(a)})}{k_c}, \tag{15}$$

which shows that the estimated value of α differs from the assigned value of α of a quantity that depends on the tip mass properties estimation error. It can be observed that such error is magnified for larger values of ω^2/k_c. This indicates that the estimation is expected to be more accurate for relatively stiff cantilevers ("SC" cases) and for shorter tips ("ST") cases, for which the tip mass m_t is small. This is in agreement with what was observed in Figure 4a,c for which, with the tip stiffness being equal, the case SCST displays higher values of k_c. A secondary effect further complicates this argument, as the quantity ω^2/k_c can be presumed to be close to the reciprocal of the system lumped mass M. Thus, the estimation is expected to be more accurate for relatively massive systems, which partially explains the better performance of the model for the LCLT case versus the SCLT case, in Figure 4b,d, respectively.

Let us now examine the case where $k \to \infty$, in other words, the high-α range, to uncover the reason for the saturation tails. In this case, $k^{(a)}$ dominates both numerator and denominator of Equation (14), which thus reduces to $k^{(e)} = k_t + \omega^2 m_t^{(e)}$. Dividing through by k_c, we obtain

$$\alpha^{(e)} = R_s^{-1} + \omega^2 m_t^{(e)}/k_c. \tag{16}$$

Since $R_s \ll 1$, this first term dominates and the asymptotic value of estimated α is equal to $\alpha^{(e)} = R_s^{-1}$. For the four cases depicted in Figure 4, the values of R_s^{-1} are approximately 1359 for the SCST case in Figure 4a, 10.9 for the SCLT case in Figure 4b, 10,858 for the LCST case in Figure 4c, and 87 for the LCLT case in Figure 4d. These values also roughly identify the starting point of the asymptotic high-α tails. Indeed, more generally, the saturation tails start occurring for a value of $\alpha^{(a)}$ roughly equal to R_s^{-1}. This observation, confirmed by the results in Figure 4, can be simply explained by observing that the horizontal asymptote should begin as $k^{(a)}$ in the denominator of Equation (14) and becomes comparable in magnitude to k_t or, equivalently, when $k^{(a)}/k_c \approx k_t/k_c$. As expected, model agreement becomes much poorer for the long tip cases (SCLT, LCLT), for which k_t is comparatively lower and R_s^{-1} is relatively large.

Finally, we examine the case where $k \to 0$, that is, the asymptotic horizontal branch in the low-α region. Once again, our point of departure is Equation (14) which, in the limit of zero sample stiffness, upon dividing through by k_c reduces to

$$\alpha^{(e)} = \frac{-\omega^2 m_t^{(a)}/k_c}{1 - R_s \omega^2 m_t^{(a)}/k_c} + \omega^2 m^{(e)}/k_c. \tag{17}$$

Note that, in Equation (17), the value of $\alpha^{(a)}$ does not appear explicitly and, therefore, the model cannot be expected to correctly estimate its value. In the hypothesis of $R_s \ll 1$, Equation (17) reduces to $\alpha^{(e)} \approx \omega^2(m^{(e)} - m^{(a)})/k_c$ and, presuming that $\omega^2/k_c \approx 1/M$ as above, $\alpha^{(e)} \approx (m^{(e)} - m^{(a)})/M$. In the "ST" cases, $(m^{(e)} - m^{(a)}) < 0$, as can be appreciated from the values for Δ and Δ' in Table 2. Indeed, for very low values of $\alpha^{(a)}$, the model yields negative values for $\alpha^{(e)}$, not displayed in Figure 4. More generally, broader accurate prediction ranges can be expected for the cases with larger k_c, as the saturation value $\alpha^{(e)}$ can take on smaller values. This is confirmed in Figure 4, where the SCST case in Figure 4a demonstrates better accuracy at low-$\alpha^{(a)}$ when compared to the LCST case in Figure 4c. The prominent low-α tails in Figure 4b,d are probably due to the massive tips causing Equation (17) to saturate for moderately large values of $\alpha^{(a)}$.

While the proposed analysis of the performance of the model is based on simplistic assumptions, our conclusions seem to be justified in view of the numerical experiments. It should also be observed that we have neglected the effect of rotational inertia and rotational stiffness embodied by K_T in the derivation of this simple argument. Including the rotational (dynamic) stiffness, however, is not expected to change the qualitative nature of the results.

In addition to the discussion above, in the low-α region, we believe that the large added mass of the tip has the effect of reducing the frequency sensitivity to changes in sample stiffness. Using a one-dimensional approximation of the system, and neglecting the tip stiffness, i.e., presuming $k_t \to \infty$, the natural frequency of the system is estimated as $\omega = \sqrt{(k + k_c)/M}$. The frequency sensitivity to changes in the system stiffness is then given by $d\omega/d(k + k_c) = 1/[2\sqrt{M}\sqrt{k + k_c}]$. Increasing the system mass or stiffness results in a decrease in frequency sensitivity. Thus, we expect the LCLT case, with lower added mass and stiffness, to outperform the SCLT case, as depicted in Figure 4b,d, and Figure 5b,d.

Importantly, model agreement for the SCLT and LCLT cases is also reduced due to the dynamic behavior of the long tip, which can no longer be treated as rigid. For long tips, new models incorporating the dynamics of the tip must be derived. This can be accomplished by considering an explicit EB-type equation for the tip to be coupled with the current governing dynamics in Equation (3). This derivation is however outside the scope of this paper, and will be tackled in subsequent work.

4.3. Performance of the Current Model Versus Traditional Models and Outlook

As discussed previously, the current method performs very well within its expected predictive range. For the long cantilever with short tip (LCST) case in Figure 4c, both the massless tip model and the no-tip model accurately predict within $\pm 10\%$ of the assigned α values for assigned α values centered around 10. In fact, their performance is nearly identical for much of the α range. This indicates that added mass and inertia are the primary effects to be considered for much of the α range. At high-α values, the massless tip model performs nearly identically with the current method. This indicates that as α is increased and the free bending mode of the cantilever tends to a highly constrained (or pinned) configuration, the primary effect in this range is due to tip length and not added mass or rotational inertia. Similar performance and behavior of the massless tip and no-tip model can be seen in the short cantilever with short tip (SCST) case in Figure 4a.

In the long tip cases (SCLT and LCLT), both the no-tip and massless tip models perform poorly. The current method performs well for the long cantilever with long tip case (LCLT) for moderate values of α and moderately well for the short cantilever with long tip (SCLT) case for moderate α values. As discussed above, we believe the prediction discrepancies are mainly due to unmodeled tip dynamics.

In the previous discussion, we have assumed perfect knowledge of the system parameters. However, uncertainties in the parameter estimation may exist in a realistic experimental setup. To assess the robustness of the proposed model, we have performed analyses to quantify the effect of uncertainty in the system parameters on the numerical predictions. Table 3 shows these results for one representative assigned α_a value, well within the detection range of the model, for the SCST case. In the analyses, we have individually varied each system parameter used in the prediction algorithm by $\pm 10\%$ and calculated the resulting estimation for α_e and compared it with the original estimate using the nominal system parameters. For low values of α_a ($\alpha_a < 10$) we see that a $\pm 10\%$ uncertainty in the system parameters has a negligible effect (less than 4%) on the prediction results. The effect of uncertainty in the system parameters increases as α_a is increased, especially in the range where model predictions with nominal parameters are already much less accurate. The largest prediction discrepancies are associated with uncertainty in the value of L. However, we do not expect difficulties in the experimental determination of this particular parameter within less than 10% uncertainty, for example, via optical microscopy.

Table 3. Effect of uncertainty in system parameters on prediction results for one assigned $\alpha_a = 0.8685$ for the SCST case. The estimation based on nominal system parameters is $\alpha_e = 0.8720$.

Parameter	Parameter +10% α_e Estimate	% Difference Predicted	Parameter −10% α_e Estimate	% Difference Predicted
b	0.8638	−0.94%	0.8787	0.77%
t	0.8675	−0.51%	0.8818	1.13%
L	0.9067	3.98%	0.8588	−1.51%
L_t	0.8709	−0.13%	0.8730	0.12%
ρ	0.8755	0.41%	0.8693	−0.31%
E	0.8695	−0.29%	0.8760	0.46%
ν	0.8713	−0.08%	0.8728	0.09%

Based on the results of this study, in cases where contact resonance microscopy will be used in conjunction with cantilevers that have tips of appreciable length and mass, it is recommended to use the current method for modeling and analysis purposes. It can be appreciated that, even for tips that introduce relatively small added mass, rotational inertia, and tip length effects (see for example the LCST case), the current method extends the predictive α range to very low α values. Thus, the proposed model will be particularly desirable when imaging soft samples in liquids, such as biological materials, using cantilevers with long tips in the trolling mode configuration.

5. Conclusions

In this work, we have introduced an updated theoretical model for contact resonance atomic force microscopy, incorporating the effects of a large, massive tip. The model employs a few geometric and material parameters, in conjunction with the knowledge of a limited number of unsprung resonance frequencies for both low-order torsional and flexural modes, for identification of some effective parameter of the system. These identified parameters are then used in the determinations of the in-contact sample stiffness from the knowledge of the in-contact resonance frequencies. The performance of the proposed model has been numerically verified using, in lieu of experimental data, results from high-fidelity finite element simulations. The updated model shows good agreement with the verification data. In general, when performing contact resonance atomic force microscopy using cantilevers with long, massive tips the bending stiffness of the tip should far exceed the bending stiffness of the cantilever. Additionally, the larger the added mass imparted to the cantilever by the tip, the larger the cantilever stiffness should be to ensure accurate measurements.

The model presented in this work has been specifically designed to be simple and easy to use, with only a minimum number of measured parameters to be obtained from an experimental campaign. However, the model is also amenable to several extensions currently in use in the field of contact resonance AFM, which may include incorporating sample viscoelasticity, including an adjustable tip position, and using multiple modes simultaneously for parameter estimation. We expect this model to be the first step in paving the way towards long tip, or trolling mode, configurations of contact resonance atomic force microscopy.

Author Contributions: Investigation, T.J.-M., M.A., and R.C.T.; methodology, T.J.-M., M.A., and R.C.T.; supervision, R.C.T.; writing–original draft, T.J.-M., M.A., and R.C.T.; writing—review and editing, T.J.-M., M.A., and R.C.T.

Funding: This material is based upon work supported by the National Science Foundation under Grant No. 1660448.

Conflicts of Interest: The authors declare no conflict of interest. The funders had no role in the design of the study; in the collection, analyses, or interpretation of data; in the writing of the manuscript; or in the decision to publish the results.

References

1. Kos, A.B.; Killgore, J.P.; Hurley, D.C. SPRITE: A modern approach to scanning probe contact resonance imaging. *Meas. Sci. Technol.* **2014**, *25*, 025405. [CrossRef]

2. Killgore, J.P.; Yablon, D.G.; Tsou, A.H.; Gannepalli, A.; Yuya, P.A.; Turner, J.A.; Proksch, R.; Hurley, D.C. Viscoelastic Property Mapping with Contact Resonance Force Microscopy. *Langmuir* **2011**, *27*, 13983–13987. [CrossRef] [PubMed]

3. Rezaei, E.; Turner, J.A. Contact resonance AFM to quantify the in-plane and out-of-plane loss tangents of polymers simultaneously. *Appl. Phys. Lett.* **2017**, *110*, 101902. [CrossRef]

4. Gonzalez-Martinez, J.F.; Kakar, E.; Erkselius, S.; Rehnberg, N.; Sotres, J. Effect of Relative Humidity on the Viscoelasticity of Thin Organic Films Studied by Contact Thermal Noise AFM. *Langmuir* **2019**, *35*, 6015–6023. [CrossRef]

5. Fiedler-Higgins, C.I.; Cox, L.M.; DelRio, F.W.; Killgore, J.P. Monitoring Fast, Voxel-Scale Cure Kinetics via Sample-Coupled-Resonance Photorheology. *Small Methods* **2019**, *3*, 1800275. [CrossRef]

6. Yip, K.; Cui, T.; Sun, Y.; Filleter, T. Investigating the detection limit of subsurface holes under graphite with atomic force acoustic microscopy. *Nanoscale* **2019**, *11*, 10961–10967. [CrossRef]

7. van Es, M.H.; Mohtashami, A.; Thijssen, R.M.; Piras, D.; van Neer, P.L.; Sadeghian, H. Mapping buried nanostructures using subsurface ultrasonic resonance force microscopy. *Ultramicroscopy* **2018**, *184*, 209–216. [CrossRef]

8. Ma, C.; Chen, Y.; Arnold, W.; Chu, J. Detection of subsurface cavity structures using contact-resonance atomic force microscopy. *J. Appl. Phys.* **2017**, *121*, 154301. [CrossRef]

9. Killgore, J.P.; Kelly, J.Y.; Stafford, C.M.; Fasolka, M.J.; Hurley, D.C. Quantitative subsurface contact resonance force microscopy of model polymer nanocomposites. *Nanotechnology* **2011**, *22*, 175706. [CrossRef]

10. Li, J.; Li, J.F.; Yu, Q.; Chen, Q.N.; Xie, S. Strain-based scanning probe microscopies for functional materials, biological structures, and electrochemical systems. *J. Mater.* **2015**, *1*, 3–21. [CrossRef]

11. Chen, Q.N.; Adler, S.B.; Li, J. Imaging space charge regions in Sm-doped ceria using electrochemical strain microscopy. *Appl. Phys. Lett.* **2014**, *105*, 201602. [CrossRef]

12. Proksch, R. Electrochemical strain microscopy of silica glasses. *J. Appl. Phys.* **2014**, *116*, 066804. [CrossRef]

13. Zhu, Q.; Pan, K.; Xie, S.; Liu, Y.; Li, J. Nanomechanics of multiferroic composite nanofibers via local excitation piezoresponse force microscopy. *J. Mech. Phys. Solids* **2019**, *126*, 76–86. [CrossRef]

14. Kofahl, C.; Güthoff, F.; Eckold, G. Direct observation of polar nanodomains in the incommensurate phase of (K0. 96Rb0. 04) 2ZnCl4 crystals using piezo force microscopy. *Ferroelectrics* **2019**, *540*, 10–17. [CrossRef]

15. Killgore, J.P.; Deolia, A.; Robins, L.; Murray, T.W. Experimental reconstruction of the contact resonance shape factor for quantification and amplification of bias-induced strain in atomic force microscopy. *Appl. Phys. Lett.* **2019**, *114*, 133108. [CrossRef]

16. Rabe, U.; Janser, K.; Arnold, W. Vibrations of free and surface-coupled atomic force microscope cantilevers: Theory and experiment. *Rev. Sci. Instrum.* **1996**, *67*, 3281–3293. [CrossRef]

17. Rabe, U.; Turner, J.; Arnold, W. Analysis of the high-frequency response of atomic force microscope cantilevers. *Appl. Phys. A Mater. Sci. Process.* **1998**, *66*, S277–S282. [CrossRef]

18. Hurley, D.C.; Turner, J.A. Measurement of Poisson's ratio with contact-resonance atomic force microscopy. *J. Appl. Phys.* **2007**, *102*, 033509. [CrossRef]

19. Yuya, P.A.; Hurley, D.C.; Turner, J.A. Contact-resonance atomic force microscopy for viscoelasticity. *J. Appl. Phys.* **2008**, *104*, 074916. [CrossRef]

20. Rezaei, E.; Turner, J.A. Contact resonances of U-shaped atomic force microscope probes. *J. Appl. Phys.* **2016**, *119*, 034303. [CrossRef]

21. Zhou, X.; Wen, P.; Li, F. Vibration analysis of atomic force microscope cantilevers in contact resonance force microscopy using Timoshenko beam model. *Acta Mech. Solida Sin.* **2017**, *30*, 520–530. [CrossRef]

22. Giessibl, F.J. The qPlus sensor, a powerful core for the atomic force microscope. *Rev. Sci. Instrum.* **2019**, *90*, 011101. [CrossRef] [PubMed]

23. Mohammadi, S.Z.; Moghaddam, M.; Pishkenari, H.N. Dynamical modeling of manipulation process in Trolling-Mode AFM. *Ultramicroscopy* **2019**, *197*, 83–94. [CrossRef] [PubMed]

24. Minary-Jolandan, M.; Tajik, A.; Wang, N.; Yu, M.F. Intrinsically high-Q dynamic AFM imaging in liquid with a significantly extended needle tip. *Nanotechnology* **2012**, *23*, 235704. [CrossRef] [PubMed]

25. Guan, D.; Barraud, C.; Charlaix, E.; Tong, P. Noncontact Viscoelastic Measurement of Polymer Thin Films in a Liquid Medium Using Long-Needle Atomic Force Microscopy. *Langmuir* **2017**, *33*, 1385–1390. [CrossRef] [PubMed]

26. Guan, D.; Charlaix, E.; Qi, R.Z.; Tong, P. Noncontact viscoelastic imaging of living cells using a long-needle atomic force microscope with dual-frequency modulation. *Phys. Rev. Appl.* **2017**, *8*, 044010. [CrossRef]

27. Zhang, Y.; Li, Y.; Song, Z.; Lin, R.; Chen, Y.; Qian, J. A High-Q AFM Sensor Using a Balanced Trolling Quartz Tuning Fork in the Liquid. *Sensors* **2018**, *18*, 1628. [CrossRef]

28. Sader, J.E. Frequency response of cantilever beams immersed in viscous fluids with applications to the atomic force microscope. *J. Appl. Phys.* **1998**, *84*, 64–76. [CrossRef]

29. Green, C.P.; Sader, J.E. Small amplitude oscillations of a thin beam immersed in a viscous fluid near a solid surface. *Phys. Fluids* **2005**, *17*, 073102. [CrossRef]

30. Green, C.P.; Sader, J.E. Torsional frequency response of cantilever beams immersed in viscous fluids with applications to the atomic force microscope. *J. Appl. Phys.* **2002**, *92*, 6262–6274. [CrossRef]

31. Tung, R.C.; Jana, A.; Raman, A. Hydrodynamic loading of microcantilevers oscillating near rigid walls. *J. Appl. Phys.* **2008**, *104*, 114905. [CrossRef]

32. Tung, R.C.; Killgore, J.P.; Hurley, D.C. Liquid contact resonance atomic force microscopy via experimental reconstruction of the hydrodynamic function. *J. Appl. Phys.* **2014**, *115*, 224904. [CrossRef]

33. Aureli, M.; Porfiri, M. Low frequency and large amplitude oscillations of cantilevers in viscous fluids. *Appl. Phys. Lett.* **2010**, *96*, 164102. [CrossRef]

34. Aureli, M.; Pagano, C.; Porfiri, M. Nonlinear finite amplitude torsional vibrations of cantilevers in viscous fluids. *J. Appl. Phys.* **2012**, *111*, 124915. [CrossRef]

35. Carrasco, C.; Ares, P.; De Pablo, P.; Gómez-Herrero, J. Cutting down the forest of peaks in acoustic dynamic atomic force microscopy in liquid. *Rev. Sci. Instrum.* **2008**, *79*, 126106. [CrossRef]

36. Meirovitch, L. *Analytical Methods in Vibrations*; MacMillan: New York, NY, USA, 1967.

37. Timoshenko, S.; Goodier, J.N. *Theory of Elasticity*; McGraw-Hill: New York, NY, USA, 1951.

38. Le Rouzic, J.; Delobelle, P.; Cretin, B.; Vairac, P.; Amiot, F. Simultaneous measurement of Young's modulus and Poisson's ratio at microscale with two-modes scanning microdeformation microscopy. *Mater. Lett.* **2012**, *68*, 370–373. [CrossRef]

39. Aureli, M.; Ahsan, S.N.; Shihab, R.H.; Tung, R.C. Plate geometries for contact resonance atomic force microscopy: Modeling, optimization, and verification. *J. Appl. Phys.* **2018**, *124*, 014503. [CrossRef]

40. Mott, P.; Roland, C. Limits to Poisson's ratio in isotropic materials—General result for arbitrary deformation. *Phys. Scripta* **2013**, *87*, 055404. [CrossRef]

41. Johnson, K.L. *Contact Mechanics*; Cambridge University Press: Cambridge, UK, 1987.

 *sensors*

Article

Atomic Force Microscopy Imaging in Turbid Liquids: A Promising Tool in Nanomedicine

Michael Leitner [1], Hannah Seferovic [1], Sarah Stainer [1], Boris Buchroithner [1], Christian H. Schwalb [2], Alexander Deutschinger [3] and Andreas Ebner [1,*]

[1] Institute of Biophysics, Johannes Kepler University, Gruberstraße 40, 4020 Linz, Austria; michael.leitner_1@jku.at (M.L.); hannah.seferovic@jku.at (H.S.); sarah.stainer@jku.at (S.S.); boris.buchroithner@jku.at (B.B.)

[2] GETec Microscopy GmbH, Seestadtstraße 27/Top 27, 1220 Vienna, Austria; chris.schwalb@getec-afm.com

[3] SCL-Sensor.Tech. Fabrication GmbH, Seestadtstraße 27/Top 27, 1220 Vienna, Austria; alexander.deutschinger@sclsensortech.com

* Correspondence: andreas.ebner@jku.at; Tel.: +43-732-2468-7637

Received: 10 June 2020; Accepted: 29 June 2020; Published: 2 July 2020

Abstract: Tracking of biological and physiological processes on the nanoscale is a central part of the growing field of nanomedicine. Although atomic force microscopy (AFM) is one of the most appropriate techniques in this area, investigations in non-transparent fluids such as human blood are not possible with conventional AFMs due to limitations caused by the optical readout. Here, we show a promising approach based on self-sensing cantilevers (SSC) as a replacement for optical readout in biological AFM imaging. Piezo-resistors, in the form of a Wheatstone bridge, are embedded into the cantilever, whereas two of them are placed at the bending edge. This enables the deflection of the cantilever to be precisely recorded by measuring the changes in resistance. Furthermore, the conventional acoustic or magnetic vibration excitation in intermittent contact mode can be replaced by a thermal excitation using a heating loop. We show further developments of existing approaches enabling stable measurements in turbid liquids. Different readout and excitation methods are compared under various environmental conditions, ranging from dry state to human blood. To demonstrate the applicability of our laser-free bio-AFM for nanomedical research, we have selected the hemostatic process of blood coagulation as well as ultra-flat red blood cells in different turbid fluids. Furthermore, the effects on noise and scanning speed of different media are compared. The technical realization is shown (1) on a conventional optical beam deflection (OBD)-based AFM, where we replaced the optical part by a new SSC nose cone, and (2) on an all-electric AFM, which we adapted for measurements in turbid liquids.

Keywords: piezoresistive cantilever; self-sensing; self-actuating; electrical readout; platelet; all electric AFM; blood; AFSEM

---

1. Introduction

Atomic force microscopy (AFM), invented in the late 1980s, has become an invaluable tool in nanoscience. Its field of application ranges from material to life science. Particularly in the life sciences, the unique possibility of molecular resolution under (almost) physiological conditions is convincing, thus justifying the importance of AFM in the study of physiological processes at the single molecule level. This unique possibility was and still is the basis for many successful research projects in which nature has been followed in its work [1–5], opening the window to physiology and nanomedicine. A prerequisite for all these studies was the possibility to perform investigations in transparent liquids (e.g., physiological buffers). A large number of important questions can be answered by measurements at such conditions, but investigations in physiologically relevant turbid liquids are not possible. Conventional AFMs work on the basis of optical readout of the cantilever deflection. For this purpose, a laser beam is focused on the backside of a reflectively coated cantilever. The position of the reflected beam is recorded by a four-segment photodiode. Although it is possible to vary the wavelength of the laser used from UV to infrared, this detection method still fails in the case of highly turbid and light-scattering physiological fluids, such as blood or milk.

The first successes in overcoming the disadvantages caused by optical beam deflection (OBD) were already achieved by Tortonese and colleagues as early as the beginning of the 1990s by implementing a p-type resistor on the surface of a cantilever [6]. Atomic resolution could be achieved by measuring the change in resistance caused by cantilever deflection [7]. Besides the piezoresistive readout based on doped silicon and polysilicon piezo-resistors [7–16], several other readout methods have been developed in the past to replace the classic OBD readout, such as: capacitive [17–21], piezoelectric [22–26], thin film metals [27] and tuning forks [28–31]. The installation of strain sensors directly into the cantilever is of particular interest as it offers several advantages over techniques with external readout. Some of the most important advantages are: (i) extremely small cantilevers far below the optical diffraction limit to increase sensitivity and so imaging speed can be realized [32,33], (ii) avoiding interference with photosensitive samples, (iii) the possibility of multi-cantilever arrays [34–36] and (iv) combining AFM with other techniques, such as scanning electron microscopy (SEM) [37–40]. In addition to the new readout method, advances in microfabrication have allowed the direct integration of actuators on the cantilever, such as piezoelectric excitation [41], Lorentz excitation [42], magnetic excitation [43–46] and thermal excitation [47]. Direct vibration excitation of the cantilever eliminates the need for an external piezo actuator and increases precision and excitation speed [48,49]. Efforts over the last 20 years to further optimize strain sensor readout have resulted in piezoresistive cantilevers that exceed standard optical beam readout in terms of low-noise imaging [15]. Furthermore, a large number of different applications in the AFM field [35,37,38,40,50–52], but also for other cantilever sensor techniques, such as torque magnetometry [53] or gas sensors [54], have been published. Although there is a great potential for the use of self-sensing AFM cantilevers in bio-applications, only very few attempts have been documented to use them in liquid [55] or for imaging biological samples in liquid [56]. To the best of our knowledge, biological AFM imaging in turbid physiological liquids has never been shown before.

In this study, we focused on the realization to perform self-sensing cantilevers (SSC)-based AFM imaging of highly interesting biological samples and to test and compare the performance with respect to different environmental and physical conditions. This was realized by upgrading a conventional AFM and by using an all-electric AFM originally developed for implementation in a SEM. We optimized both instruments for working in liquids. Initial experiments were performed in non-conductive deionized water and subsequently in physiologically relevant fluids, such as blood, blood serum and milk, without passivation of the cantilever. A passivation may change the physical parameters and thus worsen the imaging quality. In the approach used here, the entire cantilever is immersed in a sample beaker, which allows measurements in larger quantities of liquid and greatly simplifies handling compared to measurements in small sample aliquots or droplets.

2. Materials and Methods

2.1. Optical Beam Deflection Measurements

All OBD measurements were performed on a standard Keysight 5500 SPM2 (Keysight, Santa Rosa, CA, USA). Magnetically coated cantilevers (MAC Lever, Type VII) and magnetic excitation were used for the measurements in magnetic excitation mode (MAC Mode™, Keysight, Santa Rosa, CA, USA). Measurements in intermittent contact mode with acoustic excitation and contact mode measurements were performed using Bruker (Bruker, Billerica, MA, USA) MSCT cantilever E and cantilever D, respectively. This applies to all measurements under ambient conditions as well as to measurements in deionized water.

2.2. Self-Sensing Cantilevers

For all piezoresistive readout measurements, self-scanning cantilevers of the type PRSA-L300-F50-Si-PCB (300 × 100 μm) (SCL-Sensor.Tech., Vienna, Austria) were used. These cantilevers are equipped with two active piezo-resistors integrated on the cantilever and two passive piezo-resistors integrated on the chip. These four resistors are connected to a Wheatstone bridge. The resistance was about 1 kOhm each for all cantilevers used. Different supply voltages were used for the bridge depending on whether measurements were made in air or liquid. The standard supply voltage is 2.048 V for the dry state. For measurements in liquid, a supply voltage of 0.51 V was used. The cantilever itself has a size of 300 × 100 μm with a spring constant of 1–15 N/m and a resonance frequency of 30–65 kHz. Its carrier silicon chip is mounted on a Printed Circuit Board (PCB), which in turn is equipped with a 10-pin Kyocera standard connector (Kyocera, Kyoto, Japan) for easy connection to the AFM scanner. For the measurement of biological samples in liquids, cantilevers with spring constants of 1–1.3 N m^{-1} were chosen. In addition, these cantilevers are equipped with a heating loop for thermal excitation. Thermal excitation on the Keysight 5500 SPM2 was performed in magnetic AC (MAC) mode configuration, the power through the heating loop was limited to approximately 0.1 W.

2.3. Implementation of the SSC in a Commercial AFM with Optical Beam Deflection

For the implementation of the SSC into a Keysight 5500 SPM2 with OBD, two main technical developments were necessary: (i) a special nose cone (image is shown in Figure 1a left), which on the one hand supports the PCB, including the cantilever, and on the other hand, takes over the bidirectional signal transmission and amplification, and (ii) a second electronic unit between scanner and stage for additional amplification of the deflection signal and to generate signals mimicking a four-quadrant photodiode for the following hardware.

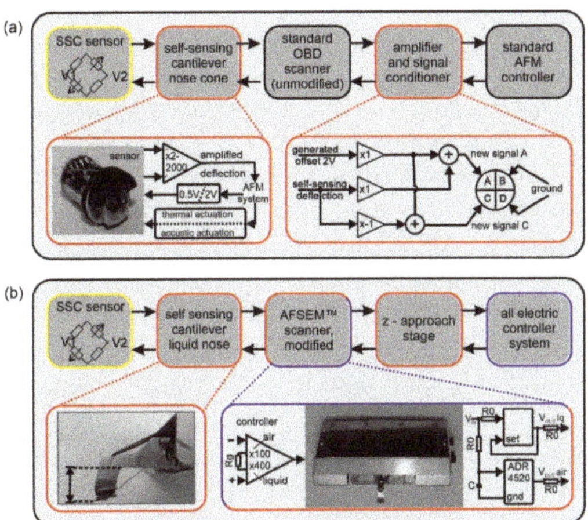

Figure 1. The flow chart in (**a**) shows the essential components for the entire electrical imaging with a commercial atomic force microscope (AFM): The self-sensing cantilever (SSC) (yellow) was mounted on the newly developed nose cone. The self-sensing signal was amplified in the nose cone, passed through an unmodified standard optical beam deflection (OBD) scanner and then subjected to additional signal amplification and processing. This is done in a new hardware between scanner and controller (new hardware components are red, unmodified are black). Part (**b**) of this figure shows a commercial self-sensing cantilever AFM adapted for liquid bio-sensing. The central upgraded parts are the extended nose ((**b**) left) and the new board design for the low-voltage sensor supply ((**b**) right) (new hardware components are red, unmodified ones are blue).

The mechanical part of the nose cone was milled from the high-temperature-resistant thermoplastic polyetheretherketone (PEEK) and adapted to carry the cantilever and the required electronics. The electronics in the nose cone consist of the following components: (i) an instrumentation amplifier with adjustable gain, (ii) power supply for the sensor, switchable between 0.51 V for measurements in liquid and 2.048 V for measurements in air, and (iii) the possibility to switch between thermal excitation via the integrated heating loop or acoustic excitation via a $2 \times 2 \times 2$ mm piezo (P.I., Lederhose, Germany). A multi-layer rigid-flexible PCB combination was developed to connect the electronics inside the nose cone with the sensor placed outside. The nose was designed in such a way that the optical path for positioning the cantilever relative to the sample is still available using the AFM's CCD camera. This design also ensures that the electronics is as close as possible to the deflection signal's origin but sealed from the cantilever in liquid. In addition to the voltage supply, the ground signal and the excitation signal (thermal/magnetic or acoustic) are taken from the scanner. The pin, originally intended for conductive measurements, is used for the deflection signal transfer. This configuration ensured that no changes had to be made to the original scanner hardware. The nose is connected to the scanner via the 6 original pins.

The second electronic hardware is located between the scanner, the stage and the photodiode connector and has two tasks: (i) additional deflection signal amplification via an adjustable instrumentation amplifier and (ii) signal conditioning to mimic four-quadrant photodiode signals for the following AFM hardware. For all self-built electronic components, a supply voltage of $+/-10$ V was taken directly from the photodiode connector. To work with the deflection signal of the self-sensing cantilever, a signal for a four-quadrant photodiode must be imitated for the controller. With optical readout, typically, four signals, A, B, C and D, are obtained. A and B represent the upper part of the four segments of the photodiode and C and D represent the lower part. From these four signals, a single deflection signal is generated using the method given in Equation (1):

$$\frac{(A+B)-(C+D)}{A+B+C+D} = \text{normalized deflection signal} \tag{1}$$

The fact that an SSC only emits a single deflection signal means that for the original AFM hardware, a four-quadrant photodiode signal has to be simulated. A scheme for generating the required signals is shown in Figure 1a (right). The original single deflection signal (DS) from the SSC is inverted and an offset voltage (OV) is added to both the original and inverted deflection signal. This results in two new signals (DS + OV) and (−DS + OV). The OV is generated with the same constant voltage source chip that is used for the sensor supply voltage (2.048 V). Now (DS + OV) is used for the photodiode segments A and B, and (−DS + OV) for the segments C and D. Using this in Equation (1) results in:

$$\frac{2*(DS+OV)-2*(-DS+OV)}{2*(DS+OV)+2*(-DS+OV)} = \frac{DS}{OV} = \text{normalized deflection signal} \tag{2}$$

The following original controller hardware restores the single deflection signal of the SSC. This signal processing is necessary in order to use SSCs on the Keysight 5500 SPM2 AFM without any technical modifications to the original hardware. In the hardware realization, A is set to (DS + OV), C to (−DS + OV), and B and D to ground. This simplifies Equation (2) to:

$$\frac{((DS+OV)+0)-((-DS+OV)+0)}{(DS+OV)+(-DS+OV)} = \frac{DS}{OV} = \text{normalized deflection signal} \tag{3}$$

All the following components use the normalized single deflection.

2.4. Software Implementation

With one exception, the software settings for SSC readout are identical to those for OBD readout: DC offset compensation is performed via the bias input field in the software. If the SSC is located away from the surface, the oscilloscope monitor in the software displays the amplified DC offset. This is compensated via the bias input field, which should result in an oscilloscope signal close to zero. For thermal excitation, the magnetic excitation is selected in the software. Here, the alternating current through the heater excites the cantilever to oscillate at its resonant frequency due to the bimetallic bimorph effect. The power consumption of the heating loop should not be higher than 0.1 W, otherwise damage may occur. Since the excitation is given as a percentage value in the software, the corresponding current I was measured and the power $P(heater)$ was calculated:

$$P(heater) = I^2 * R(heat) \leq 0.1W \tag{4}$$

The resistance $R(heat)$ of the heating loop is typically 28 Ohm. Consequently, the drive percentage has been set to a maximum of 20% to protect the cantilever heating loop.

2.5. Adaptation of the AFSEM® for Measurements in Liquids

Dry state and vacuum measurements with the AFSEM® scanner from GETec Microscopy (GETec Microscopy GmbH, Vienna, Austria) were performed without any changes to the system. For measurements in liquids, a new nose was designed and manufactured. This nose is shown in Figure 1b (left). The neck of the nose cone was extended for two reasons: (i) to allow the cantilever to be completely immersed in a sample chamber filled with liquid, and (ii) to increase the distance between the cantilever and the z-piezo. This ensures that the electronics are not damaged by liquid contact. Unlike the original nose, the piezo for acoustic excitation ($2 \times 2 \times 2$ mm P.I. Ceramics GmbH, Lederhose, Germany) was mounted through a hole from the top of the nose cone and positioned as close as possible to the sensor connector. For working in liquids, the supply voltage of the Wheatstone bridge was reduced from 2.048 to 0.51 V. To compensate for the lower electric potential of the signal, the gain has been increased by a factor of 4. Since the used scanner is designed as a tip-scanning AFM, a modified homemade sample stage was designed. The sample positioning in x and y was done by mechanical micromanipulators, while the z-axis approach was done by a fully automated z-stage with a 2″ piezo motor (Newport, Deckenpfronn, Germany). The scanner was mounted to the z-stage via a dove tail adapter and furthermore to a damped rod (Newport, Deckenpfronn, Germany). The control of the piezo motor is supported by the AFSEM® control system.

2.6. Preparation of the Cell Samples

Platelets: Thin glass slides (35×20 mm, Fischer, Austria) were cleaned by sonification in ethanol and distilled water for 5 min each and dried under a gentle nitrogen stream. A drop of fresh blood was taken with the help of an insulin lancet and incubated for 1, 3 and 5 min respectively, on the freshly cleaned glass slides. After incubation, the cell samples were carefully rinsed with deionized water and gently fixed with 0.5% formaldehyde solution for 15 min. Finally, they were again rinsed with deionized water and dried under a gentle stream of nitrogen gas. The cell samples were stored at 4 °C and used within one week. For measurements in liquids, the samples were rehydrated with the corresponding liquids of interest.

Red blood cells: The glass cover slides were cleaned as described for platelet preparation or alternatively with isopropanol instead of ethanol. In case of gold carriers, 9 MHz QCM (quartz crystal microbalance) crystals (Renlux Crystal, Shenzhen, China) were incubated in basic piranha (3:1 ammonia water and hydrogen peroxide). After rinsing with deionized water (5 times), the cleaned gold surface and the glass slides were treated identically. One drop (300 µL) of a 0.01% poly-L-lysine (PLL) solution in deionized water was incubated for 30–90 min. Blood was freshly drawn from a vein into 9 mL EDTA blood collection tubes (Greiner, Kremsmünster, Austria) in a local hospital. Immediately thereafter, physiologically intact erythrocytes were separated by centrifugation. For this purpose, 4 drops of venous blood were diluted with isotonic phosphate buffered saline (PBS) (5 mM Na_2HPO_4, 150 mM NaCl, 200 µM ethylene glycol-bis(β-aminoethylether)-N,N,N′,N′tetraacetic acid

EGTA, pH adjusted to 7.4 with HCl) and centrifuged with a microcentrifuge (Eppendorf mini spin) at 3800 rpm for 4 min. The supernatant was drained, and the pellet was resuspended in 1 mL isotonic PBS. This procedure was repeated three times. Finally, 1.5–3 µL of the pellet was dissolved in PBS. These purified erythrocytes were briefly incubated on the PLL-coated glass slides and immediately chemically fixed with 200 µL of 1% glutaraldehyde in PBS for 30–45 min. All wet samples were then washed three times with PBS for 5 min each and stored in PBS at 4 °C for a maximum of 2 days or used immediately for measurements in deionized water. Some of the samples were dried with a gentle stream of nitrogen and stored under argon for up to 4 weeks. A detailed protocol for the optimization of this procedure with regard to membrane orientation is out of the focus of this study and will be published elsewhere (Stainer et al., submitted).

2.7. Sample Preparations for Measurements in Blood, Blood Serum and Ink

After preparation of the erythrocyte sample and characterization under ambient conditions, the samples were rehydrated with freshly drawn blood or blood serum. For rehydration of erythrocyte samples with blood, the blood was diluted 1:8 with deionized water or used immediately without further dilution. For measurements in blood serum, the freshly drawn blood was centrifuged at 3800 rpm with a micro-centrifuge (Eppendorf mini spin) for 4 min to remove blood cells. The serum was taken from the centrifugation tube. These preparations were used for speed and noise measurements. For measurements in ink, a standard ink cartridge was opened, and the ink was diluted 1:1 with deionized water. All imaging solutions were used immediately after preparation.

2.8. Speed and Noise Measurements

Speed measurements were performed on a standard AFM calibration sample (HS-500MG AFM XYZ calibration standard). The imaging speed was increased stepwise from 30 to 750 µm s^{-1}. Noise and speed measurements were performed using the very same PRSA-L300-F50-Si-PCB cantilever (SCL—Sensor.Tech., Vienna, Austria). The noise measurements were done in contact mode, where the noise was calculated from both the topography and the deflection image. Before the noise measurements from the topography signal were performed, a topographical image of the calibration grid with optimized feedback parameters was taken. To ensure that there is no change in topography during the noise measurement, the x, y scan size was reduced to a minimum (~1 pm). With these parameters, an image with 1 line per second and a resolution of 256 pixels was acquired. To determine the noise from the deflection signal, the feedback parameters were then reduced to a minimum, just high enough to prevent the cantilever from drifting away from the surface. These very low feedback parameters are intended to ensure that almost the entire signal is represented in the deflection as an error signal. Since we assume that there is no topographic change, the information of the deflection image reflects the noise of the system. Using these parameters, again, an image with the same resolution was taken. The deflection images were converted into distance units by determining the contact mode sensitivity. The corresponding images were quartered using Gwyddion, and the root mean square (RMS) values of the noise as well as the standard deviation were determined.

3. Results

3.1. System Integration and Adaptation

Replacing the conventional OBD-based detection system of an AFM with an optical-free, self-sensing detection system is an essential prerequisite for performing nano-medically relevant measurements in turbid fluids, such as human blood or milk. Piezo-resistors are a promising candidate as sensing elements. These are resistors in which their resistance value changes due to mechanical expansion or geometric changes. The integration of these piezo-resistors on a cantilever is usually realized as a Wheatstone bridge of four piezo-resistors. Two of them are placed in the bending area of the cantilever, while the other two are located on the cantilever chip. Before use, a voltage is applied to zero the Wheatstone bridge (Figure 1, left, highlighted in yellow), in order to compensate for differences in the resistor branches. This configuration with four piezo-resistors leads to an increased stability regarding thermal drift or self-heating compared to single sensor probes. In order to realize an all-electric (AE) Bio-AFM with a broad applicability, we have used commercially available self-sensing cantilevers (SSCs). We followed two approaches, the implementation of the self-sensing technology in an OBD-based AFM and the use of the AFSEM® system, a SSC-based AFM, which we adapted for fluid measurements.

The optimal replacement of OBD of a commercial AFM by self-sensing cantilevers requires a design with a minimum of modifications to the existing AFM hardware and, ideally, without changing the controller software. As the AFM system, we chose a Keysight 5500 SPM2. In this system, the cantilevers are clamped by a simple mechanical spring. In contrast, an SSC requires electrical connections for the Wheatstone bridge as well as for thermal excitation, in addition to the mechanical clamping. Therefore, we have completely redesigned the nose cone (Figure 1a left).

The commercial SSCs used in this study are equipped with both a piezoresistive Wheatstone bridge for voltage-based deflection sensing, and a heating loop for thermal actuation. The implementation of these cantilevers in a commercial OBD setup requires a completely new nose design for the mounting of the self-sensing cantilever as well as the supply of the sensor and the readout of the deflection signal (Figure 1a). The new nose cone has been designed in close approximation to the original nose design, especially with regard to perfect fluid sealing and optical access via the AFM embedded CCD camera. A special printed circuit board was designed to mount and control the sensor. The flexible part of the circuit board is located on the outside of the new nose cone and is equipped with a connector for the cantilever PCB. The feed-through for the flexible part of the PCB is sealed to protect the internal electronics from fluid damage. To amplify the signal of the Wheatstone bridge as close as possible to its origin, the nose cone is equipped with an adjustable instrumentation amplifier. The deflection signal amplification can be adjusted depending on the measurement conditions (dry or liquid). The used sensor supply voltage of 2.048 V for imaging in dry conditions and 0.51 V for imaging in liquid environment is also realized in the nose cone. Since the space for the PCB is very limited and the smallest surface mounted device (SMD) standard had to be used, individual nose cones were manufactured for measurements in liquid and dry environment. Furthermore, the excitation pins of the AFM system were directly connected, either to the heating loop for thermal excitation or to the excitation piezo. For acoustic excitation, a piezo was embedded in the nose cone as close as possible to the cantilever. The Keysight 5500 SPM2 AFM used is designed in such a way that optical access is possible for both the OBD laser and CCD camera. Thus, all parts for all-electric readout have been placed outside the optical axis, which still allows the use of the CCD camera. The standard OBD scanner has 6 pins to cover the different operating modes with different nose cones (e.g., scanning tunneling microscopy and conductive AFM). The two pins, which are intended for cantilever excitation, are still used, while the other pins, which are not required for standard AFM imaging, are used to supply the new nose cone electronics with electricity/power and to transfer the amplified deflection signal to the subsequent hardware. This configuration ensured that no changes to the original scanner hardware were necessary.

In order to utilize the original OBD signal processing path for all-electric readout, the single deflection signal needs to be converted into an equivalent of the OBD signals. This must be achieved by mimicking the signal that the controller hardware typically receives from a four-segment photodiode. A second hardware has been developed for this purpose, including: (i) a signal conditioning hardware and (ii) an additional adjustable instrumentation amplifier. Here, the four signals of a four-segment photodiode are simulated from the single SSC deflection. The scheme of this signal processing is shown in Figure 1a (right), and the detailed description is given in the experimental part in Section 2.3. In short, the original deflection signal was inverted, and an offset was added to both the original and the inverted signal. Using the non-inverted deflection signal plus offset for the upper two segments and the inverted deflection signal plus offset for the two lower segments of the four-quadrant photodiode allows the SSC to be connected to the original controller hardware.

In contrast to the implementation of SSC in an OBD AFM, the upgrade of a commercial all-electric AFM (AE-AFM) for physiological measurements in (turbid) liquids requires an alternative approach. The used AFSEM® was originally developed to perform measurements in vacuum (e.g., in the chamber of a SEM) or in air. Using the original configuration would damage the nose of the scanner by immersion in water, as open electrical connections would be short-circuited. As shown in Figure 1b (left), we have designed a nose with a significantly longer distance between cantilever and electronics (Figure 1b (left, arrow)). This, on the one hand, allows for safe measurements in liquids like none-conductive buffers and, on the other hand, the small nose shape allows experiments in Petri dishes and smaller chambers. In addition, the voltage supply for the Wheatstone bridge was reduced from 2.048 to 0.51 V to avoid the risk of undesired electrochemical reactions. The reduction of the supply voltage thus also causes a proportional change in signal intensity, resulting in a reduced signal-to-noise ratio. In contrast to the lower deflection signal due to the reduced supply voltage, the gain of the instrumentation amplifier has been changed from 100- to 400-fold. No further modifications of the system were necessary.

All these changes aimed to allow measurements in physiological fluids ranging from buffer and blood serum to pure blood or milk. We see a high potential for AE-AFM investigations in blood. Therefore, we have chosen blood compartments as the ideal feasibility test system to compare the performance of AE-AFM with OBD-based AFMs. For the direct comparison of the optical and the electrical readout, the ability to image simple calibration grids for all possible combinations was tested. These ranged from contact mode to three different excitations for intermittent contact mode with optical and electrical readout. As a result, we successfully performed measurements of a standard AFM grid using electrical readout in air, deionized water and freshly drawn blood. All these measurements were done in contact mode and in intermittent contact mode (acoustic as well as thermal excitation). Magnetic excitation for AE-AFM imaging was not realized due to a lack of magnetically coated SSCs. In addition, the same measurements were performed with optical beam deflection. Here, the thermal excitation was not shown as there was a lack of optical cantilever equipped with a heating loop. Data are shown in supplementary information (Supplementary Figure S1). It is obvious that OBD fails in optically non-transparent liquids. In contrast, AE-AFM allowed such investigations in turbid liquids, including blood and bovine milk.

3.2. Imaging of Biological Samples in Dry State

Besides the comparison of different readout methods and excitation modes, a direct comparison of the image quality on highly relevant physiological samples is of interest. First, images of biologically relevant samples using the self-sensing cantilever readout on the OBD-based AFM were taken of human platelets in a dry state. Platelets are one of the major players in hemostasis. The conventional OBD-based AFM has been successfully used to study the activation mechanism at the single-cell [57] and single-molecule level using imaging, recognition imaging and single-molecule force spectroscopy [58]. The latter study was based on investigations on platelets that were partially fixed and dried in different activation states. In contrast to this earlier study, we focused here on fast (clot formation) processes that occur within the first minutes. Thus, fresh blood was incubated for 1, 3 or 5 min on a cleaned solid surface, followed by careful washing, gentle fixation and drying. In this study, however, the main purpose was to compare the performance of the AE-AFM with OBD readout in terms of topography. The images shown in Figure 2 are taken in air and show typical and expected platelet activation, including filopodia formation.

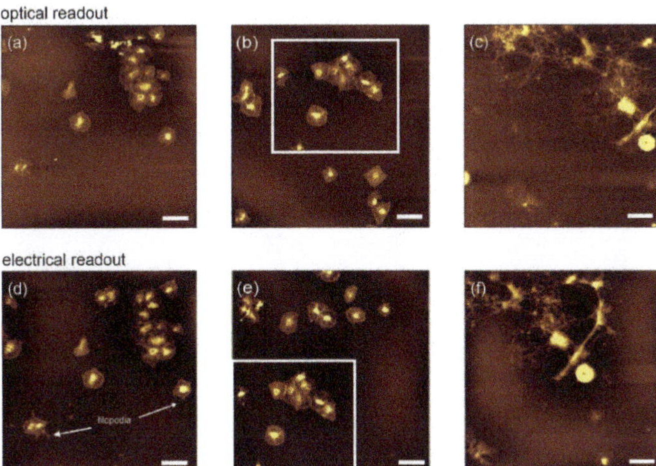

Figure 2. Comparison of optical and electrical readout on physiologically relevant samples in air: The upper part shows platelets after one (**a**), three (**b**) and five minutes (**c**) incubation, imaged with optical cantilevers and magnetic excitation. In comparison, the same sample positions were imaged with self-sensing cantilevers (**d–f**), yielding equal results (lower part). The scale bar is 10 μm for all images. Z-scale: (**a,d**): 191 nm, (**b,e**): 200 nm, (**c,f**): 900 nm.

All three samples (1, 3 and 5 min) were initially imaged with a standard OBD readout system and a magnetically coated optical cantilever in intermittent contact mode (Figure 2a–c, respectively). After changing to our AE-AFM scanner nose and to a self-sensing cantilever, the same positions (guided by the AFM's CCD camera) were imaged (Figure 2d–f). AE-AFM imaging was also performed in intermittent contact mode by means of acoustic excitation. Apart from the different levels of activation, it is shown that the imaging quality and capabilities are clearly comparable, and AE-AFM is perfectly suited for imaging biological samples. Even in the dry state, the advantages of using AE-AFM are obvious, ranging from (i) simplified usage due to the lack of optical adjustment, (ii) higher stability due to complete electrical readout [10] and (iii) the possibility of easy combination with other techniques (e.g., SEM).

3.3. Combined AFM–SEM Investigations and Measurements in Different Environments

As a test system for combined simultaneous measurements with AFM and SEM, we have selected another physiologically highly relevant blood cell, the erythrocytes (RBC—red blood cells). In order to achieve optimal AFM and SEM resolution and to avoid artifacts caused by differences in cell stiffness, a new preparation of ultra-flat RBC ghosts was used. The erythrocytes were immobilized on a flat gold surface (e.g., a cleaned QCM), which allowed an adjustable membrane orientation. These samples were characterized by correlated AFM–SEM imaging. Areas of interest were determined by SEM imaging, followed by positioning of the SSC and acquisition of three-dimensional (3D) topographic images in the vacuum chamber of the SEM. The results are shown in Figure 3a,b. Most of the examined cells appeared extremely flat and represented their outer membrane. A detailed analysis showed that 35–55% of the gold surface was covered with ultra-flat erythrocyte ghost cells. Finally, we performed AE-AFM measurements in various fluids, ranging from deionized water to physiological media, such as blood serum, to human blood and bovine milk. As shown in Figure 3 (right side), we first started by imaging a calibration grid in deionized water, followed by the cell sample (Figure 3c). The same was done for blood serum (Figure 3d) and freshly drawn human blood (Figure 3e). On the left side in Figure 3c–e, the liquid nose of the AE-AFM is shown in the respective media. In the center, an AFM calibration grid measured in this liquid is presented. On the right side, a typical AFM image of ultra-flat erythrocyte ghosts in the corresponding medium is shown. It should be mentioned that the grid for blood was recorded in a 1:8 diluted blood:water solution to reduce contamination of the grid by products of blood clotting. In contrast to blood serum, native blood exhibits relatively rapid clotting phenomena induced by shear stress or compartments of the intrinsic or extrinsic coagulation cascade. Therefore, the ultra-flat RBC membranes in pure native blood showed some differences in their topographical appearance, which are strongly dependent on the clotting state of the blood (unaffected RBC images taken with 1:8 diluted blood are shown in the Supplementary Materials, Figure S1, lower right). Although artifacts can be caused, e.g., by undesired adhesion of platelets or other blood compartments to the sensor surface, technically, stable imaging over a sufficiently long time is possible. We were able to perform AFM imaging of such a preparation of ultra-flat RBC ghosts for more than four hours without any loss of performance in the readout system.

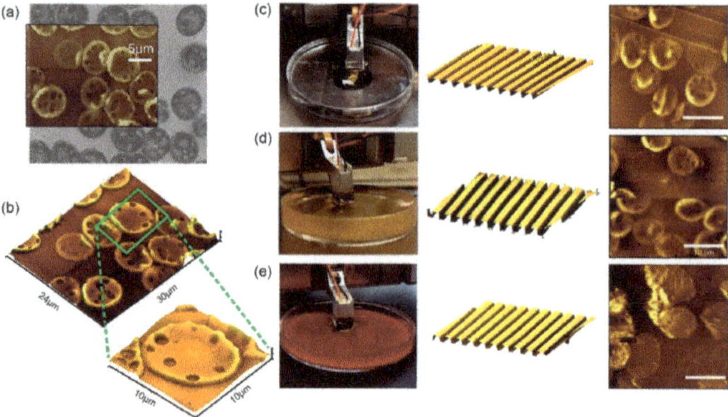

Figure 3. (left) Correlated atomic force microscopy—scanning electron microscopy (AFM—SEM): (**a**) Ultra-flat erythrocyte ghosts on gold were first recorded with the SEM and imaged with the AFSEM® inside the SEM at the same position. (**b**) Overview and three-dimensional (3D) single erythrocyte AFM image at the same position. Images of ultra-flat erythrocyte ghost cells were taken in various liquids, such as deionized water (**c**), blood serum (**d**) and freshly drawn human blood (**e**). For each liquid, a proof-of-principle measurement of a calibration grid was performed ((**c–e**), middle), followed by imaging of erythrocyte ghost cells ((**c–e**), right). The grid in (**e**) was recorded in 1:8 diluted blood. Scale bars are 10 μm for AFM images in (**c–e**).

In summary, we were able to show comparable results on both systems, the adapted OBD AFM and the AFSEM® adapted for working in liquids (AE-AFM). Furthermore, combined SEM and AFM images of the same sample position in vacuum were shown. More importantly, measurements of blood compartments such as platelets or erythrocytes are possible and were shown—for the first time ever—in their real native environment (i.e., undiluted human blood). Nevertheless, it must be mentioned that long-term experiments in native blood require further physiological and technical optimizations with respect to imaging conditions, ranging from adjustable blood flow to controlled sedimentation and passivation of the cantilever surface, but the technical limitations in general have now been demonstrably overcome.

3.4. Detailed System Characterization

In addition to the technical realization of all-electric AFM imaging, including first measurements of cell samples in turbid liquids like blood, we investigated and compared the imaging performance in different media, like deionized water, ink, blood serum and pure blood. A key parameter is the z-noise of the system. This z-noise is composed of three main parts: noise caused by the cantilever movement, noise caused by the measuring principle and noise of the readout electronics [15]. In both readout methods, OBD or SSC, the combination of these three noise sources yield the limit for the minimum detectable deflection. The latter equals to the root mean square (RMS) voltage noise [15]. The measured z-noise can be strongly influenced by external noise sources such as acoustic noise, mechanically transmitted vibrations or electronic noise. Thus, it is necessary to minimize external sources of interference before determination of the z-noise caused by the setup itself. To optimize the noise level, measurements were performed with a mechanically fixed sample stage and minimized mechanical loop in a closed glove box on a passively damped table. To exclude deviation caused by different cantilevers, all noise measurements were always performed with the same cantilever. The RMS AFM imaging noise was characterized by so-called two-dimensional (2D) noise images in contact mode. For noise determination, we used two different approaches, (i) the determination of the noise from a topographical image, and (ii) from the deflection image. Before determining the z-noise from the topography, an image was taken from a calibration grid with optimized feedback parameters. In order to obtain a 2D noise image without topographic changes, the x, y scan size was set to a minimum value. To determine the noise from the deflection image, the feedback parameters were minimized to avoid any shift of information from the deflection/error into the topographical image. To convert the voltage signal into a length measurement, the contact mode sensitivity was determined.

As a reference, the system z-noise was determined in dry conditions (2.91 Å (topography method)). The noise values in deionized water, diluted ink, blood serum, as well as in diluted and pure human blood, are shown in Table 1. In liquid, the z-noise is somewhat higher, whereby, for example, a value of 3.52 Å (topography method) was measured for blood. The noise values from the deflection images are slightly lower compared to those from the topography method but show a similar trend (2.63 Å for air, 3.1 Å for blood). Compared to previous noise measurements with self-sensing cantilevers of this type, focusing on the minimal reachable z-noise [15], we are about a factor 3 higher at these typical measuring conditions. This may be caused by the additional mechanical loop of the highly flexible AE-AFM z-stage and the different setup design (tip scanning). We could show that changing from dry state to different solutions, including highly turbid blood, yield very minor increases in the z-noise. In summary, it can be said that even with measurements in pure blood, a sufficient z-resolution for physiological measurements (i.e., 0.30 ± 0.15 nm) can be achieved.

Table 1. Comparison of z-noise values recorded in different media.

Noise in Different Media		Dry State [Å]	Deionized Water [Å]	Diluted Ink [Å]	Blood Sera [Å]	Diluted Blood [Å]	Pure Blood [Å]
Topography	RMS *	2.91	3.64	3.36	3.62	3.03	3.52
	SD #	0.024	0.02	0.043	0.15	0.077	0.069
Deflection	RMS *	2.63	3.00	2.67	2.50	2.31	3.10
	SD #	0.026	0.042	0.023	0.07	0.026	0.15

* root mean square; # standard deviation.

We further investigated the influence of the imaging speed, especially in turbid liquids, using our optimized AE-Bio AFM. Although the used AE-AFM was not developed as a high-speed AFM system, we investigated and analyzed the influence of different fluids on the imaging performance of moderate and higher scanning rates. For all these measurements, the same PRSA-L300-F50-Si-CB 300 × 100 cantilever was used to allow an unbiased comparison. As a sample, we used an AFM XYZ calibration grating and a scan size of 30 μm. The tip velocity was varied from 30 to 750 μm s^{-1}. First, reference measurements were taken in contact mode in air, followed by the different liquids, like deionized water, diluted ink, serum, as well as diluted and pure human blood. Figure 4 shows a comparison of the performance in deionized water with the most challenging environmental condition, pure fresh human blood.

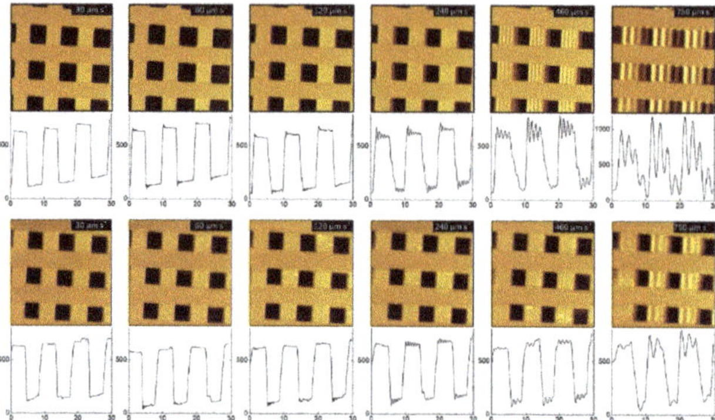

Figure 4. Comparison of speed performance in different media. The upper part shows images acquired in deionized water, and the lower part in pure human blood. The speed has been increased from 30 μm/s (1 s/line) to 750 μm/s (0.04 s/line). The corresponding horizontal cross-sections have been extracted from the very same position in the middle row of the image. Measurements have been performed on a HS-500MG AFM XYZ calibration standard. x, y scan range is 30 μm for all images.

As can be seen from the topography as well as from the corresponding cross-section (Figure 4, upper part), imaging in deionized water works fine at standard speed (30 μm s^{-1}) and up to an imaging speed of about 60 μm s^{-1}. At a speed of 60 μm s^{-1}, the first artifacts appear and can be seen in the cross-section (pits of the calibration grid, Figure 4, upper part, second image). With a further stepwise increase of the speed to 240 μm s^{-1}, the feedback artifacts increase and are clearly visible in the topography. Especially when using imaging velocities of 750 μm s^{-1} or higher, the quality of the topographical images is strongly distorted. The same results are also obtained in pure blood (Figure 4, lower part), showing first feedback artefacts at similar scanning speed. Speed measurements at all other environmental conditions are shown in Supplementary Figure S2. By comparing the speed measurements at all different imaging conditions, it becomes clear that the scanning speed is not significantly influenced by measuring in (turbid) liquids. In summary, we proved that the z-noise in blood increases by about 35% compared to deionized water and no limitations in scanning speed appear by measuring in turbid viscous blood compared to the reference measurement at dry conditions. In contrast, we can conclude from the images recorded in Figure 4 that the speed performance in blood is even slightly superior to the measurements in deionized water. The scan speed is determined by the spring constant of the cantilever, its effective mass, the damping constant of the cantilever in the surrounding medium and the stiffness of the sample [59]. The viscosity of human blood depends on different factors, especially from the shear rate. It is in the range of 3.26 ± 0.43 mPa s at a shear rate of 100 s^{-1} to 5.46 ± 0.84 mPa s at a shear rate of 1 s^{-1}. In contrast, viscosity of water at 20 °C is 1 mPa s [60]. In blood, the damping constant and the effective mass of the cantilever are also increased. At our moderate imaging velocities, the viscous drag effect still allows correct data but the increased damping constant and effective mass explain the minor shift of the imaging speed limit compared to measurements in deionized water.

4. Discussion

The technical challenge to perform biological AFM imaging in turbid liquids is now solved by combining commercially available and sufficiently soft SSCs with upgrades in AFM design and electronics. Although we are able—for the first time ever—to present images of blood cells in highly turbid fresh human blood, the all-electric AFM as a tool for in-situ investigation of clot formation is still at a very early stage. As mentioned above, mimicking arterial blood flow requires precise control of physical, biochemical and medical parameters. In addition, anti-adhesive coatings of the surrounding surfaces may be required. In any case, we are convinced that all these requirements are feasible, making the AE-AFM to a key tool in hemostatic research. By constructing a suitable sensor nose design, AE-AFM also has the potential to observe passivation and undesired coagulation in stents and can therefore play a crucial role in the development of stents and stent coatings. In addition, our biosensing all-electric AFM technically allows further characterization methods, including high-resolution elasticity mapping, gaining mechanical data, nano-indentation and functional investigation. Simultaneous topography and recognition imaging is currently not implemented but is in the planning stage. In contrast to conventional OBD-based AFM, our AE-AFM developments simplify the usage of cantilever arrays for the investigation of biological samples (e.g., in blood). This may open new great opportunities like multi-information topography and recognition imaging using different bio-functionalized cantilever on the same sample in a parallel manner. Finally, it should not be overlooked that AE-AFM measurements could be used as a powerful (nano-) investigation tool in completely different areas, such as industrial applications (e.g., milk tank inspection) or biofilm formation in optically non-transparent liquids. We believe that SSC-based bio-AFMs will become a promising tool, especially in the biological/physiological/medical sciences.

Supplementary Materials: The following are available online at http://www.mdpi.com/1424-8220/20/13/3715/s1, Figure S1: Comparison of electrical and optical readout, Figure S2: Comparison of AE-AFM speed measurements at different environmental conditions.

Author Contributions: M.L., B.B., H.S. performed AE-AFM measurements, S.S. prepared erythrocyte samples, C.H.S., A.D. and M.L. worked on the technical development. A.E. planned and supervised the entire work. M.L. and A.E. wrote the manuscript. All authors have read and agreed to the published version of the manuscript.

Funding: This work was supported by the European Comission and the state of Upper Austria by the EFRE Project IWB 2014-2020, 2018-98292, Open Access Funding by the University of Linz.

Conflicts of Interest: The authors declare no conflicts of interest.

References

1. Ando, T. High-speed atomic force microscopy and its future prospects. *Biophys. Rev.* **2018**, *10*, 285–292. [CrossRef]
2. Li, M.; Xi, N.; Wang, Y.; Liu, L. Atomic Force Microscopy in Probing Tumor Physics for Nanomedicine. *IEEE Trans. Nanotechnol.* **2019**, *18*, 83–113. [CrossRef]
3. Santos, N.C.; Carvalho, F.A. *Atomic Force Microscopy*; Springer: Berlin, Germany, 2019.
4. Lamprecht, C.; Hinterdorfer, P.; Ebner, A. Applications of biosensing atomic force microscopy in monitoring drug and nanoparticle delivery. *Expert Opin. Drug Deliv.* **2014**, *11*, 1237–1253. [CrossRef]
5. Dufrêne, Y.F.; Ando, T.; Garcia, R.; Alsteens, D.; Martinez-Martin, D.; Engel, A.; Gerber, C.; Müller, D.J. Imaging modes of atomic force microscopy for application in molecular and cell biology. *Nat. Nanotechnol.* **2017**, *12*, 295–307. [CrossRef]
6. Tortonese, M.; Yamada, H.; Barrett, R.C.; Quate, C.F. Atomic force microscopy using a piezoresistive cantilever. In Proceedings of the TRANSDUCERS '91: 1991 International Conference on Solid-State Sensors and Actuators. Digest of Technical Papers, San Francisco, CA, USA, 24–27 June 1991.
7. Tortonese, M. Atomic resolution with an atomic force microscope using piezoresistive detection. *Appl. Phys. Lett.* **1993**, *62*, 834–836. [CrossRef]
8. Linnemann, R.; Gotszalk, T.; Hadjiiski, L.; Rangelow, I.W. Characterization of a cantilever with an integrated deflection sensor. *Thin Solid Films* **1995**, *264*, 159–164. [CrossRef]
9. Jumpertz, R.; Hart, A.V.D.; Ohlsson, O.; Saurenbach, F.; Schelten, J. Piezoresistive sensors on AFM cantilevers with atomic resolution. *Microelectr. Eng.* **1998**, *41–42*, 441–444. [CrossRef]
10. Thaysen, J.; Boisen, A.; Hansen, O.; Bouwstra, S. Atomic force microscopy probe with piezoresistive read-out and a highly symmetrical Wheatstone bridge arrangement. *Sens. Actuators A Phys.* **2000**, *83*, 47–53. [CrossRef]
11. Yu, X.; Thaysen, J.; Hansen, O.; Boisen, A. Optimization of sensitivity and noise in piezoresistive cantilevers. *J. Appl. Phys.* **2002**, *92*, 6296–6301. [CrossRef]
12. Doll, J.C.; Pruitt, B.L. High-bandwidth piezoresistive force probes with integrated thermal actuation. *J. Micromech. Microeng.* **2012**, *22*, 095012. [CrossRef]

13. Tosolini, G.; Scarponi, F.; Cannistraro, S.; Bausells, J. Biomolecule recognition using piezoresistive nanomechanical force probes. *Appl. Phys. Lett.* **2013**, *102*, 253701. [CrossRef]
14. Bausells, J. Piezoresistive cantilevers for nanomechanical sensing. *Microelectr. Eng.* **2015**, *145*, 9–20. [CrossRef]
15. Dukic, M.; Adams, J.D.; Fantner, G.E. Piezoresistive AFM cantilevers surpassing standard optical beam deflection in low noise topography imaging. *Sci. Rep.* **2015**, *5*, 16393. [CrossRef]
16. Kunicki, P.; Angelov, T.; Ivanov, T.; Gotszalk, T.; Rangelow, I. Sensitivity Improvement to Active Piezoresistive AFM Probes Using Focused Ion Beam Processing. *Sensors* **2019**, *19*, 4429. [CrossRef]
17. Brugger, J.; Buser, R.; de Rooij, N. Micromachined atomic force microprobe with integrated capacitive read-out. *J. Micromech. Microeng.* **1992**, *2*, 218. [CrossRef]
18. Blanc, N.; Brugger, J.; De Rooij, N.F.; Dürig, U. Scanning force microscopy in the dynamic mode using microfabricated capacitive sensors. *J. Vac. Sci. Technol. B Microelectr. Nanometer Struct. Process. Meas. Phenom.* **1996**, *14*, 901–905. [CrossRef]
19. Forsen, E.; Abadal, G.; Ghatnekar-Nilsson, S.; Teva, J.; Verd, J.; Sandberg, R.; Svendsen, W.; Perez-Murano, F.; Esteve, J.; Figueras, E.; et al. Ultrasensitive mass sensor fully integrated with complementary metal-oxide-semiconductor circuitry. *Appl. Phys. Lett.* **2005**, *87*, 043507. [CrossRef]
20. Kim, S.-J.; Ono, T.; Esashi, M. Capacitive resonant mass sensor with frequency demodulation detection based on resonant circuit. *Appl. Phys. Lett.* **2006**, *88*, 053116. [CrossRef]
21. Verd, J.; Uranga, A.; Abadal, G.; Teva, J.; Torres, F.; Pérez-Murano, F.; Fraxedas, J.; Esteve, J.; Barniol, N. Monolithic mass sensor fabricated using a conventional technology with attogram resolution in air conditions. *Appl. Phys. Lett.* **2007**, *91*, 013501. [CrossRef]
22. Itoh, T.; Suga, T. Development of a force sensor for atomic force microscopy using piezoelectric thin films. *Nanotechnology* **1993**, *4*, 218. [CrossRef]
23. Lee, J.H.; Hwang, K.S.; Park, J.; Yoon, K.H.; Yoon, D.S.; Kim, T.S. Immunoassay of prostate-specific antigen (PSA) using resonant frequency shift of piezoelectric nanomechanical microcantilever. *Biosens. Bioelectr.* **2005**, *20*, 2157–2162. [CrossRef] [PubMed]
24. Karabalin, R.B.; Matheny, M.H.; Feng, X.L.; Defaÿ, E.; Le Rhun, G.; Marcoux, C.; Hentz, S.; Andreucci, P.; Roukes, M.L. Piezoelectric nanoelectromechanical resonators based on aluminum nitride thin films. *Appl. Phys. Lett.* **2009**, *95*, 103111. [CrossRef]
25. Ivaldi, P.; Abergel, J.; Matheny, M.H.; Villanueva, L.G.; Karabalin, R.B.; Roukes, M.L.; Andreucci, P.; Hentz, S.; Defaÿ, E. 50 nm thick AlN film-based piezoelectric cantilevers for gravimetric detection. *J. Micromech. Microeng.* **2011**, *21*, 085023. [CrossRef]
26. Moore, S.I.; Ruppert, M.G.; Yong, Y.K. An optimization framework for the design of piezoelectric AFM cantilevers. *Precis. Eng.* **2019**, *60*, 130–142. [CrossRef]
27. Choi, Y.-S.; Gwak, M.-J.; Lee, D.-W. Polymeric cantilever integrated with PDMS/graphene composite strain sensor. *Rev. Sci. Instrum.* **2016**, *87*, 105004. [CrossRef]
28. Akiyama, T.; Staufer, U.; De Rooij, N.F.; Frederix, P.L.T.M.; Engel, A. Symmetrically arranged quartz tuning fork with soft cantilever for intermittent contact mode atomic force microscopy. *Rev. Sci. Instrum.* **2003**, *74*, 112–117. [CrossRef]
29. Gonzalez, L.; Martínez-Martín, D.; Otero, J.; De Pablo, P.J.; Puig-Vidal, M.; Gómez-Herrero, J. Improving the lateral resolution of quartz tuning fork-based sensors in liquid by integrating commercial AFM tips into the fiber end. *Sensors* **2015**, *15*, 1601–1610. [CrossRef]
30. Saitoh, K.; Taguchi, A.; Kawata, S. Tip-Enhanced Raman Scattering Microscope Using Quartz-Tuning-Fork AFM Probe. In *JSAP-OSA Joint Symposia*; Optical Society of America: Washington, DC, USA, 2017.
31. Voigtländer, B. Quartz Sensors in Atomic Force Microscopy. In *Atomic Force Microscopy*; Springer: Berlin, Germany, 2019; pp. 301–307.
32. Li, M.; Tang, H.X.; Roukes, M.L. Ultra-sensitive NEMS-based cantilevers for sensing, scanned probe and very high-frequency applications. *Nat. Nanotechnol.* **2007**, *2*, 114. [CrossRef]
33. Dukic, M.; Winhold, M.; Schwalb, C.H.; Adams, J.D.; Stavrov, V.; Huth, M.; Fantner, G.E. Direct-write nanoscale printing of nanogranular tunnelling strain sensors for sub-micrometre cantilevers. *Nat. Commun.* **2016**, *7*, 12487. [CrossRef]
34. Yoshikawa, G.; Lang, H.P.; Akiyama, T.; Aeschimann, L.; Staufer, U.; Vettiger, P.; Aono, M.; Sakurai, T.; Gerber, C. Sub-ppm detection of vapors using piezoresistive microcantilever array sensors. *Nanotechnology* **2008**, *20*, 015501. [CrossRef]

35. Holz, M.; Reuter, C.; Ahmad, A.; Reum, A.; Ivanov, T.; Guliyev, E.; Rangelow, I.W.; Lee, H.S. *Parallel Active Cantilever AFM Tool for High-Throughput Inspection and Metrology*; SPIE Advanced Lithography; SPIE: Washington, DC, USA, 2019; Volume 10959.

36. Somnath, S.; Kim, H.J.; Hu, H.; King, W.P. Parallel nanoimaging and nanolithography using a heated microcantilever array. *Nanotechnology* **2013**, *25*, 014001. [CrossRef] [PubMed]

37. Kreith, J.; Strunz, T.; Fantner, E.J.; Fantner, G.E.; Cordill, M.J. A versatile atomic force microscope integrated with a scanning electron microscope. *Rev. Sci. Instrum.* **2017**, *88*, 053704. [CrossRef] [PubMed]

38. Winhold, M.; Leitner, M.; Lieb, A.; Frederix, P.; Hofbauer, F.; Strunz, T.; Sattelkov, J.; Plank, H.; Schwalb, C.H. Correlative In-Situ AFM & SEM & EDX Analysis of Nanostructured Materials. *Microsc. Microanal.* **2017**, *23*, 26–27.

39. Angelov, T.; Ahmad, A.; Guliyev, E.; Reum, A.; Atanasov, I.; Ivanov, T.; Ishchuk, V.; Kaestner, M.; Krivoshapkina, Y.; Lenk, S.; et al. Six-axis AFM in SEM with self-sensing and self-transduced cantilever for high speed analysis and nanolithography. *J. Vac. Sci. Technol. B Microelectr. Nanometer Struct. Process. Meas. Phenom.* **2016**, *34*, 06KB01. [CrossRef]

40. Rangelow, I.W.; Kaestner, M.; Ivanov, T.; Ahmad, A.; Lenk, S.; Lenk, C.; Guliyev, E.; Reum, A.; Hofmann, M.; Reuter, C.; et al. Atomic force microscope integrated with a scanning electron microscope for correlative nanofabrication and microscopy. *J. Vac. Sci. Technol. B* **2018**, *36*, 06J102. [CrossRef]

41. Rogers, B.; York, D.; Whisman, N.; Jones, M.; Murray, K.; Adams, J.D.; Sulchek, T.; Minne, S.C. Tapping mode atomic force microscopy in liquid with an insulated piezoelectric microactuator. *Rev. Sci. Instrum.* **2002**, *73*, 3242–3244. [CrossRef]

42. Buguin, A.; Roure, O.D.; Silberzan, P. Active atomic force microscopy cantilevers for imaging in liquids. *Appl. Phys. Lett.* **2001**, *78*, 2982–2984. [CrossRef]

43. Florin, E.L.; Radmacher, M.; Fleck, B.; Gaub, H.E. Atomic force microscope with magnetic force modulation. *Rev. Sci. Instrum.* **1994**, *65*, 639–643. [CrossRef]

44. Han, W.; Lindsay, S.; Jing, T. A magnetically driven oscillating probe microscope for operation in liquids. *Appl. Phys. Lett.* **1996**, *69*, 4111–4113. [CrossRef]

45. Lantz, M.; O'shea, S.; Welland, M. Force microscopy imaging in liquids using ac techniques. *Appl. Phys. Lett.* **1994**, *65*, 409–411. [CrossRef]

46. Revenko, I.; Proksch, R. Magnetic and acoustic tapping mode microscopy of liquid phase phospholipid bilayers and DNA molecules. *J. Appl. Phys.* **2000**, *87*, 526–533. [CrossRef]

47. Pedrak, R.; Ivanov, T.; Ivanova, K.; Gotszalk, T.; Abedinov, N.; Rangelow, I.W.; Edinger, K.; Tomerov, E.; Schenkel, T.; Hudek, P. Micromachined atomic force microscopy sensor with integrated piezoresistive sensor and thermal bimorph actuator for high-speed tapping-mode atomic force microscopy phase-imaging in higher eigenmodes. *J. Vac. Sci. Technol. B Microelectr. Nanometer Struct. Process. Meas. Phenom.* **2003**, *21*, 3102–3107. [CrossRef]

48. Fantner, G.E.; Burns, D.J.; Belcher, A.M.; Rangelow, I.W.; Youcef-Toumi, K. DMCMN: In depth characterization and control of AFM cantilevers with integrated sensing and actuation. *J. Dyn. Syst. Meas. Control* **2009**, *131*, 061104. [CrossRef]

49. Herruzo, E.T.; Garcia, R. Frequency response of an atomic force microscope in liquids and air: Magnetic versus acoustic excitation. *Appl. Phys. Lett.* **2007**, *91*, 143113. [CrossRef]

50. Winhold, M.; Leitner, M.; Frank, P.; Hosseini, N.; Sattelkov, J.; Fantner, G.E.; Plank, H.; Schwalb, C.H. Correlative In-Situ Analysis on the Nanoscale by combination of AFM and SEM. *Microsc. Microanal.* **2018**, *24*, 1922–1923. [CrossRef]

51. Frank, P.; Leitner, M.; Hummel, S.; Hosseini, N.; Wang, Y.; Winkler, R.; Fantner, G.E.; Plank, H.; Zeng, Y.; Schwalb, C.H. Crystallographic and Nanomechanical Analysis by Correlative In-situ AFM & SEM. *Microsc. Microanal.* **2018**, *24*, 2280–2281.

52. Sattelkow, J.; Fröch, J.E.; Winkler, R.; Hummel, S.; Schwalb, C.; Plank, H. Three-Dimensional Nanothermistors for Thermal Probing. *ACS Appl. Mater. Interfaces* **2019**, *11*, 22655–22667. [CrossRef]

53. Semenenko, B.; Esquinazi, P. Diamagnetism of bulk graphite revised. *Magnetochemistry* **2018**, *4*, 52. [CrossRef]

54. Huber, C.; Reith, P.; Badarlis, A. Gas Density and Viscosity Measurement with a Microcantilever for Online Combustion Gas Monitoring. In Proceedings of the Sensors and Measuring Systems, 19th ITG/GMA-Symposium, Nuremberg, Germany, 26–27 June 2018; VDE: Berlin, Germany, 2018.

55. Xia, F.; Yang, C.; Wang, Y.; Youcef-Toumi, K.; Reuter, C.; Ivanov, T.; Holz, M.; Rangelow, I.W. Lights Out! Nano-Scale Topography Imaging of Sample Surface in Opaque Liquid Environments with Coated Active Cantilever Probes. *Nanomaterials* **2019**, *9*, 1013.

56. Fantner, G.E.; Schumann, W.; Barbero, R.J.; Deutschinger, A.; Todorov, V.; Gray, D.S.; Belcher, A.M.; Rangelow, I.W.; Youcef-Toumi, K. Use of self-actuating and self-sensing cantilevers for imaging biological samples in fluid. *Nanotechnology* **2009**, *20*, 434003. [CrossRef]

57. Karagkiozaki, V.; Pappa, F.; Arvaniti, D.; Moumkas, A.; Konstantinou, D.; Logothetidis, S. The melding of nanomedicine in thrombosis imaging and treatment: A review. *Future Sci.* **2016**, *2*, FSO113. [CrossRef] [PubMed]

58. Posch, S.; Neundlinger, I.; Leitner, M.; Siostrzonek, P.; Panzer, S.; Hinterdorfer, P.; Ebner, A. Activation induced morphological changes and integrin $\alpha IIb\beta3$ activity of living platelets. *Methods* **2013**, *60*, 179–185. [CrossRef] [PubMed]

59. Butt, H.J.; Siedle, P.; Seifert, K.; Fendler, K.; Seeger, T.; Bamberg, E.; Weisenhorn, A.L.; Goldie, K.; Engel, A. Scan speed limit in Atomic force microscopy. *J. Microsc.* **1993**, *169*, 75–84. [CrossRef]

60. Rosenson, R.S.; Mccormick, A.; Uretz, E.F. Distribution of blood viscosity values and biochemical correlates in healthy adults. *Clin. Chem.* **1996**, *42*, 1189–1195. [CrossRef] [PubMed]

Article

Nano-Scale Stiffness and Collagen Fibril Deterioration: Probing the Cornea Following Enzymatic Degradation Using Peakforce-QNM AFM

Ahmed Kazaili [1,2], Hayder Abdul-Amir Al-Hindy [3], Jillian Madine [4] and Riaz Akhtar [1,*]

1 Department of Mechanical, Materials and Aerospace Engineering, School of Engineering, University of Liverpool, Liverpool L69 3GH, UK; ahmed_mohammed381@yahoo.com
2 Department of Biomedical Engineering, College of Engineering, University of Babylon, Babylon, Hillah 51002, Iraq
3 College of Pharmacy, University of Babylon, Babylon, Hillah 51002, Iraq; phar.hayder.abdul@uobabylon.edu.iq
4 Department of Biochemistry and Systems Biology, Institute of Systems, Molecular and Integrative Biology, Faculty of Health and Life Sciences, University of Liverpool, Liverpool L69 7BE, UK; j.madine@liverpool.ac.uk
* Correspondence: r.akhtar@liverpool.ac.uk; Tel.: +44-151-794-5770

Abstract: Under physiological conditions, the cornea is exposed to various enzymes, some of them have digestive actions, such as amylase and collagenase that may change the ultrastructure (collagen morphology) and sequentially change the mechanical response of the cornea and distort vision, such as in keratoconus. This study investigates the ultrastructure and nanomechanical properties of porcine cornea following incubation with α-amylase and collagenase. Atomic force microscopy (AFM) was used to capture nanoscale topographical details of stromal collagen fibrils (diameter and D-periodicity) and calculate their elastic modulus. Samples were incubated with varying concentrations of α-amylase and collagenase (crude and purified). Dimethylmethylene blue (DMMB) assay was utilised to detect depleted glycosaminoglycans (GAGs) following incubation with amylase. Collagen fibril diameters were decreased following incubation with amylase, but not D-periodicity. Elastic modulus was gradually decreased with enzyme concentration in amylase-treated samples. Elastic modulus, diameter, and D-periodicity were greatly reduced in collagenase-treated samples. The effect of crude collagenase on corneal samples was more pronounced than purified collagenase. Amylase was found to deplete GAGs from the samples. This enzymatic treatment may help in answering some questions related to keratoconus, and possibly be used to build an empirical animal model of keratoconic corneas with different progression levels.

Keywords: Peakforce-QNM; AFM; cornea; collagenase; amylase; nanomechanics; collagen; collagen fibril morphology; keratoconus; collagen fibril diameter

Citation: Kazaili, A.; Abdul-Amir Al-Hindy, H.; Madine, J.; Akhtar, R. Nano-Scale Stiffness and Collagen Fibril Deterioration: Probing the Cornea Following Enzymatic Degradation Using Peakforce-QNM AFM. *Sensors* **2021**, *21*, 1629. https://doi.org/10.3390/s21051629

Academic Editors: Maria Lepore and Bruno Tiribilli

Received: 30 December 2020
Accepted: 18 February 2021
Published: 26 February 2021

Publisher's Note: MDPI stays neutral with regard to jurisdictional claims in published maps and institutional affiliations.

1. Introduction

There is a need to investigate and develop a better understanding of corneal ultrastructure and biomechanics at the nano-level. Understanding corneal ultrastructure and its response to chemicals is important for a number of ocular disorders, such as keratoconus that is characterised by a significant deterioration in the collagenous network resulting in a cone-shape cornea.

The human cornea consists of five layers, in which the stroma represents about 90% of its thickness. The stroma is composed of many lamellae that are parallel to the corneal surface. The stroma mainly consists of Type I collagen fibrils that are arranged in a high degree of lateral order and run in the same direction as their lamellae. Collagen fibrils consist of several collagen molecules that are aligned together in staggered arrangement to form a banding pattern, which is referred to as D-periodicity. Collagen fibrils in human corneas have relatively uniform diameters of 32.5 ± 1.5 nm and a D-periodicity of 65 nm [1,2], which may be slightly higher in porcine corneas [3]. Collagen fibrils of the stroma are associated with proteoglycans that keep them aligned and give support to provide the overall shape and strength of the cornea. Proteoglycans contain chains of glycosaminoglycans (GAGs), which are polysaccharide molecules that attract water and are thought to provide the extracellular matrix with additional physical properties not provided by collagen fibrils alone [4].

The cornea is exposed to a number of enzymes that are either secreted by lacrimal gland or produced by the corneal cells, such as keratocytes. Some of these enzymes, such as amylase [5] and collagenase [6] are believed to have digestive actions in the stromal layer of the corneas. It was also found that these enzymes cause a reduction in collagenous tissue stiffness; therefore, they may contribute in the progression of keratoconus [7].

Alpha-amylase is an active enzyme in the tear fluid that is thought to increase in patients with keratoconus [8], where it was found that it can decrease corneal stiffness [9]. Previous studies have used extensometers to examine the effect of alpha-amylase on corneal sections [9,10]. They found that corneal stiffness decreases with alpha-amylase incubation. The corresponding alteration in tissue ultrastructure has not been investigated previously.

The other digestive enzyme in the cornea is collagenase, which attacks the peptide bonds of the triple helix region on collagen. It is found in epithelial and stromal cells and can be released into the stromal layer in response to trauma for biomechanical modulation of the collagenous network [6,11–13]. Collagenase has been approved by United States Food and Drug Administration (FDA) as an injection to improve the range of motion in joints affected by advanced Dupuytren's disease [14]. In the eye, collagenase was found to increase in patients with keratoconus [12,15,16] due to reduction of its inhibitor [17,18]. However, no ultrastructural or nanomechanical investigation has been conducted to show its activity on corneal tissues.

The investigation of ultrastructural topography and biomechanical properties of the cornea requires a technique that is less destructive to the samples, such as the atomic force microscope (AFM). Other techniques such as scanning electron microscopy (SEM) and transmission electron microscopy (TEM) may provide ultrastructural details of the cornea, but samples need to be sputter-coated and dried extensively, affecting measurement of diameters and axial D-periodicity of collagen fibrils [19] as they are altered due to significant dehydration [20]. Therefore, AFM has been selected in this paper to capture in vitro ultrastructural topography and nanomechanical properties without extensive dehydration and sputter-coating which leads to alteration of the ultrastructure through extensive dehydration [3,19,20].

The use of AFM in the investigation of corneal diseases can help in addressing many questions. Early AFM studies focused on the collagen fibrils characterisation in mammalian corneas [21]. Other studies yielded elastic properties of corneal layers [22,23] which is useful for understanding mechanical properties in healthy and diseased corneas. For example, AFM was used to analyse photoablated stromal corneas in comparison to untreated samples [24]. The study showed undulations and granule-like features on the ablated stromal surface when a 193 nm excimer laser was used, and confirmed the precision of laser surgery in removing submicrometric amounts of the stroma. AFM was also utilised to investigate the ultrastructural topography in stromal layer of human corneas following collagen cross-linking treatment with riboflavin and ultraviolet-A light [25]. However, to date, AFM has not been used to examine ultrastructural changes in either keratoconic corneas or corneas exposed to enzymatic degradation in vitro.

This study takes advantage of recent advances in AFM such as PFQNM-AFM (Peak Force Quantitative Nanomechanical atomic force microscopy), which enables fast acquisition and mapping of a sample's mechanical properties [26]. PFQNM-AFM has demonstrated utility for characterising localised mechanical properties of collagen-rich tissues such as the sclera [27] and arterial tissue [28]. Rapid acquisition of tissue ultrastructure is achievable with comparable high-resolution mechanical property maps.

With PFQNM-AFM, in this study, we investigate ultrastructural and nanomechanical changes following enzymatic incubation of porcine corneal tissue with amylase and collagenase. We assess GAG depletion following amylase incubation, and also assess the difference between crude collagenase and purified collagenase on corneal degradation. We exploit the ability of the AFM microcantilever to act as a "force sensor" at the nano-scale for this novel application to characterise mechanical degradation in the cornea following in vitro degradation.

2. Materials and Methods

2.1. Tissue Samples

Eighteen porcine eyes were sourced from 5 to 6 month old pigs from a local abattoir shortly after slaughter. They were divided into three main groups (6 corneas each group); amylase-group, crude collagenase group, and purified collagenase group. The corneas were dissected immediately on arrival (within an hour from slaughter) at the University of Liverpool. Corneal samples were chosen from the apex (3 mm diameter) after desquamating the epithelial layer using a cotton-tipped applicator and tweezers. The preparation procedure included snap freezing and cryosectioning to make the corneal sample ready for AFM experiments.

After dissecting, the samples were carefully rinsed with Phosphate Buffered Saline (PBS) (Sigma-Aldrich, Dorset, UK), placed in a cryomould with appropriate orientation and embedded in the non-infiltrating optimum cutting temperature resin (Tissue-Tek, CellPath, Powys, UK).

Cryosectioning was performed by utilising a Leica cryostat (model CM1850, Leica Microsystems Ltd., Milton Keynes, UK) to section the frozen corneal samples to a thickness of 5 μm. In the amylase group (6 corneas), eleven sections of 5 μm thickness were taken from the first third of each cornea. The first third was defined as the anterior stromal layer of 150 μm after the epithelium. In the crude collagenase group (6 corneas), five sections of 5 μm thickness were taken from the first third of each cornea. A similar number of sections were prepared from the purified collagenase group. In all corneas, sections were taken after a depth of 20 μm from the anterior surface of the stroma. Sections were then stored at $-80\,^{\circ}$C until they were tested.

2.2. Enzymatic Treatment

For the amylase group, the cryosectioned tissue (n = 66) was treated with α-amylase (type Aspergillus oryzae, Sigma-Aldrich, Dorset, UK) for 40 min. Amylase was diluted in PBS to varying concentrations (0.2, 0.4, 0.6, 0.8, 1, 1.2, 1.4, 1.6, 1.8, and 2 mg/mL). The control sections in this group (n = 6) were incubated in PBS only, for the same period. One drop (40 mL) of amylase at 37.5 $^{\circ}$C was applied on the amylase treated group and kept at 37.5 $\pm$ 1.1 $^{\circ}$C for 40 min. Afterwards, the corneal sections were washed with cold PBS (4 $^{\circ}$C) and left in air for 18 min in preparation for AFM testing.

For the crude collagenase group, the samples (n = 30) were incubated with crude collagenase (type *Clostridium histolyticum*, Sigma-Aldrich, Dorset, UK) for 15 min at 37.4 $\pm$ 1.8 $^{\circ}$C. The treated samples of this group (n = 24) were incubated with one drop (40 mL) of crude collagenase (0.05, 0.1, 0.15, and 0.2 mg/mL) at 37.5 $^{\circ}$C that were diluted in PBS. Those samples were snap-washed firstly by a cold (4 $^{\circ}$C) aqueous solution of dichloromethylene diphosphonic acid disodium (NaDTA) at a concentration of 1 mg/mL to inhibit the activity of the collagenase; and subsequently by PBS twice and tested after 18 min. The control samples of this group (n = 6) were incubated in PBS for 15 min and snap-washed with NaDTA and PBS.

For the purified collagenase group, the samples (n = 30) were treated with varying concentrations of purified collagenase (type *Clostridium histolyticum*, Sigma-Aldrich, Dorset, UK) at 37.5 $^{\circ}$C (0.05, 0.1, and 0.15 mg/mL) for 15 min. Six samples served as controls, which were incubated in PBS only, for 15 min to investigate the effect of snap-washing the sections with cold (4 °C) NaDTA. The treated samples (n = 24) of this group were also snap-washed with cold NaDTA and PBS following the enzymatic incubation and tested after 18 min. Table 1 summarises these groups and treatment solutions.

Table 1. A summary of the groups and treatment parameters.

Samples	Amylase Group	Crude Collagenase Group	Purified Collagenase Group
No. of Porcine Eyes	6	6	6
No. of Sections	66	66	66
Incubation time (min)	40	15	15
Enzyme concentrations (mg/mL)	0.2, 0.4, 0.6, 0.8, 1, 1.2, 1.4, 1.6, 1.8, and 2	0.05, 0.1, 0.15, and 0.2	0.05, 0.1, 0.15, and 0.2
Washing solution	PBS	NaDTA	NaDTA

2.3. PFQNM-AFM Method

A Bruker MultiMode 8 AFM with E-piezoelectric scanner (Bruker Nano Inc., Nano Surfaces Division, Tucson, AZ, USA) was utilised to investigate the topographical and elastic property changes of the cryosectioned samples following enzymatic treatment. The PFQNM mode in air was used, which is most suited for biological samples with structural heterogeneity [29,30]. This mode is characterised by its ability to control the applied forces to the sample (or the peak force), which allows indentations to be limited to several nanometres that both maintains resolution and prevents sample damage. In addition, Peak Force QNM mode allows measurements at an extremely wide range of elastic moduli (1 MPa to 50 GPa) [30]. This mode uses the Derjaguin–Muller–Toporov (DMT) model and a curve fitting process of the unloading portion of the force to calculate elastic modulus, as outlined in other papers [30,31].

The AFM experiments were conducted with a silicon probe with a rectangular tip, type RTESPA-300 (Bruker Nano Inc., CA, USA). It was used due to its capabilities of capturing high resolution topographical images and its ability to measure a wide elastic modulus range of the samples being tested (200 MPa–5000 MPa). The nominal tip radius was 10 nm, and spring constant of the cantilever was 22 N/m with a resonance frequency of 300 kHz.

Relative calibration of the AFM was performed before every test. The calibration procedure was used to define the parameters of the PF-QNM mode. These parameters were measured relative to a known reference sample, Vishay Photostress PS1 Polymer (Vishay; Wendell, NC, USA). This reference sample had a known elastic modulus of 2.7 ± 0.1 GPa, which was utilised to calibrate the AFM and estimate the tip radius. A direct method of thermal tuning was carried out to measure the spring constant of the cantilever. Finally, deflection sensitivity was calibrated to convert Volts measured on the photodetector to nanometres of motion, which was performed by measuring a force curve on an "infinitely stiff" surface relative to the chosen cantilever. Therefore, a sapphire sample (Sapphire-12M; Bruker Nano Inc., Nano Surfaces Division, CA, USA) was utilised, which is stiff enough that the cantilever does not indent it during the force curve measurement.

With the use of the optical microscope integrated with the AFM, images of the cryosectioned corneas were captured for 3 different locations on each sample. Topographical images were collected at 5 × 5 μm and also 1 × 1 μm. The 1 × 1 μm images were suitable to visualise individual collagen fibrils. Peak force error images, which are referred to as the derivative of topography image [32], were also captured to visualise collagen fibrils orientation. The peak force frequency and amplitude were set to 2 kHz and 150 nm respectively. A scan rate of 0.799 Hz was utilised for all samples. The 1 × 1 μm images were captured with 256 horizontal lines, and each line was scanned at a resolution of 256 pixel/line. The 5 × 5 μm images were scanned with 512 horizontal lines, and each line was scanned at a resolution of 512 pixel/line. These setting were chosen following many trials to obtain good quality images whilst minimising artefacts. In addition, these settings were recommended by the manufacturer for biological tissues for optimum results.

2.4. Collagen Fibril Analysis

Collagen fibril diameter and D-periodicity were measured with NanoScope Analysis 1.7 software (Bruker Nano Inc., Nano Surfaces Division, CA, USA). Collagen fibrils that were straight with high contrast and at approximately zero inclination angle were manually selected from each height image to measure their diameter and D-periodicity. The recorded values then were averaged for each image. Figure 1 shows an example of the collagen fibril surface profile that was collected.

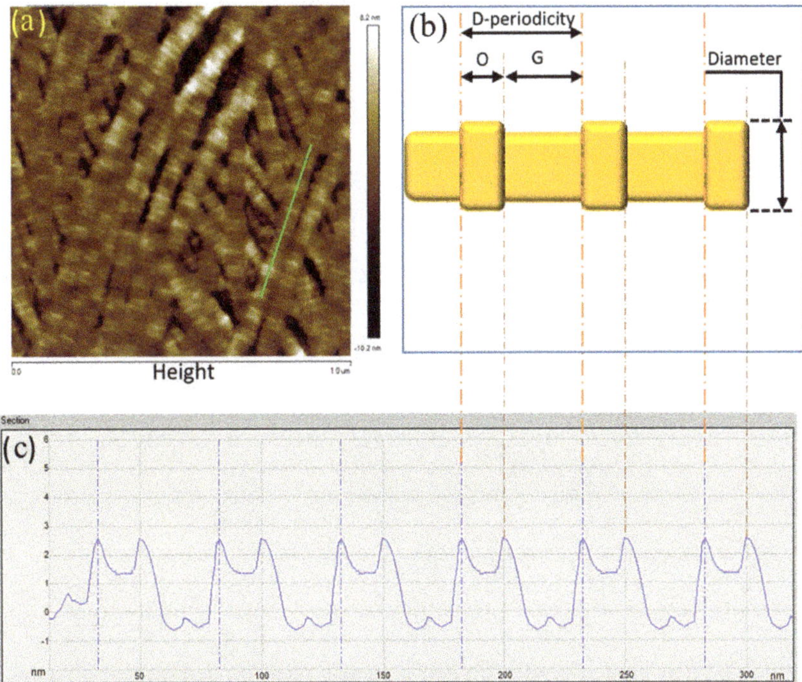

Figure 1. Collagen fibril surface profile. (**a**) Height image of the anterior lamella of a porcine cornea obtained with Atomic force microscopy (AFM). The green line in image (**a**) shows the area selected for analysis. (**b**) Schematic diagram showing collagen fibril morphology. "O" and "G" refer to overlap and gap zones. (**c**). Line profile generated from the green line shown in (**a**) with the corresponding O, G and periodicity shown with reference to (**b**). The peak-to-peak distance was measured manually to, which represents the D-periodicity. This analysis process was carried out 3 times as a minimum to calculate the average value of collagen fibril diameter and D-periodicity for each image. For this image, collagen fibril diameter and D-periodicity were 55.5 ± 2.4 nm and 67.8 ± 1.1 nm, respectively.

2.5. Glycosaminoglycans Quantification

Eight fresh porcine corneas were utilised for these experiments. The epithelium of the corneas was removed. Four portions of 6.5 mg in weight were cut from the anterior central region of the stroma of each cornea and organised into four groups: tissue culture (TC) group (n = 8), control group (n = 8) and amylase groups (n = 16) that was subdivided into two subgroups. Each sample of the first amylase subgroup (n = 8) were incubated with 1 mL of α-amylase (2 mg/mL) for 60 min. Samples in the second amylase subgroup (n = 8) were incubated with the same amount and concentration of the α-amylase for 120 min.

Control samples were incubated with 1 mL of PBS for 120 min. TC group samples were incubated with 1 mL of TC (CARRY-C, Alchimia, Italy) for the same period. All sample were slightly shaken and incubated at 37.5 °C. These treatment solutions were assessed with the Dimethylmethylene Blue (DMMB) assay to quantify the depleted proteoglycan in the solutions [33].

2.6. Data analysis and Statistics

All statistical analysis were performed using OriginPro 2016 version 9.3 (OriginLab Northampton, MA, USA). Data are expressed as mean values and standard deviation (mean ± standard deviation) unless otherwise stated. The two-sample t-test was used to test the statistical difference, with the significance level (α) set as 0.05 for all tests.

3. Results

3.1. Amylase Group

Figure 2 shows representative topographical images of control samples of the amylase group. In control samples, collagen fibrils appeared to be aligned and packed together Collagen fibrils had diameters of 55.5 ± 2.4 nm and axial D-periodicity of 67.46 ± 2.3 nm

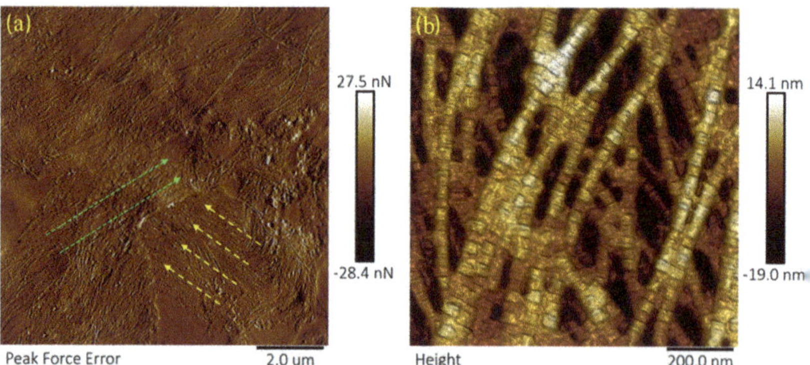

Figure 2. Typical topography (height) and peak force error images showing collagen fibrils of the anterior lamella of a porcine cornea (control samples). (**a**) A peak force error image of a control sample with a scan size of 10 μm. The yellow arrows show the direction of a lamella that runs at a right angle to another bundle of collagen fibrils (the green arrows). In each bundle, collagen fibrils are packed together (**b**) A height image of collagen fibrils of a control sample. Scan size was 1 μm "Height" represents the type of AFM image. A "3D effect" filter was applied for enhancement.

AFM images at different amylase concentrations are shown in Figure 3. It was also noticed in amylase treated samples that collagen fibril "splitting up" or fusion was observed The alignment and regularity of collagen fibrils in the same lamella gradually decreased with increased amylase concentrations, see Figure 3a,c,e. Some collagen fibrils seemed to fuse with another adjacent fibril, as shown in Figure 3b,f.

Amylase treatment was associated with collagen fibril dimeter reduction, which was significant with high amylase concentrations ($p < 0.0498$), as shown in Figure 4a The maximum reduction in collagen fibril diameter was found in samples treated with 2 mg/mL amylase concentration, with a significant reduction of approximately 26% relative to the control group ($p < 0.0001$). No changes in D-periodicity were observed following incubation with varying concentrations of amylase, ($p > 0.05$), Figure 4b. The maximum and minimum values of D-periodicity of amylase treated samples were 70.8 nm and 63.1 nm respectively.

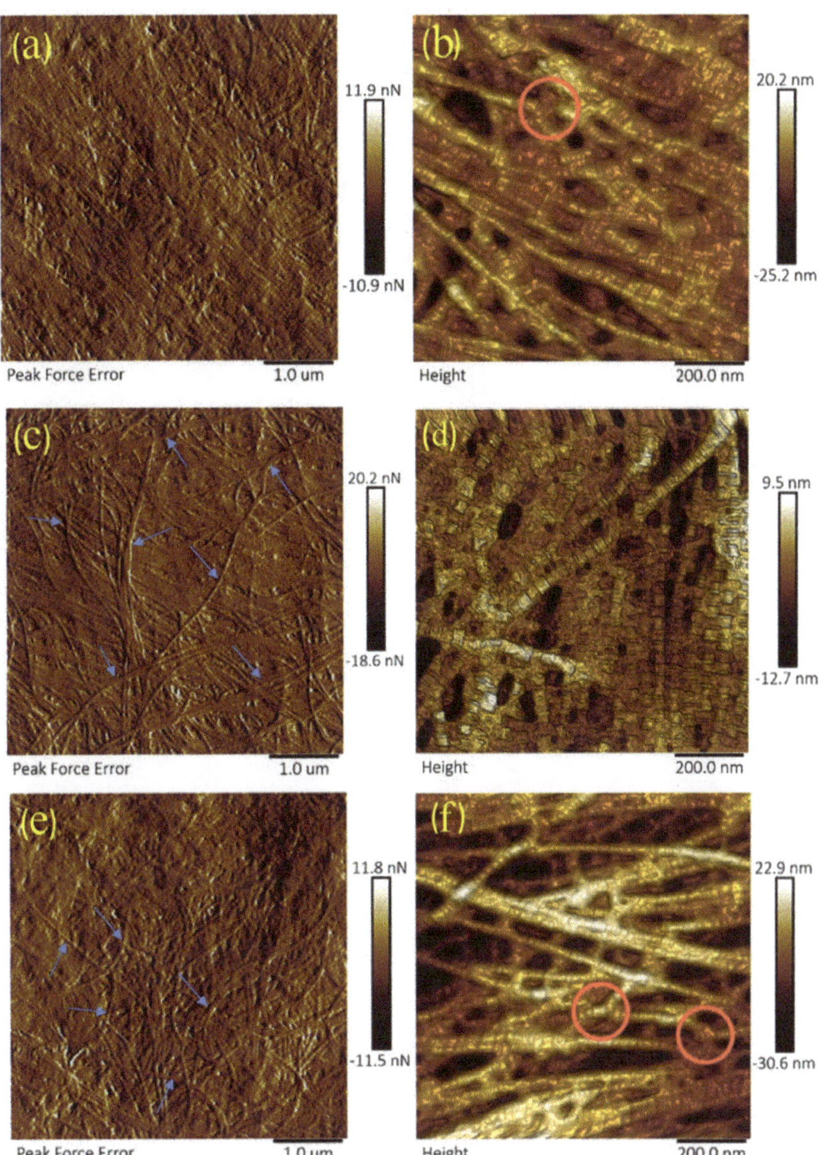

Figure 3. Topography and peak force error images showing collagen fibrils of the anterior lamella of porcine corneas following incubation with varying concentration of amylase. The arrows mark collagen fibrils that ran in irregular directions to other collagen fibrils following amylase incubation. Circles mark the places where the collagen fibril underwent fusion and splitting up. Sections (**a,b**) were incubated with 0.2 mg/mL amylase. The samples in (**c,d**) were incubated with 1 mg/mL amylase. Samples in (**e,f**) were incubated with 2 mg/mL amylase. The scan size for (**a,c,e**) was 5 μm. The scan size for (**b,d,f**) was 1 μm. A "3D effect" filter was applied for enhancement on the height images.

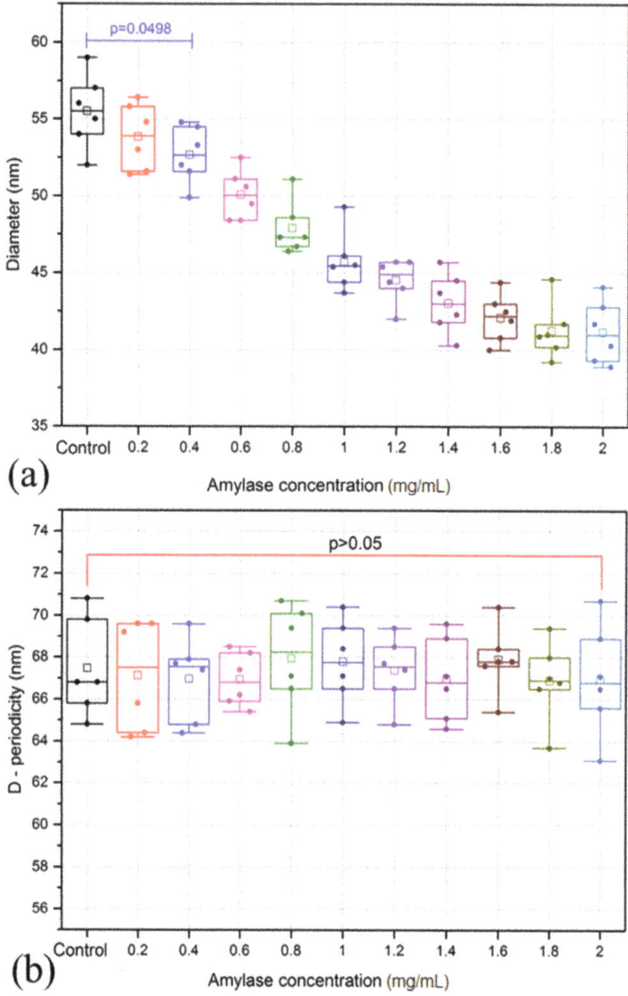

Figure 4. Collagen fibril diameters (**a**) and D-periodicity (**b**) of the anterior lamella of porcine cornea samples following incubation with varying concentrations of amylase. All data is represented as box plots and data overlaid with lower and upper borders of the box to represent the lower and upper quartiles, and the middle horizontal line to represent the median. The upper and lower whiskers represent 5th and 95th percentile of the data. n = 6 porcine eyes/group.

The elastic modulus (E) was found to decrease with amylase incubation. Example E maps following amylase treatment are shown in Figure 5. For the control samples, a mean E of 2.27 ± 0.15 GPa was recorded. E of treated corneas was decreased by 2.2% with incubation in 0.2 mg/mL amylase solution. The greatest reduction in elastic modulus of 50.2% was following incubation in 2 mg/mL amylase solution ($p < 0.0001$).

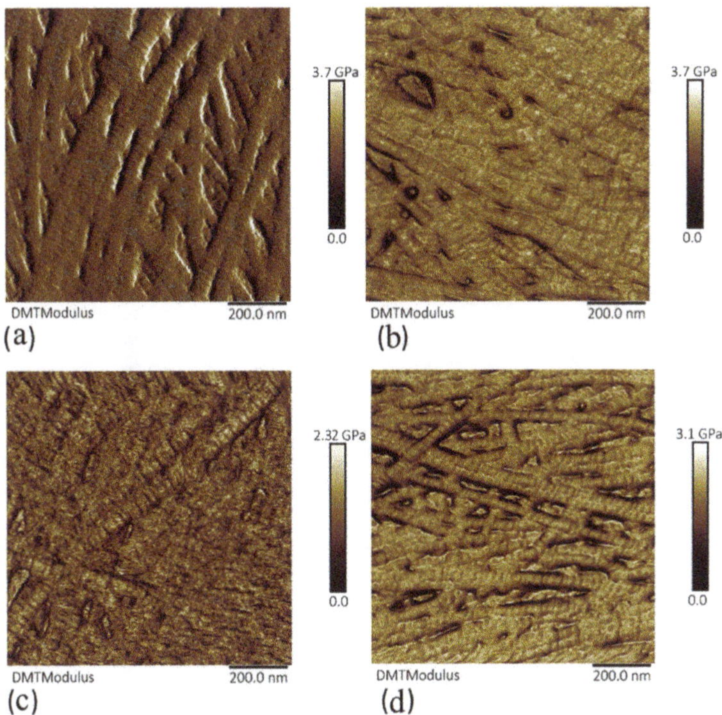

Figure 5. Elastic modulus maps of the anterior lamella of porcine corneas of control and amylase-treated samples. (**a**) Control sample. Amylase-treated samples incubated with (**b**) 0.2 mg/mL, (**c**) 1 mg/mL, and (**d**) 2 mg/mL. A scan size of 1 µm was selected for each image.

Figure 6 shows the trend with amylase concentration. E was decreased in a negative logarithmic relation with increasing amylase concentration. However, no significance differences were detected in reduction of E when the corneas were incubated with amylase of 1.8 and 2.0 mg/mL ($p > 0.086$).

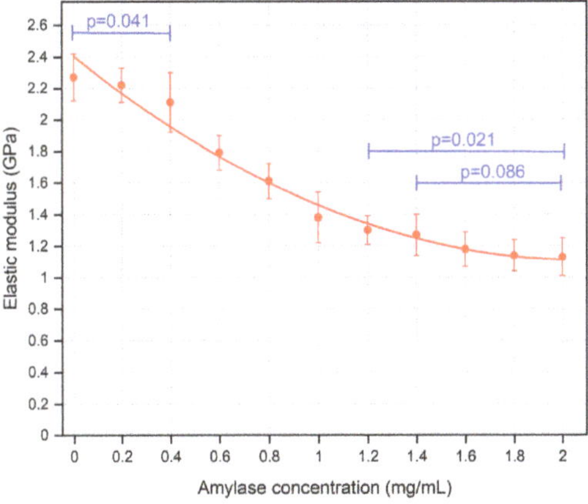

Figure 6. Mean values of elastic modulus for the corneal samples treated with varying concentrations of amylase. The curve was fit with a second order equation. Zero amylase concentration refers to control samples. n = 6 porcine eyes/group. Error bars represent the standard deviation.

3.2. Crude and Purified Collagenase Groups

Topographical images of sections treated with crude (Figure 7) and purified (Figure 8) collagenase revealed a significant deterioration of collagen fibrils with increasing concentration of the enzymes from 0.05 to 0.2 mg/mL. Topographical details (collagen fibril diameter and D-periodicity) were not easily identified in sections that were incubated with crude and purified collagenases of concentration 0.2 mg/mL, as shown in Figures 7d and 8d. The sections treated with crude collagenase (0.2 mg/mL) showed traces of degraded collagen fibrils; however, it was possible to identify non-degraded collagen fibrils in samples that were treated with the lower concentrations of the enzymes.

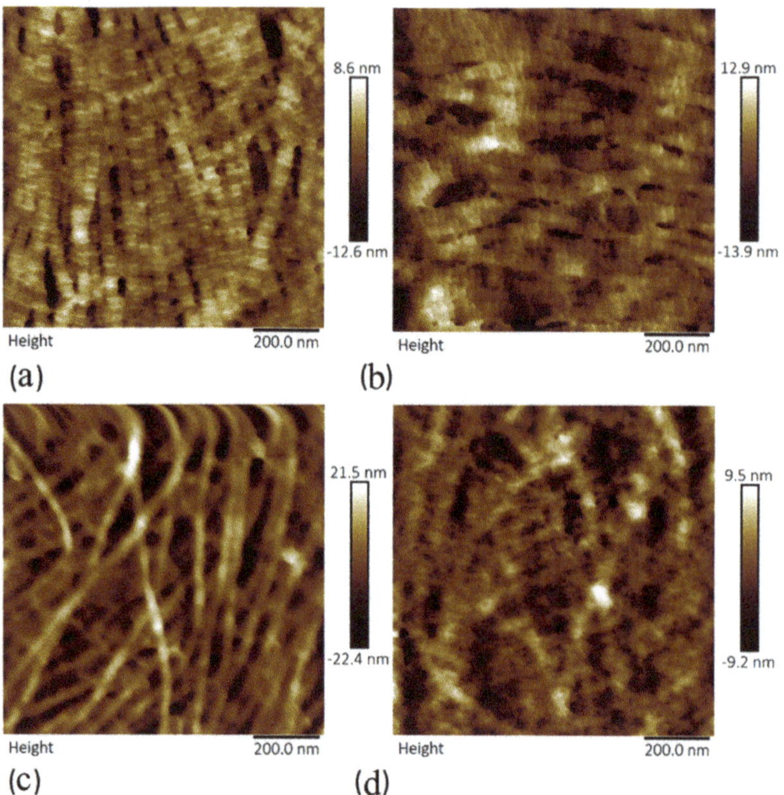

Figure 7. Topography images showing collagen fibrils of anterior lamella of porcine corneas following incubation with varying concentration of crude collagenase: (**a**) 0.05 mg/mL, (**b**) 0.1 mg/mL, (**c**) 0.15 mg/mL, and (**d**) 0.2 mg/mL. The scan size was 1 μm for each image.

A significant reduction in collagen fibril diameter was found with both crude and purified collagenase treatment (Figure 9). The reduction in collagen fibril diameter was higher with increased concentrations of the enzymes ($p < 0.0001$). The mean collagen fibril diameter for the control samples for the crude collagenase group was 57.63 ± 2.12 nm and for the controls in the purified collagenase group, the mean collagen fibril diameter was 58.71 ± 2.26 nm. No significant difference was found between the control sections of both groups ($p = 0.41$). The minimum collagen fibril diameter was 39.16 ± 2.1 nm, observed in purified collagenase treated sections of 0.2 mg/mL, whilst no collagen fibrils were clearly visible in tissue sections where the highest crude collagenase was used.

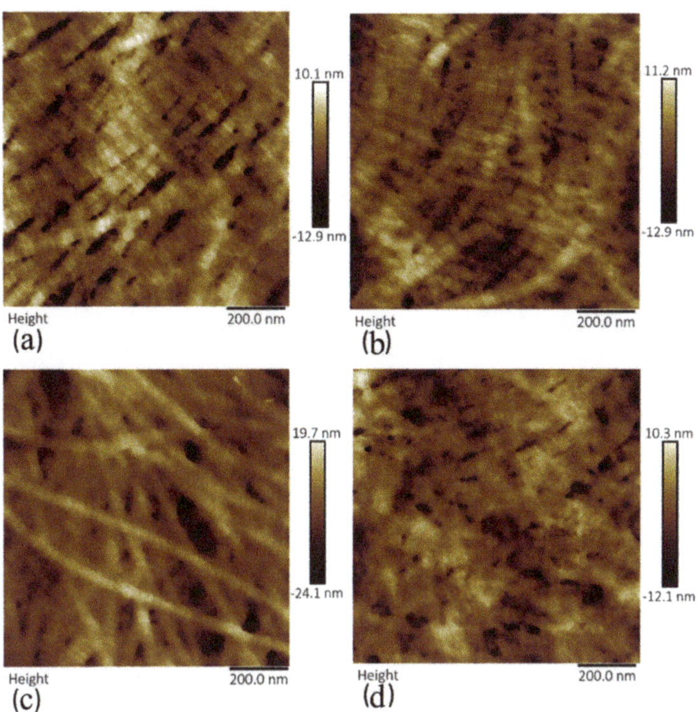

Figure 8. Topography images showing collagen fibrils of anterior lamella of porcine corneas following incubation with varying concentration of purified collagenase: (**a**) 0.05 mg/mL, (**b**) 0.1 mg/mL, (**c**) 0.15 mg/mL, and (**d**) 0.2 mg/mL. The scan size was 1 µm for each image.

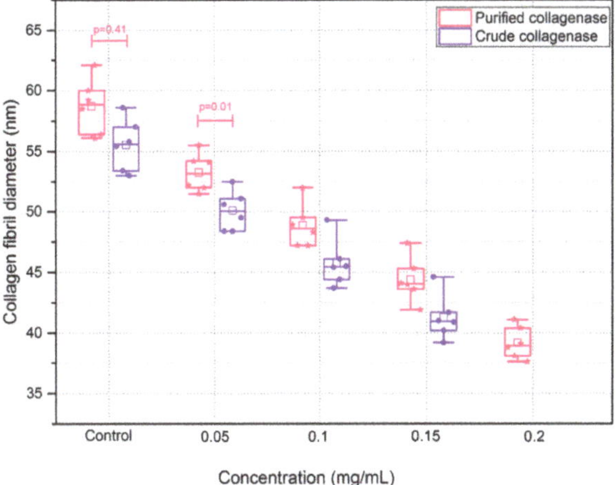

Figure 9. Collagen fibril diameter following incubation with varying concentrations of crude and purified collagenase (n = 6 porcine eyes/group). All data are represented as box plots and data overlaid with lower and upper borders of the box to represent the lower and upper quartiles, and the middle horizontal line to represent the median. The upper and lower whiskers represent 5th and 95th percentile of the data.

The results also showed that PBS slightly decreased collagen fibril diameter. Figure 10 shows that the sections that were incubated in PBS (control sections for 40 min) had collagen fibril diameters (55.5 ± 2.4 nm) significantly less than those tissue sections that were incubated for 15 min in PBS (control sections of crude and purified collagenase groups).

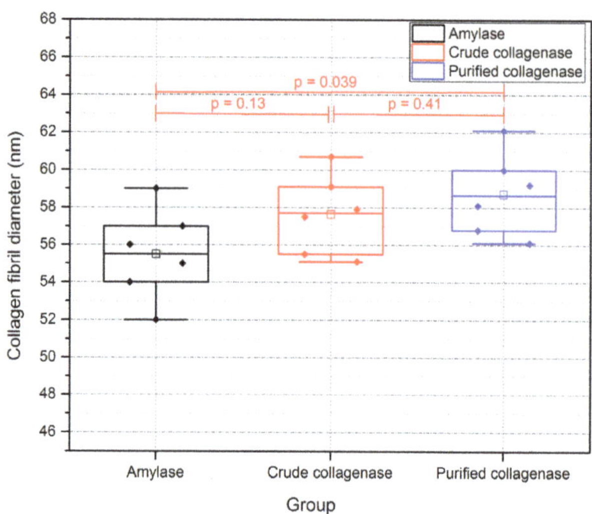

Figure 10. Collagen fibril diameters of control sections following incubation with PBS only for 15 and 40 min. (n = 6 porcine eyes/group.) Control sections of amylase group were incubated with PBS for 40 min. Controls in the crude and purified collagenase groups were incubated with PBS for 15 min.

Collagen fibril D-periodicity was significantly reduced following incubation with crude and purified collagenases (p = 0.03) (Figure 11). No significant difference was found in D-periodicity of collagen fibrils in the control groups for crude and purified collagenase, 67.6 ± 2.3 nm and 67.4 ± 2.6 nm, respectively. The reduction in D-periodicity was approximately 8.2% in 0.05 mg/mL crude collagenase group as compared to the control sections (p < 0.0001), whilst at the same concentration for the purified collagenase group, the D-periodicity was decreased by approximately 4.2% in contrast to the controls. The D-periodicity was significantly decreased with increased concentration of the collagenases (p < 0.0001), where it was reduced to 45.1 ± 2.4 nm (approximately 33.1% lower relative to the control) at the highest concentration of purified collagenase. It was not possible to measure D-periodicity of collagen fibrils in samples that were treated with 0.2 mg/mL crude collagenase because individual collagen fibrils could not be identified.

The elastic modulus significantly decreased in sections that were incubated with crude collagenase for 15 min, and it decreased more with increasing concentration (p < 0.001), as shown in Figure 12. For the control samples, the mean E was 2.21 ± 0.16 GPa, with a 27.6% decrease following incubation with 0.05 mg/mL crude collagenase (p < 0.001). The maximum reduction of the elastic modulus was 76.5%, which was observed following incubation with 0.2 mg/mL crude collagenase. Similar trends in terms of E were found with the purified collagenase-treated corneas, with a significant decrease following incubation with varying concentrations of the enzyme (p < 0.05), Figure 12. For the controls, mean E was 2.16 ± 0.18 GPa. E appeared to decrease linearly with an increase in purified collagenase concentration. A greater reduction in E was found in the crude collagenase group as compared to the purified collagenase group. This difference was statistically significant at each concentration (p < 0.001). No significant difference was found in E for the controls when comparing the crude and purified collagenase groups (p = 0.92).

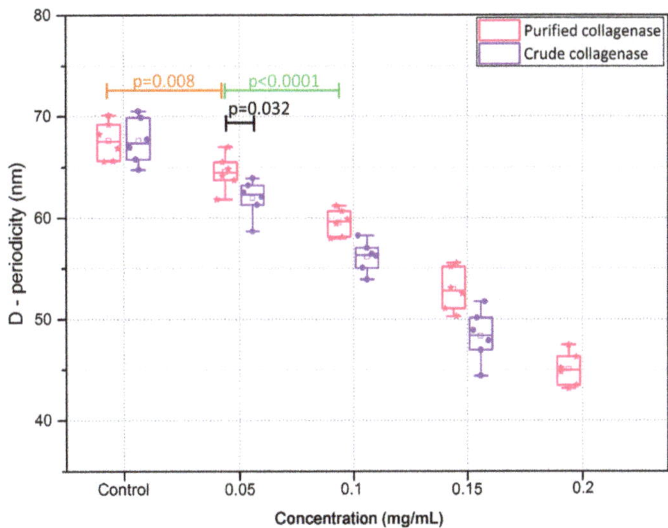

Figure 11. D-periodicity of collagen fibrils of porcine cornea sections following incubation with varying concentrations of crude and purified collagenase (n = 6 porcine eyes/group).

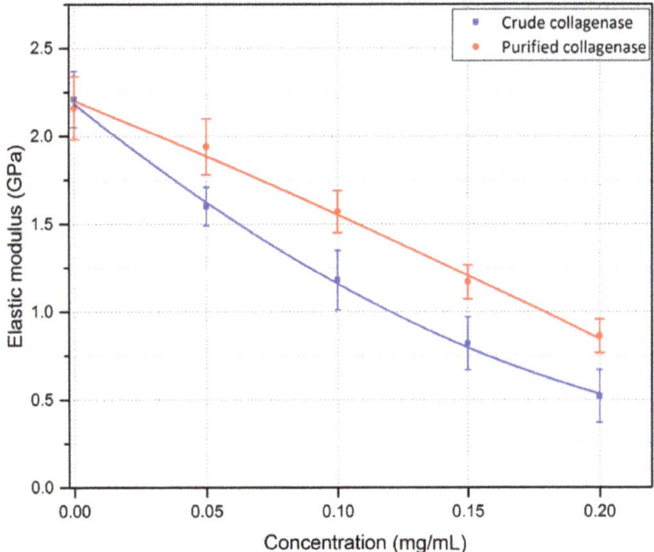

Figure 12. Mean values of elastic modulus of corneal samples treated with varying concentrations of crude and purified collagenase. Zero concentration refers to the control samples (n = 6 porcine eyes/group). Error bars represent the standard deviation.

3.3. GAG Quantification

Figure 13 shows the quantities of GAGs in treatment solutions. The quantities of GAGs released in treatment solutions after 60 min and 120 min in amylase were 47.3 ± 8.4 µg/mL and 73.3 ± 7.5 µg/mL, respectively. GAG quantities released in treatment solutions of the amylase group were significantly higher than in other groups (control and tissue culture groups), $p < 0.0001$. A small quantity of GAGs (14.2 ± 3.1 µg/mL) were released in treatment solutions of the PBS group. There was little or no GAGs released in the TC group. Statistically significant differences were found among the groups ($p < 0.01$).

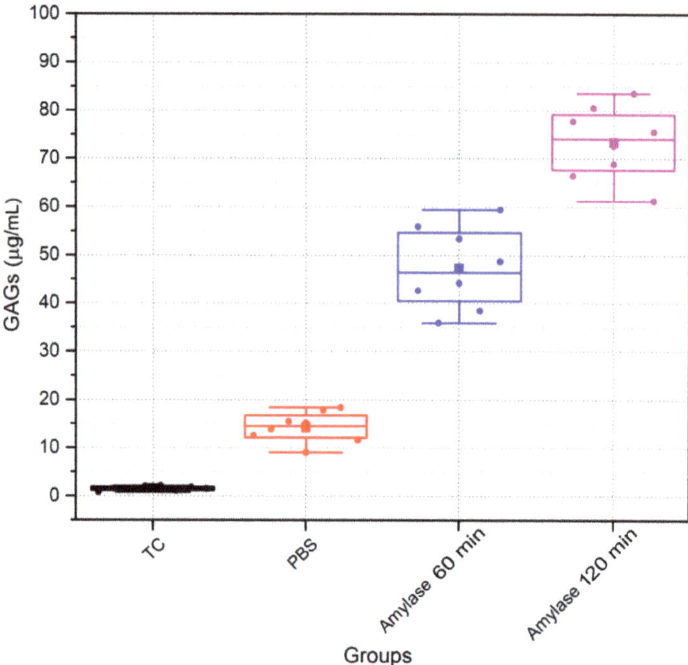

Figure 13. Depleted sulphated glycosaminoglycans (GAGs) of porcine corneal samples following incubation with tissue culture (TC), Phosphate Buffered Saline (PBS), and amylase. (n = 8 porcine eyes/group).

4. Discussion

This paper utilises the PFQNM-AFM mode to investigate nano-scale alterations in the mechanical properties and collagen fibril ultrastructure of the porcine cornea following in vitro enzymatic degradation. The paper presents a novel application of this fast data acquisition AFM mode, utilising the AFM cantilever as a sensor for detecting alterations in the cornea following enzymatic degradation with amylase, crude collagenase, and purified collagenase.

The use of AFM has significant advantages in investigating nanomechanical and ultrastructural details over other nano-imaging techniques (such as SEM) and over conventional mechanical testing techniques such as extensometry [34]. AFM has made it possible to observe differences in stiffness of a composite material (soft tissues) and visually distinguish between hard and soft regions. With regard to imaging, collagen fibrils can be visualised without the need for special treatments (such as metal/carbon coating) that would irreversibly alter or damage the samples [34]. Clearly, an advantage of AFM is the ability to not only record ultrastructure but also to quantify the nanomechanical properties of samples [19,35]. With the use of PFQNM mode, AFM provides a link between samples topography and its mechanical properties in nanoscale, with rapid acquisition of high-resolution mechanical property maps [30].

4.1. Corneal Degradation with Amylase

Collagen fibril diameters of the corneal sections were decreased following incubation with amylase. In an earlier AFM study, it was found that the collagen fibril diameter in the porcine corneal stroma is 55.6 ± 5.2 nm for hydrated sections [3], which is consistent with collagen fibril diameters of the air-dried control sections in the current study. The reduction of collagen fibril diameters following incubation with amylase suggests the important role of proteoglycans in maintaining the collagen fibril diameter. It was found that the sclera (white of the eye) of highly myopic human eyes is associated with reduction of proteoglycans and contained an increased number of smaller diameter collagen fibrils in comparison with normal human sclera, which leads to an increase in sclera elasticity [36]. That finding might justify the currently presented change in collagen fibril diameters following the proposed depletion of proteoglycans with amylase.

We found that the axial collagen fibril D-periodicity did not change following treatment with amylase. The axial periodicity of Type I collagen fibrils in normal human corneal stroma has been reported as 65 nm, with X-ray diffraction [37] and 67 nm with AFM [20], which is close to the mean values of collagen fibril D-periodicity in the control samples in this study. Given that we found non-significant changes in collagen fibril D-periodicity in the amylase treated samples, our findings suggest that collagen fibrils were not digested by the amylase.

Loss of collagen fibril orientation, splitting-up and fusion of collagen fibrils was observed following enzymatic treatment with amylase. A similar finding was also reported in sclera samples following incubation with amylase [38], suggesting that depletion of proteoglycans which are located between collagen fibrils results in this alteration to collagen fibrils.

GAGs were depleted from the stroma of the cornea samples following incubation with α-amylase. GAG depletion appears to weaken the collagenous network of the tissue. The quantity of depleted GAGs from the samples was a function of incubation time. A small quantity of GAGs was released after incubation with PBS, which could have occurred as a result of tissue swelling which may damage the tissue surface and thereby lead to GAG release.

The elastic modulus was significantly decreased in amylase treated sections. Proteoglycans act as cross-links between the collagen fibrils, which together lead to the mechanical properties of the tissue. By analogy, the depletion or break down of these cross-links leads to deterioration of the normal ultrastructural organisation of the tissue, and subsequently reduction of tissue stiffness [39]. The reduction of stiffness in collagenous tissue following amylase incubation was also stated in a number of studies [9,10,40], where amylase was utilised to deplete proteoglycans of collagenous tissues.

We found that as the concentration of amylase increased, there was a greater reduction in E, which implies that high concentrations of amylase increases the number of cleavages on α-1,4 glycosidic bonds. These bonds link the GAGs to the core protein of the proteoglycans [41]. GAGs fill the space between the collagen fibrillar network in the extracellular matrix. Depletion of these GAGs with amylase appears to disrupt the organised network of collagen fibrils, leading to a weakening of the physical properties of the collagenous tissue, as has been suggested previously [4,39].

4.2. Corneal Degradation with Collagenase

Both crude and purified collagenase significantly degraded corneal samples, with significant changes in the ultrastructure which were manifested by alterations in collagen fibril diameters and their structural organisation. Collagen fibrils in the collagenase treated samples (crude and purified) exhibited a reduction in diameter with increased enzyme concentration. The reduction in diameter is mainly attributed to digestion of collagen fibrils by collagenases as is the established function of this enzyme [42–45]. The current finding agrees with a previous study that used AFM to measure the adhesion force and ultrastructure of the collagen fibrils on the Achilles tendons of rats following injection of collagenase [46].

It was found that collagen fibril D-periodicity significantly decreased following incubation with collagenases (crude and purified). Hence, significant digestion of the collagen fibrils occurred with these enzymes. In support of the current results, Lee and colleagues

(2011) found that D-periodicity of collagen fibrils in the Achilles tendon of rats was significantly reduced after collagenase incubation [46]. Therefore, it can be said the collagenase did not fragment the collagen fibrils but instead digested them along the fibril axis. This suggestion has been supported by previous studies, where it was reported the collagenase did result in fibril fragmentation but decreased their diameter and D-periodicity [46,47].

Collagenase decreased the stiffness of the corneal tissue, and this reduction increased with enzyme concentration. These results agree with previous studies in which bacterial collagenase reduced tissue stiffness [47–49]. It is well known that collagen fibrils provide structural integrity to tissues and their deterioration causes a reduction in stiffness and subsequent geometrical modifications in the tissue [37,50,51]. This deterioration can be interpreted as a reduction in collagen fibril diameters, which was found to be correlated with their stiffness [27].

Interestingly, we found that crude collagenase has a stronger effect on ultrastructural details and mechanical property deterioration than purified collagenase. This result agrees with Gaul et al. (2018) who found that arterial tissue incubated with crude collagenase showed more degradation responses to strain than those incubated with purified collagenase [42]. This is attributed to contamination of crude collagenase with other proteolytic enzymes [43]. Therefore, the ultrastructural details were not possible to be observed in samples where high concentration of crude collagenase was used.

Elastic modulus values of porcine corneas, reported in previous studies [52,53], is less than those measured by AFM (nano-scale). E of control corneas (whole thickness) tested by nanoindentation with a 100 μm flat punch [52] and inflation testing [53] were in the range of 40 to 700 kPa depending on testing conditions such as internal pressure and hydration. However, E measured on corneal sections (5 μm thick) in this study were in the range of 2 to 2.45 GPa. The difference between the testing length scale is one of the reasons for higher E measured by AFM, where elastic modulus at the macroscale reflects the bulk elastic response of the corneas in the tangential direction. E measured by AFM reflects the mechanical properties of the tissue ultrastructure. It has been reported that differences in testing scales from macro- to the micro-scale leads to higher E, which can be attributed to the tissue components being probed at each length scale [54,55]. Another reason for high values is due to lack of hydration. The tissue sections were air-dried to obtain higher resolution images of the collagen fibrils than is possible under liquid. Sample thickness also affects E where the thicker the sample, the less effect there is of the substrate on sample stiffness [56]. However, increasing the thickness of the AFM sections reduces the quality of the topographical images. Therefore, an optimal tissue thickness was required to balance image quality and reduce substrate effects.

5. Limitations

It was hypothesised that the thickness of the tissue sections after the enzymatic treatment would decrease however, it was not possible to measure cryosection thickness following enzymatic degradation. Reduction in tissue thickness may have impacted E. Another limitation includes testing the samples in air that resulted in elevated values of corneal elastic modulus.

6. Conclusions

PFQNM-AFM served as a suitable method for examining nanomechanical and ultrastructural changes in the cornea following incubation with amylase and collagenase. Amylase treatment reduces collagen fibril diameter and corneal stiffness, but not D-periodicity of collagen fibrils. The reduction in corneal stiffness and collagen fibril diameter increased with amylase concentration due to GAG depletion, which is thought to break-down proteoglycan linkages with collagen fibrils that leads to deterioration in corneal stiffness.

Collagenase treatment gradually deteriorates the ultrastructure and the stiffness of the cornea. These changes are significantly higher than the changes obtained with amylase treatment. The deterioration of the ultrastructure includes reduction in both the diameter and the D-periodicity of collagen fibrils. Incubation with crude collagenase has a greater effect on corneal samples than purified collagenase. The disruption of collagen fibril morphology results in reduction of nano-scale stiffness.

Author Contributions: A.K. conceived and planned the experiments, conducted all the experimental work, analysed the data, and edited and wrote the manuscript. H.A.-A.A.-H. verified the analytical method and edited the manuscript. J.M. planned the DMMB assay experiments, verified the analytical method, and edited the manuscript. R.A. supervised all the work, conceived and planned the experiments, analysed the data, and edited and wrote the manuscript. All authors have read and agreed to the published version of the manuscript.

Funding: This project has been funded with a PhD scholarship fund awarded to AK from High Committee of Education Development in Iraq.

Institutional Review Board Statement: Not applicable.

Informed Consent Statement: Not applicable.

Data Availability Statement: The data presented in this study are available on request from the corresponding author.

Acknowledgments: The authors would like to acknowledge Professor Ahmed Elsheikh, Brendan Geraghty, Zhuola, Ya Hua, Chim, and Zhuo Chang for their valuable advice and support, and the wider Biomechanical Engineering Group at University of Liverpool. We also thank Dave Atkinson and Derek Neary for their technical support.

Conflicts of Interest: The authors declare no conflict of interest.

References

1. Meek, K. Corneal collagen—its role in maintaining corneal shape and transparency. *Biophys. Rev.* **2009**, *1*, 83–93. [CrossRef]
2. Meek, K.M.; Knupp, C. Corneal structure and transparency. *Prog. Retin. Eye Res.* **2015**, *49*, 1–16. [CrossRef] [PubMed]
3. Xia, D.; Zhang, S.; Hjortdal, J.; Li, Q.; Thomsen, K.; Chevallier, J.; Besenbacher, F.; Dong, M. Hydrated Human Corneal Stroma Revealed by Quantitative Dynamic Atomic Force Microscopy at Nanoscale. *ACS Nano* **2014**, *8*, 6873–6882. [CrossRef] [PubMed]
4. Zhang, F.; Zhang, Z.; Linhsrdt, R.J. Glycosaminoglycans. In *Handbook of Glycomics*, 1st ed.; Cummings Elsevier, R.D., Pierce, M.J., Eds.; Academic Press: Cambridge, MA, USA, 2010; Volume 3, pp. 59–80.
5. Van Haeringen, N.J.; Ensink, F.; Glasius, E. Amylase in human tear fluid: Origin and characteristics, compared with salivary and urinary amylases. *Exp. Eye Res.* **1975**, *21*, 395–403. [CrossRef]
6. Kaiserman, I.; Sella, S. Chronic ocular inflammation and keratoconus. In *Controversies in the Management of Keratoconus*, 1st ed.; Barbara, A., Ed.; Springer: Cham, Germany, 2019; Volume 2, pp. 17–27.
7. Galvis, V.; Sherwin, T.; Tello, A.; Merayo, J. Keratoconus: An inflammatory disorder? *Eye* **2015**, *29*, 843–852. [CrossRef]
8. Legkikh, L.; Koledintsev, M.; Semenova, A.; Okuyama, K. Biochemical Investigations of Lacrima in Early Diagnosis of Keratoconus. (Abstract) Sangubashi Eye Clinic, Japan. 2017. Available online: www.sangubashi.com/English/report/syanhai.htm (accessed on 12 January 2020).
9. Spoerl, E.; Teral, N.; Raiskup, F.; Pillunat, L. Amylase reduces the biomechanical stiffness of the cornea. *Investig. Ophthalmol. Vis. Sci.* **2012**, *53*, 1531.
10. Wollensak, G.; Spörl, E.; Mazzotta, C.; Kalinski, T.; Sel, S. Interlamellar cohesion after corneal crosslinking using riboflavin and ultraviolet A light. *Br. J. Ophthalmol.* **2011**, *95*, 876–880. [CrossRef]
11. Balasubramanian, S.; Pye, D.; Willcox, M. Are proteinases the reason for keratoconus? *Curr. Eye Res.* **2010**, *35*, 185–191. [CrossRef]
12. Collier, S.; Madigan, M.; Penfold, P. Expression of membrane-type 1 matrix metalloproteinase (MT1-MMP) and MMP-2 in normal and keratoconus corneas. *Curr. Eye Res.* **2000**, *21*, 662–668. [CrossRef]
13. Collier, S.A. Is the corneal degradation in keratoconus caused by matrix-metalloproteinases? *Clin. Exp. Ophthalmol.* **2001**, *29*, 340–344. [CrossRef] [PubMed]
14. Hurst, L.C.; Badalamente, M.A.; Hentz, V.R.; Hotchkiss, R.N.; Kaplan, F.T.; Meals, R.A.; Smith, T.M.; Rodzvilla, J. Injectable collagenase clostridium histolyticum for Dupuytren's contracture. *N. Engl. J. Med.* **2012**, *361*, 968–979. [CrossRef]
15. Mackiewicz, Z.; Määttä, M.; Stenman, M.; Konttinen, L.; Tervo, T.; Konttinen, Y. Collagenolytic proteinases in keratoconus. *Cornea* **2006**, *25*, 603–610. [CrossRef]
16. Seppälä, H.; Määttä, M.; Rautia, M.; Mackiewicz, Z.; Tuisku, I.; Tervo, T.; Konttinen, Y. EMMPRIN and MMP-1 in keratoconus. *Cornea* **2006**, *25*, 325–330. [CrossRef]
17. Romero-Jiménez, M.; Santodomingo-Rubido, J.; Wolffsohn, J. Keratoconus: A review. *Contact Lens Anterior Eye* **2010**, *33*, 157–166. [CrossRef] [PubMed]

18. Volatier, T.L.; Figueiredo, F.C.; Connon, C.J. Keratoconus at a molecular level: A review. *Anat. Record* **2019**, *1*, 1–5. [CrossRef] [PubMed]

19. Last, J.; Russell, P.; Nealey, P.; Murphy, C. The applications of atomic force microscopy to vision science. *Investig. Ophthalmol. Vi Sci.* **2010**, *51*, 6083–6094. [CrossRef]

20. Jastrzebska, M.; Tarnawska, D.; Wrzalik, R.; Chrobak, A.; Grelowski, M.; Wylegala, E.; Zygadlo, D.; Ratuszna, A. New insight int the shortening of the collagen fibril D-period in human cornea. *J. Biomol. Struct. Dyn.* **2016**, *35*, 551–563. [CrossRef] [PubMed]

21. Fullwood, N.J.; Hammiche, A.; Pollock, H.M.; Hourston, D.J.; Song, M. Atomic force microscopy of the cornea and scler *Curr. Eye Res.* **1996**, *14*, 529–535. [CrossRef] [PubMed]

22. Last, J.; Thomasy, S.; Croasdale, C.; Russell, P.; Murphy, C. Compliance profile of the human cornea as measured by atomic forc microscopy. *Micron* **2012**, *43*, 1293–1298. [CrossRef]

23. Lombardo, M.; Lombardo, G.; Carbone, G.; De Santo, M.P.; Barberi, R.; Serrao, S. Biomechanics of the anterior human cornea tissue investigated with atomic force microscopy. *Investig. Ophthalmol. Vis. Sci.* **2012**, *53*, 1050–1057. [CrossRef]

24. Lombardo, M.; Santo, M.; Lombardo, G.; Barberi, R.; Serrao, S. Atomic force microscopy analysis of normal and photoablate porcine corneas. *J. Biomech.* **2006**, *39*, 2719–2724. [CrossRef] [PubMed]

25. Choi, S.; Lee, S.C.; Lee, H.J.; Cheong, Y.; Jung, G.B.; Jin, K.H.; Park, H.K. Structural response of human corneal and scleral tissue to collagen cross-linking treatment with riboflavin and ultraviolet A light. *Lasers Med. Sci.* **2013**, *28*, 1289–1296. [CrossRef]

26. Hu, J.; Chen, S.; Huang, D.; Zhang, Y.; Lü, S.; Long, M. Global mapping of live cell mechanical features using PeakForce QNM AFM. *Biophys. Rep.* **2020**, *6*, 1–10. [CrossRef]

27. Papi, M.; Paoletti, P.; Geraghty, B.; Akhtar, R. Nanoscale characterization of the biomechanical properties of collagen fibrils in th sclera. *Appl. Phys. Lett.* **2014**, *104*, 103703. [CrossRef]

28. Chang, Z.; Paoletti, P.; Barrett, S.; Chim, Y.; Caamaño-Gutiérrez, E.; Hansen, M.; Beck, H.; Rasmussen, L.; Akhtar, R. Nanomechan ics and ultrastructure of the internal mammary artery adventitia in patients with low and high pulse wave velocity. *Acta Biomate* **2018**, *73*, 437–448. [CrossRef]

29. Dokukin, M.; Sokolov, I. Quantitative Mapping of the Elastic Modulus of Soft Materials with HarmoniX and PeakForce QNM AFM Modes. *Langmuir* **2012**, *28*, 16060–16071. [CrossRef] [PubMed]

30. Pittenger, B.; Erina, N.; Su, C. *Quantitative Mechanical Property Mapping at the Nanoscale with Peak Force QNM*; Application Not Veeco Instruments Inc., 2010; pp. 1–12. Available online: https://www.bruker.com/fileadmin/user_upload/8-PDF-Docs SurfaceAnalysis/AFM/ApplicationNotes/AN128-RevB0Quantitative_Mechanical_Property_Mapping_at_the_Nanoscale_ with_PeakForceQNM-AppNote.pdf (accessed on 12 January 2020).

31. Young, T.; Monclus, M.; Burnett, T.; Broughton, W.; Ogin, S.; Smith, P. The use of the PeakForceTM quantitative nanomechanica mapping AFM-based method for high-resolution Young's modulus measurement of polymers. *Meas. Sci. Technol.* **2011**, *22* 125703. [CrossRef]

32. Heu, C.; Berquand, A.; Elie-Caille, C.; Nicod, L. Glyphosate-induced stiffening of HaCaT keratinocytes, a Peak Force Tapping study on living cells. *J. Struct. Biol.* **2012**, *178*, 1–7. [CrossRef]

33. Farndale, R.W.; Sayers, C.A.; Barrett, A.J. A direct spectrophotometric microassay for sulfated glycosaminoglycans in cartilage cultures. *Connect. Tissue Res.* **1982**, *9*, 247–248. [CrossRef]

34. Jalili, N.; Laxminarayana, K.K. A review of atomic force microscopy imaging systems: Application to molecular metrology anc biological sciences. *Mechatronics* **2004**, *14*, 907–945. [CrossRef]

35. Alessandrini, A.; Facci, P. AFM: A versatile tool in biophysics. *Meas. Sci. Technol.* **1982**, *16*, R65. [CrossRef]

36. Harper, A.R.; Summers, J.A. The dynamic sclera: Extracellular matrix remodeling in normal ocular growth and myopi development. *Exp. Eye Res.* **2015**, *133*, 100–111. [CrossRef] [PubMed]

37. Meek, K.; Boote, C. The organization of collagen in the corneal stroma. *Exp. Eye Res.* **2004**, *78*, 503–512. [CrossRef]

38. Akhtar, R. The Role of Proteoglycans in the Ultrastructure and Mechanical Properties of the Sclera. Ph.D. Thesis, University o Liverpool, Liverpool, UK, 2018.

39. Murienne, B.; Jefferys, J.; Quigley, H.; Nguyen, T. The effects of glycosaminoglycan degradation on the mechanical behavior o the posterior porcine sclera. *Acta Biomater.* **2015**, *12*, 195–206. [CrossRef]

40. Tanaka, E.; Aoyama, J.; Tanaka, M.; Van Eijden, T.; Sugiyama, M.; Hanaoka, K.; Watanabe, M.; Tanne, K. The proteoglycar contents of the temporomandibular joint disc influence its dynamic viscoelastic properties. *J. Biomed. Mater. Res.* **2003**, *65*, 386–392 [CrossRef] [PubMed]

41. Quintarelli, G.; Dellovo, M.C.; Balduini, C.; Castellani, A.A. The effects of alpha amylase on collagen-proteoglycans and collagen-glycoprotein complexes in connective tissue matrices. *Histochemie* **1969**, *18*, 373–375. [CrossRef]

42. Gaul, R.T.; Nolan, D.R.; Ristori, T.; Bouten, C.V.C.; Loerakker, S.; Lally, C. Strain mediated Enzymatic Degradation of arteria tissue: Insights into the role of the non-collagenous tissue matrix and collagen crimp. *Acta Biomater.* **2018**, *77*, 301–310. [CrossRef

43. Hanoune, J.; Stengel, D.; Lacombe, M.L.; Feldmann, G.; Coudrier, E. Proteolytic activation of rat liver adenylate cyclase by a contaminant of crude collagenase from Clostridium histolyticum. *J. Biol. Chem.* **1977**, *252*, 2039–2045. [CrossRef]

44. Jayes, F.; Liu, B.; Moutos, F.; Kuchibhatla, M.; Guilak, F.; Leppert, P. Loss of stiffness in collagen-rich uterine fibroids after digestion with purified collagenase Clostridium histolyticum. *Am. J. Obstet. Gynecol.* **2016**, *215*, 596.e1–596.e8. [CrossRef]

45. Van der Kraan, P.M.; Vitters, E.L.; van Beuningen, H.M.; van de Putte, L.B.; van den Berg, W.B. Degenerative knee joint lesions in mice after a single intra-articular collagenase injection. A new model of osteoarthritis. *J. Exp. Pathol.* **1990**, *71*, 19–31.

46. Lee, G.J.J.; Choi, S.; Chon, J.; Yoo, S.; Cho, I.; Park, H.K.K. Changes in collagen fibril pattern and adhesion force with collagenase-induced injury in rat Achilles tendon observed via AFM. *J. Nanosci. Nanotechnol.* **2011**, *11*, 773–777. [CrossRef]
47. Långsjö, T.; Rieppo, J.; Pelttari, A.; Oksala, N.; Kovanen, V.; Helminen, H. Collagenase-induced changes in articular cartilage as detected by electron-microscopic stereology, quantitative polarized light microscopy and biochemical assays. *Cells Tissues Organs* **2002**, *172*, 265–275. [CrossRef] [PubMed]
48. Lyyra, T.; Arokoski, J.P.; Oksala, N.; Vihko, A.; Hyttinen, M.; Jurvelin, J.S.; Kiviranta, I. Experimental validation of arthroscopic cartilage stiffness measurement using enzymatically degraded cartilage samples. *Phys. Med. Biol.* **1999**, *44*, 525–535. [CrossRef] [PubMed]
49. Stolz, M.; Raiteri, R.; Daniels, A.U.; VanLandingham, M.; Baschong, W.; Aebi, U. Dynamic elastic modulus of porcine articular cartilage determined at two different levels of tissue organization by indentation-type atomic force microscopy. *Biophys. J.* **2004**, *86*, 3269–3283. [CrossRef]
50. Fratzl, P. Collagen: Structure and mechanics, an introduction. In *Collagen*, 1st ed.; Fratzl, P., Ed.; Springer: Boston, MA, USA, 2008; Volume 1, pp. 1–13.
51. Sherman, V.R.; Yang, W.; Meyers, M.A. The materials science of collagen. *J. Mech. Behav. Biomed. Mater.* **2015**, *52*, 22–50. [CrossRef]
52. Kazaili, A.; Geraghty, B.; Akhtar, R. Microscale assessment of corneal viscoelastic properties under physiological pressures. *J. Mech. Behav. Biomed. Mater.* **2019**, *100*, 103375. [CrossRef]
53. Kazaili, A.; Lawman, S.; Geraghty, B.; Eliasy, A.; Zheng, Y.; Shen, Y.; Akhtar, R. Line-Field Optical Coherence Tomography as a tool for In vitro characterization of corneal biomechanics under physiological pressures. *Sci. Rep.* **2019**, *9*, 1–13. [CrossRef]
54. Akhtar, R.; Sherratt, M.J.; Cruickshank, J.K.; Derby, B. Characterizing the elastic properties of tissues. *Mater. Today* **2011**, *14*, 96–105. [CrossRef]
55. Crichton, M.; Chen, X.; Huang, H.; Kendall, M. Elastic modulus and viscoelastic properties of full thickness skin characterised at micro scales. *Biomaterials* **2013**, *34*, 2087–2097. [CrossRef]
56. Picas, L.; Milhiet, P.E.E.; Hernández-Borrell, J. Atomic force microscopy: A versatile tool to probe the physical and chemical properties of supported membranes at the nanoscale. *Chem. Phys. Lipids* **2012**, *165*, 845–860. [CrossRef] [PubMed]

Article

Impact of Experimental Parameters on Cell–Cell Force Spectroscopy Signature

Reinier Oropesa-Nuñez [1,†], Andrea Mescola [2,†], Massimo Vassalli [3,*] and Claudio Canale [4]

1   Department of Materials Science and Engineering, Uppsala University, Ångströmlaboratoriet, Box 35, SE-751 03 Uppsala, Sweden; reinier.oropesa@angstrom.uu.se
2   CNR-Nanoscience Institute-S3, Via Campi 213/A, 41125 Modena, Italy; andrea.mescola@nano.cnr.it
3   James Watt School of Engineering, University of Glasgow, Glasgow G128LT, UK
4   Department of Physics, University of Genoa, via Dodecaneso 33, 16146 Genoa, Italy; canale@fisica.unige.it
*   Correspondence: massimo.vassalli@glasgow.ac.uk
†   These authors contributed equally to the paper.

Abstract: Atomic force microscopy is an extremely versatile technique, featuring atomic-scale imaging resolution, and also offering the possibility to probe interaction forces down to few pN. Recently, this technique has been specialized to study the interaction between single living cells, one on the substrate, and a second being adhered on the cantilever. Cell–cell force spectroscopy offers a unique tool to investigate in fine detail intra-cellular interactions, and it holds great promise to elucidate elusive phenomena in physiology and pathology. Here we present a systematic study of the effect of the main measurement parameters on cell–cell curves, showing the importance of controlling the experimental conditions. Moreover, a simple theoretical interpretation is proposed, based on the number of contacts formed between the two interacting cells. The results show that single cell–cell force spectroscopy experiments carry a wealth of information that can be exploited to understand the inner dynamics of the interaction of living cells at the molecular level.

Keywords: cell-cell interaction; force spectroscopy; atomic force microscopy; cell mechanics; mechanobiology

Citation: Oropesa-Nuñez, R.; Mescola, A.; Vassalli, M.; Canale, C. Impact of Experimental Parameters on Cell–Cell Force Spectroscopy Signature. *Sensors* **2021**, *21*, 1069. https://doi.org/10.3390/s21041069

Academic Editor: Petr Skládal
Received: 28 December 2020
Accepted: 1 February 2021
Published: 4 February 2021

Publisher's Note: MDPI stays neutral with regard to jurisdictional claims in published maps and institutional affiliations.

1. Introduction

Cells have several evolved mechanisms to sense and respond to mechanical stimuli in their environment. Mechanical forces transmitted through cell–matrix and cell–cell interactions play a pivotal role in the organization, growth, maturation, and function of living tissues [1–5]. Cell–cell interactions are not only crucial to maintaining tissue morphogenesis and homeostasis, but they also activate signaling pathways important for the regulation of different cellular processes including, cell survival, cell migration, and differentiation [6,7]. Alterations in the plasma membrane composition, and consequently, its nanomechanical properties and the nanoscale forces arising from the cell–cell interactions, can impair cellular mechanosensitivity [8,9] and eventually lead to the onset of several human pathologies. Although cell–cell connections are commonly represented as two phospholipid bilayers tethered by a few receptors, the compartments created possess properties that are distinct from and more complex than other parts of the plasma membrane. Indeed, cell mechanics is affected during the pathological mechanisms in breast cancer diseases by the alteration of the expression of cell membrane components [10]. The interaction of Aβ42 oligomers was also found to negatively influence the membrane's biophysics of hippocampal neurons [11]. Therefore, the understanding of the underlying molecular pathways of cell–cell interactions is a crucial aspect for a better comprehension of human pathologies.

In this context, mechanobiology has emerged as an active field to quantify the mechanics of cell–cell and cell–matrix interactions integrating biophysical measurements and technique developments ranging from the molecular to cellular level. Among the biophysical techniques in this field, atomic force microscopy (AFM) is an exciting analytical tool to measure the binding mechanics of cell–cell molecules. The AFM was originally used to obtain surface topography. Moreover, it precisely measures the interaction force between the probe tip and the sample surface with pico-nano Newton resolution. Different AFM approaches have been successfully applied to study cellular systems. The use of large colloidal or flat probes in indentation experiments allows the determination of the mechanical properties of a cell, mediating the result on a large contact area [12]. At the same time, the use of a sharp tip as indenter allows detection of local changes of mechanical properties [13,14] as well as investigation of single-molecule unfolding events [15]. The stiffer cytoskeletal filaments network [16] can be characterized by AFM, as well as the softer nuclear compartment [17].

Functional probes in molecular recognition force microscopy mode allowed for detecting specific ligand–receptor interaction forces [18] on cells. Although the size and shape can change significantly, the probe is always made by a rigid material, generally silicon or silicon nitride, in some cases coated/functionalized with a single molecular species.

A more advanced methodology for single cell force spectroscopy has been further established to quantify cell–substrate adhesion [19]. The idea is to bind a living cell to an AFM tipless cantilever, using it as an extraordinarily powerful, but at the same time complex, probe. The presence of many specific binding sites and different non-specific interaction sources makes the analysis of force spectroscopy curves acquired in this modality a challenging task. For this reason, the technique has often been applied on simple and well-controlled samples; material substrates [20], substrates coated/functionalized with a single molecular species [21], or multifunctional substrates of very well-known molecules [22].

The cellular probe can be used to extend the analysis beyond the cell–substrate interaction, and studying with great detail cell–cell adhesion. Cell–cell force spectroscopy (CCFS) is a technique in which a cell attached to the tipless AFM cantilever is brought in contact with another living cell, while the interaction force is collected [23]. The quantitative evaluation of cell–cell interaction offers a powerful tool for biomedical research and it paves the way for future diagnostic translation. In fact, this experimental procedure can be further extended to probe cell–tissue interaction. Nowadays, the challenging idea to consider the adhesion between a particular cell with cells derived from a tissue possibly involved in a pathological state, such as cancer, as a diagnostic marker of the pathology is not far to be realized.

Besides its great potential, CCFS has been the focus of only a relatively low number of papers published in the field. The complexity of cell–cell curves requires a particular attention in the analysis. In this context, it is fundamental to control the experimental conditions, having a clear understanding of how particular acquisition parameters can influence the results. Clear and standard methodologies are still not defined. In this work we present a systematic approach to CCFS. Using Chinese Hamster Ovary (CHO) cells and testing cell-cell interaction varying different experimental settings, aiming to disclose how these settings are influencing the results.

2. Materials and Methods

2.1. Cell Culture

CHO (CCL-61T; ATCC, Teddington, UK) cell lines were cultured on Petri dishes (Techno Plastic Products, Neuchâtel, Switzerland), coated with poly-D-lysine (PDL; Sigma Aldrich, Milano, Italy), in Dulbecco's modified Eagle's medium (Gibco, Paisley, UK) containing 4.5% glutamine and glucose, 10% inactivated fetal bovine serum, 1.0% penicillin streptomycin, and 1.0% nonessential amino acids (Gibco) at 37 °C in 5.0% CO_2. The cells were split every 4–5 days before reaching a confluency rate of <80%.

2.2. Cantilever Functionalization and Cell Capture

Single beam silicon tipless cantilevers TL1-50 (NanoWorld, Neuchâtel, Switzerland), with a nominal spring constant of 0.03 N/m, were irradiated in an ultraviolet/ozone cleaner (ProCleaner; Bioforce Nanosciences, Ames, IA, USA) for 15 min before functionalization. The cantilevers were functionalized with concanavalin A (ConA, Sigma-Aldrich, Milano, Italy) as previously described [24,25].

For cell attachment, CHO cells (density of 3×10^3 cells mL^{-1}) were removed from the Petri dish via trypsinization. Briefly, the culture medium was removed from the Petri containing the confluent cells, and the cells were first washed with sterile phosphate-buffered saline (PBS) and, subsequently, incubated with 0.5% trypsin-EDTA 0.05% (Gibco, Thermofisher, Milano, Italy) for 2 min at 37 °C. The trypsinized cells were resuspended in 1 mL of PBS buffer and centrifuged for 5 min at $200 \times g$. After centrifugation, the cells were resuspended in PBS and gently agitated. A few detached cells were injected into a standard sterile Petri dish where CHO cells were cultured at a density of 3×10^3 cells mL^{-1}. Before cell seeding, a small part of the coverslip was coated with agarose, by spreading a 20 μL drop of 0.15% w/w agarose solution (Sigma-Aldrich, Milano, Italy) until gelification. The lack of adhesion between the cell and the repulsive agarose surface significantly increased the efficiency of the cell capture procedure. A single cell was captured by pressing for 30 s the functionalized cantilever onto a cell lying on the agarose spot with a controlled force of 2.0 nN, and then by lifting the cantilever. The system was left for 10 min to get a stable cell-cantilever contact before the acquisition of force–distance (F–D) curves.

2.3. Cell–Cell Force Spectroscopy (CCFS)

CCFS experiments were carried out using a Nanowizard III system (Bruker, JPK Instruments, Berlin, Germany), coupled with an AxioObserver D1 (Zeiss, Oberkochen, Germany) inverted optical microscope. A CellHesion module (Bruker, JPK Instruments) was used to extend to 100 μm the vertical displacement range of the AFM. All experiments were carried out at 37 °C in PBS containing 2.0 mM CaCl$_2$ and 2.0 mM MgCl$_2$ and setting the force-curve length at 80 μm to achieve complete detachment of the cell probe from the target cell. For each set of experiments, only the measurement condition in consideration was changed while keeping the rest of the parameters unaltered. All the acquired F–D curves were processed with the JPK Data Processing software to correct for the bending of the cantilever and to remove the baseline offset and exported in txt format for further analysis.

Approach and retract speeds: the approach and retract speeds were also analyzed. For all the cases, the contact was kept for 45 s using the constant-height as the delay mode (see below). The setpoint force, hence the maximum force exerted between the two cells, was set at 1.0 nN. For the evaluation of the approach speed influence in the cell–cell interaction, the cantilever was lifted at a constant velocity of 10.0 μm/s. Different approach speeds (1, 2.5, 10, 25, and 50 μm/s) were investigated. To understand how the retract speed influences the measurements, the approach speed was set at a constant velocity of 10.0 μm/s. After the contact, the cell probe was lifted at 1, 2.5, 10, 25, and 50 μm/s. Over 10 F–D curves per speed were acquired for every experiment, for a total of 53 and 354 curves for approach and retract speeds, respectively.

Delay time: different extended pauses (1, 15, 30, 45, 60, and 120 s) were studied. The cell probe was lowered at a constant speed of 10.0 μm/s until the cell probe contacted the target cell and the preset force (setpoint force) of 1.0 nN was reached. The constant height was used as the delay mode. The cell probe was then retracted, lifting the cantilever at a constant velocity of 10.0 μm/s. A total of 211 curves were acquired.

Setpoint force: three setpoint forces (1, 10, and 30 nN) were studied. The cell probe was lowered at a constant speed of 10.0 μm/s until the cell probe contacted the target cell and the preset force was reached. The contact was kept for 45 s using the constant height as the delay mode. Then the cell probe was retracted, lifting the cantilever at a constant velocity of 10.0 μm/s to register the F–D curve. At least 33 F–D curves per each setpoint were acquired for a total of 100 F–D curves.

Delay mode: the contact between the two cells can be maintained in two different modalities, constant-force and constant-height. In constant-force, after reaching the setpoint force, the piezo actuator compensates for any change of the interaction force associated with the cell shape's adaption and remodeling under an applied load. In constant height mode, the piezo actuator maintains a fixed position after reaching the setpoint force. In this second case, the cell position is fixed, but the interaction force can change due to the cell's ability to remodel its shape. The cell probe was lowered and retracted at a constant speed of 10.0 μm/s. The cell probe was kept in contact with the target cell for 45 s after the preset force of 1.0 nN was reached. A total of 52 F–D curves for constant force mode and 41 F–D curves for the constant height mode were analyzed.

2.4. Data Analysis

Force spectroscopy curves were exported from the JPK software to text format, and further analyzed using a custom software developed using Python 3 and the scientific libraries offered by NumPy/SciPy [26]. The source code of the software is currently available through github [27]. The software is designed to analyze the retract segment of each force–distance curve, batch processing a folder based on a selected set of parameters and exporting the results in a comma-separated values (CSV) text file for further statistics (see below). Each curve is processed to identify the baseline (based on the part of the curve far from the sample) and the origin of the Z is placed where the retract curve first crosses the baseline value; all distances are calculated with respect to this point. The curve is thus segmented, using peaks in the first derivative to identify discontinuity points (based on a Savitzky–Golay filter [28]. The final detachment point (Z_{det}, F_{det}) is thus identified as the last discontinuity point. The detachment work W (see Figure 1) is calculated as the area under the curve from Z = 0 to the detachment point $Z = Z_{det}$.

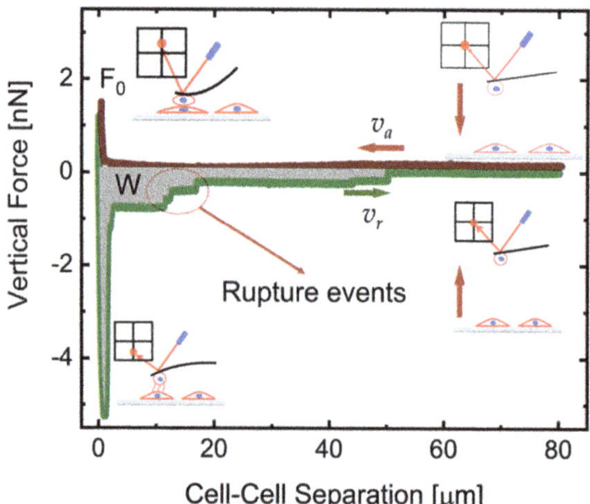

Figure 1. Typical force–distance (F–D) curve acquired in cell–cell force spectroscopy (CCFS). The curve starts when the CHO cell attached to a tipless cantilever is positioned at 80 μm on top of a selected target CHO cell. This is represented as the cartoon in the top right. The atomic force microscope (AFM) cantilever then begins to move towards the target cell with a constant approach speed v_a (red line) reaching the preset setpoint force value F_0 (see cartoon in the top left). At this point, cells are kept in contact having constant either the applied force or the position (delay mode). After the contact time (delay time, τ), the AFM cantilever is pushed-away from the target cell and

cells begin their detachment (cartoon in the lower left). The detachment process (green line) occurs with a constant retract speed v_r and it is characterized by several rupture events. The latter are generally associated with specific cell-cell interactions. The F–D curve ends when a final plateau at force ~0 nN is reached corresponding with the full detachment of the cells.

3. Results

Cell–cell interactions were investigated by AFM-based force spectroscopy (hereafter called cell–cell force spectroscopy, CCFS). Generally, these experiments consist of force versus displacement curves obtained when a cell adhered to the tipless cantilever is brought in contact with a second cell seeded in the culture dish, as represented in Figure 1. The measure starts when the tipless cantilever, functionalized with the CHO cell, is settled above the target CHO cell but far from the sample. Then it is moved towards the sample with a constant approach speed v_a (red line in Figure 1) until the preset force value F_s (setpoint) is reached. It is relevant to keep this force under a few nN, in order to restrict the interaction between the cells to the membrane and cortical region, without causing any damage to the cells. Cells are kept in contact for different preset times (contact time τ) while keeping constant either the applied force or the position (delay mode). Subsequently, the AFM cantilever is pushed away from the surface with a constant retract speed v_r, independent of the approach one. The retract segment (green line in Figure 1) shows many steps, associated with the rupture of specific bonds between the cells, and the experiment ends when the cells are fully detached (observed in the curve when a final plateau at force ~0 nN is reached). Each curve is then segmented, and a set of relevant parameters is extracted (see Section 2.4), including the detachment (or adhesion) work W that was further used to primarily quantify the cell-cell interactions. All these experimental parameters impact the corresponding detachment curve, and they can disclose different aspects of the cell–cell interaction.

The influence of the retraction speed was analyzed by varying v_r while keeping constant all other parameters: approach speed v_a = 10.0 µm/s, setpoint force F_0 = 1.0 nN, contact time τ = 45 s, delay mode: constant height. The results are reported in Figure 2a where the detachment work W is plotted as a function of the retract speed. The distribution of measured W clearly shows a trend towards higher average values, accompanied by larger distributions and the appearance of a tail for large values in the distribution (Figure 2a). While the retraction speed is expected to impact the adhesion work largely, another less obvious and often neglected parameter was also investigated: the approach speed. A second set of experiments where the approach speed v_a has been varied is reported in panel b) of Figure 2. All the other parameters have been kept constant: retract speed v_r = 10.0 µm/s, contact time τ = 45 s, delay mode: constant height. Interestingly, the trend in this case suggests a decrease of the detachment work with increasing approach speed (Figure 2b) and no broadening of the distribution is apparent for this experimental configuration.

Other parameters that can influence cell–cell adhesion are the delay time τ and the setpoint force F_0 (see Figure 2c,d). The impact of the delay time was studied in an experimental set acquired with the same approach and retract speed of 10.0 µm/s, and keeping the height constant while in contact, after having reached the sample with a setpoint force F_0 = 1.0 nN. The average detachment work in this condition increases with the delay time over the full range of 120 s (Figure 2c). Conversely, the standard deviation of the data remains almost constant for all points but the last one, which shows a broader distribution of values (Figure 2c). A similar analysis was performed for the setpoint force F_0, while keeping the speeds and the delay time constant, at 10.0 µm/s and 45 s respectively. In this case, only 3 values were acquired, as larger forces were clearly associated with permanent deformations/damage of the cells. The results reported in Figure 2d show that the adhesion work grows with the indentation force, and higher forces are associated with broader distributions.

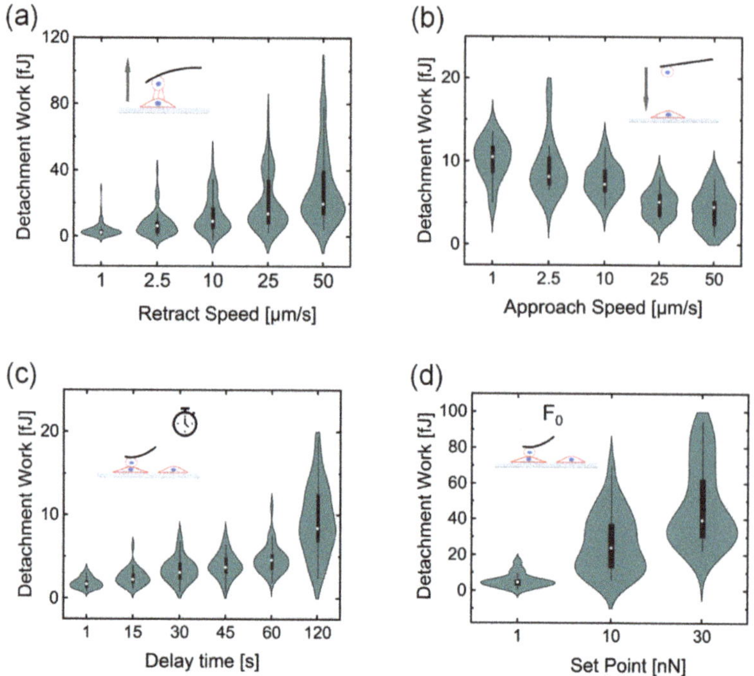

Figure 2. Influence of the experimental settings on cell-cell interactions. (**a**,**b**) show the distribution of the work spent to detach the probe cell from the target cell when the retraction speed (**a**) and the approach speed (**b**) were modified. (**c**,**d**) display the effect of the delay time τ (**c**) and setpoint force (**d**) in the distribution of the adhesion work.

Furthermore, the impact of two different delay modes, constant-force and constant height, was also investigated (Figure 3). The cell probe was lowered and retracted at a constant speed of 10.0 µm/s and kept in contact with the target cell for 45 s after the preset force of 1.0 nN was reached. The vertical deflection variation as well as the relative piezo displacement over the time for the two selected delay modes are reported in Figure 3a,b. The results show that, when the setpoint force is kept constant (Figure 3a, blue line) during all the time the cells are in contact, the relative piezo displacement (Figure 3a, green line) decreases due to the dynamic rearrangement of the cells able to remodel their structure when subjected to an external force. On the other hand, when the constant height mode was employed (Figure 3b), after an initial and rapid increase of the vertical deflection to reach the setpoint force of 1.0 nN, the force decreases settling down to lower values (Figure 3b, blue line) being the height to be kept constant as shown by the relative piezo displacement signal (Figure 3b, green line). The distribution of the detachment work while using these two modes is presented in Figure 3c. Results reveal that work needed to detach both cells in the case of constant-force mode is significantly higher than in the case of constant height mode and higher forces are associated with broader distributions.

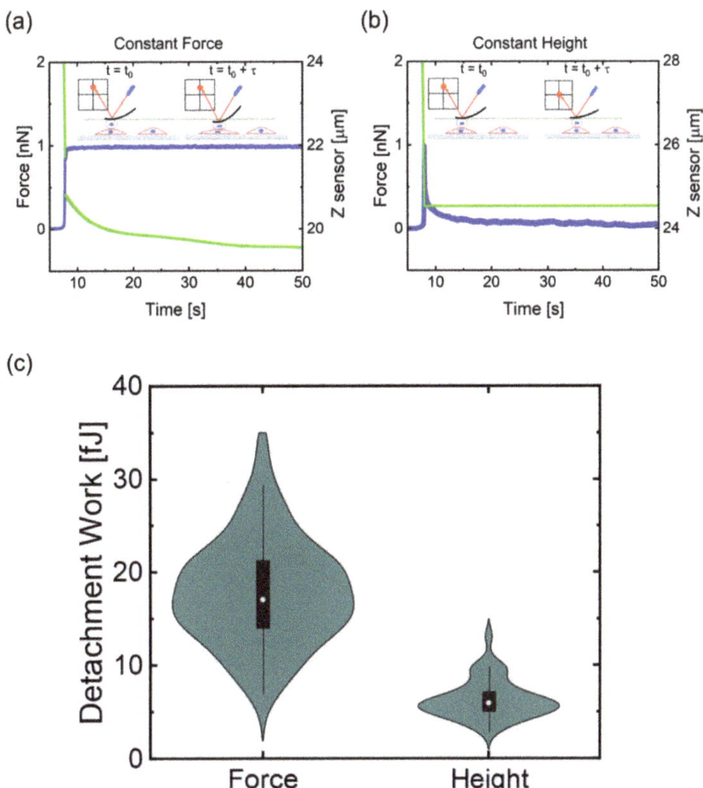

Figure 3. Influence of the two different delay modes on cell–cell interactions. When cells are kept under a constant applied force during the contact time (blue line in (**a**)), they reshape their structure resulting in a compensatory decrease in the relative piezo displacement signal (green line). In contrast, when it is kept constant the separation between the AFM cantilever and the sample (green line in (**b**)), the vertical deflection signal in the photodiode initially increases to reach the preset force setpoint (blue line) to, subsequently, decrease during the contact time. (**c**) shows the distribution of the work spent to detach the probe cell from the target cell when both delay modes are investigated.

4. Discussion

The pattern of interaction measured with CCFS is extremely rich, and the fine details are expected to depend on the specific cellular system and the characteristics of the molecular players involved in the adhesion [29]. Nevertheless, the general behavior observed in Figure 2 can be described at the first order in terms of simplified components. To the simplest approximation, the CCFS experiment can be modeled as two viscoelastic (semi-)spheres that are brought in contact with constant velocity v_a till the interaction force reaches a maximum value F_0, where the motion is halted for a time τ During the contact, we can assume that a set of N surface bonds are formed, and they contribute, together with the intrinsic properties of the cell, to the definition of the detachment work W measured while the cells are being pulled apart with constant speed v_r.

We assume that the extraction work W is proportional to the number N of bonds being formed during the interaction, and that this number in turn simply depends on the area of interaction between the cells and the time they spend in contact:

$$W \propto N \propto \int_{T\_i} S(t)dt$$

where the integral is made over the interaction time T_i and $S(t)$ indicates the instantaneous interaction surface. Under this simple approximation we can revisit the results presented in the previous section, where only one experimental parameter at a time was modified, keeping all the other conditions constant.

For the experiments of Figure 2c, only the delay time was changed, while approach and retract speed were kept constant. Given that the setpoint force is the same for all these curves, we do not expect a major change of the contact area, and the number N (and so the extraction work W) is expected to be directly proportional to the time spent in contact, plus a small offset associated to the indentation phase (from $F = 0$ to $F = F_0$). The experiments in Figure 2c show this trend, as demonstrated in Figure 4a where the average values of W for this experiment, with the corresponding linear fit, is shown.

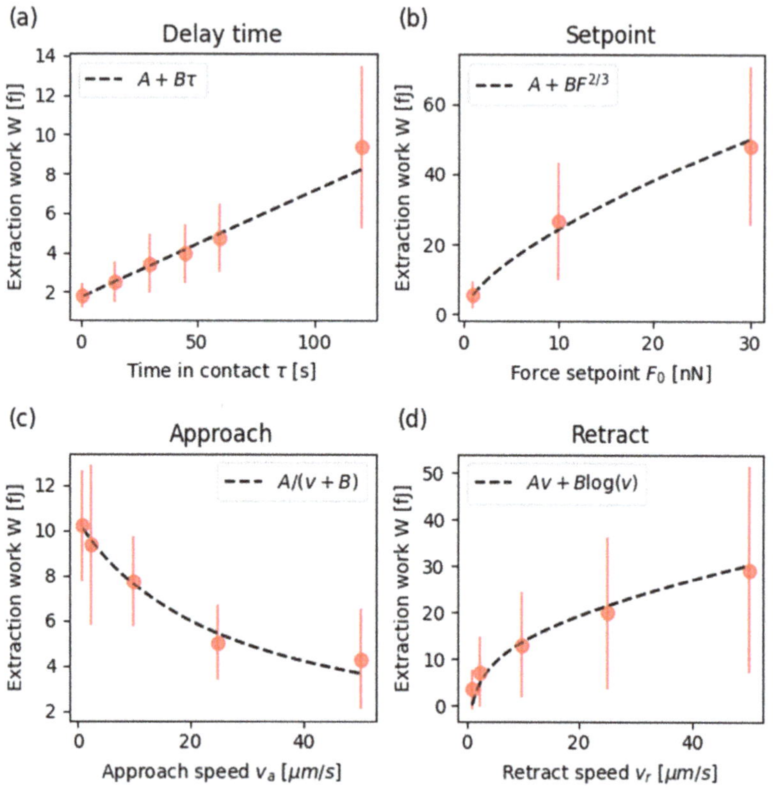

Figure 4. Average extraction work W as a function of (**a**) delay time, (**b**) force setpoint, (**c**) approach speed and (**d**) retract speed. The experimental data were fitted with the equation reported in the corresponding legend, obtaining the following parameters: (**a**) A = 0.05 fJ and a slope B = 1.72 fJ/s; (**b**) offset A = 0.42 fJ and a coefficient B = 5.13 fJ $N^{3/2}$; (**c**) A = 282 zJ/s and B = 27.0 μm/s; (**d**) A = 0.21 nPa s and B = 5.0 fJ.

The same simplified consideration can be applied to guess the trend of W as a function of the setpoint force. The maximum force directly affects the area of interaction between the cells. To have an approximation, we can use the Hertz contact mechanics theory [30] that gives an estimate for the area between the surfaces:

$$A = \pi a^2 = \pi R \, \delta_0$$

where a is the contact radius, R the effective radius of the cells, δ_0 the indentation corresponding to the maximum force. The same theory connects this indentation to the force F_0:

$$F_0 \propto \delta^{3/2} \rightarrow \delta_0 \propto F_0^{2/3}$$

and this suggests that the area should be proportional to:

$$A \propto F_0^{2/3}$$

and so the adhesion work W. In Figure 4b we present the best fit of $W(F_0)$ with this power law.

Interestingly, the extraction work appears to be influenced by the approach speed as well (see Figure 2b). In particular, W slowly decays with the speed v_a, indicating that the number of established bonds is lower for higher speed. In fact, the time the two cells remain in contact is inversely proportional to the approach speed, as the final (hold) position is the same for all configurations. This behaviour is directly fitted to the data in Figure 4c, where an additional coefficient accounts for constant contributions in addition to the variable one.

To understand the different values of extraction work measured as a function of the retract speed, the number of bonds cannot be the dominating factor, as this is expected to be the same for all conditions. Instead, the origin of the different values has to be found in the dynamical response of the cell–cell complex. While being pulled, this structure will present two main contributions. The first is a proper viscous drag, associated to the nature of the cell, which is expected to be linearly proportional to the pulling speed v_r. While this effect dominates at higher speed, the second mechanism contributes to deviate the curve from the simple linear drag for low speeds. In fact, the two cells are connected by soft bonds, whose resistance is expected to follow a kinetic activation mechanism, which results in a logarithmic dependence on the speed [31]. Putting these two terms together, it is possible to fit the experimental data well with only two parameters (see Figure 4d).

Moreover, the difference between delay modes has been compared and the results are reported in Figure 3. This experiment introduces a different aspect to the analysis, related to the active nature of the adhesion process. Cells are exquisitely mechanosensitive; they are able to feel the external force [32,33] and react by rearranging their shape and reducing the stress. This mechanism cannot be simply described in terms of physical components, but it should include the biological mechanism of mechanosensing, and its main molecular players [34]. Nevertheless, it is crucial to control the contact mode to obtain reliable results in CCFS, as demonstrated in Figure 3.

5. Conclusions

This work shows how the choice of the experimental conditions can influence the results derived from CCFS experiments. The specific shape of the CCFS curve depends on a plethora of events associated with the nanoscale interaction between the cells, convolving physical and biological contributions such as the roughness of the membrane, the activation of specific receptors on the surface, or the presence or invaginations and water pockets within the adhesion surface. Although it is not trivial to deconvolve the contribution of any single component contributing to the inherent complexity of the physical contact between the two living cells, we demonstrated that a first-order understanding of the trends observed at different conditions can be deduced by simple theoretical considerations, where the main parameter is the number of bonds being formed between the cells during the contact phase.

Furthermore, this work also suggests that the choice of experimental conditions that are generally not considered, such as the approach speed and the delay mode, can bring significantly, and in some cases unexpected, variations in the results. The hints suggested by our work represent a step forward towards the definition of the best experimental practices.

Author Contributions: Conceptualization, R.O.-N., C.C. and M.V.; methodology, R.O.-N. and C.C.; software, M.V.; validation; formal analysis, R.O.-N. and A.M.; investigation, R.O.-N. and A.M.; writing—original draft preparation, M.V., R.O.-N., A.M. and C.C.; supervision, C.C. and M.V. All authors have read and agreed to the published version of the manuscript.

Funding: R.O.-N. acknowledges the economic support of the grant Olle Engkvist (project number 194–0644).

Institutional Review Board Statement: Not applicable.

Informed Consent Statement: Not applicable.

Data Availability Statement: The data presented in this study are openly available in the Enlighten Research Data repository of the University of Glasgow at doi:10.5525/gla.researchdata.1103.

Acknowledgments: M.V. would like to acknowledge Ettore Landini for supporting the development of the Python code used to open the JPK text files.

Conflicts of Interest: The authors declare no conflict of interest.

References

1. Engler, A.J.; Sen, S.; Sweeney, H.L.; Discher, D.E. Matrix Elasticity Directs Stem Cell Lineage Specification. *Cell* **2006**, *126*, 677–689. [CrossRef]
2. Keller, R.; Davidson, L.A.; Shook, D.R. How We Are Shaped: The Biomechanics of Gastrulation. *Differentiation* **2003**, *71*, 171–205. [CrossRef] [PubMed]
3. Nelson, C.M.; Jean, R.P.; Tan, J.L.; Liu, W.F.; Sniadecki, N.J.; Spector, A.A.; Chen, C.S. Emergent Patterns of Growth Controlled by Multicellular Form and Mechanics. *Proc. Natl. Acad. Sci. USA* **2005**, *102*, 11594–11599. [CrossRef]
4. Paszek, M.J.; Zahir, N.; Johnson, K.R.; Lakins, J.N.; Rozenberg, G.I.; Gefen, A.; Reinhart-King, C.A.; Margulies, S.S.; Dembo, M.; Boettiger, D.; et al. Tensional Homeostasis and the Malignant Phenotype. *Cancer Cell* **2005**, *8*, 241–254. [CrossRef]
5. Levental, K.R.; Yu, H.; Kass, L.; Lakins, J.N.; Egeblad, M.; Erler, J.T.; Fong, S.F.T.; Csiszar, K.; Giaccia, A.; Weninger, W.; et al. Matrix Crosslinking Forces Tumor Progression by Enhancing Integrin Signaling. *Cell* **2009**, *139*, 891–906. [CrossRef]
6. Kashef, J.; Franz, C.M. Quantitative Methods for Analyzing Cell–Cell Adhesion in Development. *Dev. Biol.* **2015**, *401*, 165–174. [CrossRef]
7. Belardi, B.; Son, S.; Felce, J.H.; Dustin, M.L.; Fletcher, D.A. Cell–Cell Interfaces as Specialized Compartments Directing Cell Function. *Nat. Rev. Mol. Cell Biol.* **2020**, *21*, 750–764. [CrossRef] [PubMed]
8. Ridone, P.; Pandzic, E.; Vassalli, M.; Cox, C.D.; Macmillan, A.; Gottlieb, P.A.; Martinac, B. Disruption of Membrane Cholesterol Organization Impairs the Activity of PIEZO1 Channel Clusters. *J. Gen. Physiol.* **2020**, *152*, e201912515. [CrossRef] [PubMed]
9. Romero, L.O.; Massey, A.E.; Mata-Daboin, A.D.; Sierra-Valdez, F.J.; Chauhan, S.C.; Cordero-Morales, J.F.; Vásquez, V. Dietary Fatty Acids Fine-Tune Piezo1 Mechanical Response. *Nat. Commun.* **2019**, *10*, 1200. [CrossRef]
10. Suresh, S. Biomechanics and Biophysics of Cancer Cells. *Acta Biomater.* **2007**, *3*, 413–438. [CrossRef] [PubMed]
11. Ungureanu, A.-A.; Benilova, I.; Krylychkina, O.; Braeken, D.; De Strooper, B.; Van Haesendonck, C.; Dotti, C.G.; Bartic, C. Amyloid Beta Oligomers Induce Neuronal Elasticity Changes in Age-Dependent Manner: A Force Spectroscopy Study on Living Hippocampal Neurons. *Sci. Rep.* **2016**, *6*, 25841. [CrossRef] [PubMed]
12. Mescola, A.; Vella, S.; Scotto, M.; Gavazzo, P.; Canale, C.; Diaspro, A.; Pagano, A.; Vassalli, M. Probing Cytoskeleton Organisation of Neuroblastoma Cells with Single-Cell Force Spectroscopy: Probing Cytoskeleton of Neuroblastoma Cells with scfs. *J. Mol. Recognit.* **2012**, *25*, 270–277. [CrossRef] [PubMed]
13. Radmacher, M. Measuring the Elastic Properties of Living Cells by the Atomic Force Microscope. In *Methods in Cell Biology*; Elsevier Science: San Diego, CA, USA, 2002; Volume 68, pp. 67–90. ISBN 978-0-12-544171-1.
14. Mescola, A.; Dauvin, M.; Amoroso, A.; Duwez, A.-S.; Joris, B. Single-Molecule Force Spectroscopy to Decipher the Early Signalling Step in Membrane-Bound Penicillin Receptors Embedded into a Lipid Bilayer. *Nanoscale* **2019**, *11*, 12275–12284. [CrossRef]
15. Mazzoni, E.; Mazziotta, C.; Iaquinta, M.R.; Lanzillotti, C.; Fortini, F.; D'Agostino, A.; Trevisiol, L.; Nocini, R.; Barbanti-Brodano, G.; Mescola, A.; et al. Enhanced Osteogenic Differentiation of Human Bone Marrow-Derived Mesenchymal Stem Cells by a Hybrid Hydroxylapatite/Collagen Scaffold. *Front. Cell Dev. Biol.* **2021**, *8*, 1658. [CrossRef]
16. Doss, B.L.; Pan, M.; Gupta, M.; Grenci, G.; Mège, R.-M.; Lim, C.T.; Sheetz, M.P.; Voituriez, R.; Ladoux, B. Cell Response to Substrate Rigidity Is Regulated by Active and Passive Cytoskeletal Stress. *Proc. Natl. Acad. Sci. USA* **2020**, *117*, 12817–12825. [CrossRef]
17. Ferrera, D.; Canale, C.; Marotta, R.; Mazzaro, N.; Gritti, M.; Mazzanti, M.; Capellari, S.; Cortelli, P.; Gasparini, L. Lamin B1 Overexpression Increases Nuclear Rigidity in Autosomal Dominant Leukodystrophy Fibroblasts. *FASEB J.* **2014**, *28*, 3906–3918. [CrossRef]
18. Bellani, S.; Mescola, A.; Ronzitti, G.; Tsushima, H.; Tilve, S.; Canale, C.; Valtorta, F.; Chieregatti, E. GRP78 Clustering at the Cell Surface of Neurons Transduces the Action of Exogenous Alpha-Synuclein. *Cell Death Differ.* **2014**, *21*, 1971–1983. [CrossRef] [PubMed]

19. Helenius, J.; Heisenberg, C.-P.; Gaub, H.E.; Muller, D.J. Single-Cell Force Spectroscopy. *J. Cell Sci.* **2008**, *121*, 1785–1791. [CrossRef]
20. Taubenberger, A.V.; Hutmacher, D.W.; Muller, D.J. Single-Cell Force Spectroscopy, an Emerging Tool to Quantify Cell Adhesion to Biomaterials. *Tissue Eng. Part B Rev.* **2014**, *20*, 40–55. [CrossRef]
21. Schubert, R.; Strohmeyer, N.; Bharadwaj, M.; Ramanathan, S.P.; Krieg, M.; Friedrichs, J.; Franz, C.M.; Muller, D.J. Assay for Characterizing the Recovery of Vertebrate Cells for Adhesion Measurements by Single-Cell Force Spectroscopy. *FEBS Lett.* **2014**, *588*, 3639–3648. [CrossRef]
22. Canale, C.; Petrelli, A.; Salerno, M.; Diaspro, A.; Dante, S. A New Quantitative Experimental Approach to Investigate Single Cell Adhesion on Multifunctional Substrates. *Biosens. Bioelectron.* **2013**, *48*, 172–179. [CrossRef] [PubMed]
23. Puech, P.-H.; Poole, K.; Knebel, D.; Muller, D.J. A New Technical Approach to Quantify Cell–Cell Adhesion Forces by AFM. *Ultramicroscopy* **2006**, *106*, 637–644. [CrossRef]
24. Keshavan, S.; Oropesa-Nuñez, R.; Diaspro, A.; Canale, C.; Dante, S. Adhesion and migration of CHO cells on micropatterned single layer graphene. *2D Mater.* **2017**, *4*, 025022. [CrossRef]
25. Oropesa-Nuñez, R.; Keshavan, S.; Dante, S.; Diaspro, A.; Mannini, B.; Capitini, C.; Cecchi, C.; Stefani, M.; Chiti, F.; Canale, C. Toxic HypF-N Oligomers Selectively Bind the Plasma Membrane to Impair Cell Adhesion Capability. *Biophys. J.* **2018**, *114*, 1357–1367. [CrossRef]
26. Virtanen, P.; Gommers, R.; Oliphant, T.E.; Haberland, M.; Reddy, T.; Cournapeau, D.; Burovski, E.; Peterson, P.; Weckesser, W.; Bright, J.; et al. SciPy 1.0: Fundamental Algorithms for Scientific Computing in Python. *Nat. Methods* **2020**, *17*, 261–272. [CrossRef] [PubMed]
27. Savitzky, A.; Golay, M.J.E. Smoothing and Differentiation of Data by Simplified Least Squares Procedures. *Anal. Chem.* **1964**, *36*, 1627–1639. [CrossRef]
28. Vassalli, M. massimovassalli/CellForceSpectroscopy: Alpha Version (Version 0.1). Zenodo. 2021. Available online: http://doi.org/10.5281/zenodo.4469887 (accessed on 3 February 2021).
29. Rakshit, S.; Zhang, Y.; Manibog, K.; Shafraz, O.; Sivasankar, S. Ideal, Catch, and Slip Bonds in Cadherin Adhesion. *Proc. Natl. Acad. Sci. USA* **2012**, *109*, 18815–18820. [CrossRef] [PubMed]
30. Kontomaris, S.V.; Malamou, A. Hertz Model or Oliver & Pharr Analysis? Tutorial Regarding AFM Nanoindentation Experiments on Biological Samples. *Mater. Res. Express* **2020**, *7*, 033001. [CrossRef]
31. Evans, E.; Ritchie, K. Strength of a Weak Bond Connecting Flexible Polymer Chains. *Biophys. J.* **1999**, *76*, 2439–2447. [CrossRef]
32. Ridone, P.; Vassalli, M.; Martinac, B. Piezo1 Mechanosensitive Channels: What Are They and Why Are They Important. *Biophys. Rev.* **2019**, *11*, 795–805. [CrossRef]
33. Gaub, B.M.; Müller, D.J. Mechanical Stimulation of Piezo1 Receptors Depends on Extracellular Matrix Proteins and Directionality of Force. *Nano Lett.* **2017**, *17*, 2064–2072. [CrossRef] [PubMed]
34. Cheng, B.; Lin, M.; Huang, G.; Li, Y.; Ji, B.; Genin, G.M.; Deshpande, V.S.; Lu, T.J.; Xu, F. Cellular Mechanosensing of the Biophysical Microenvironment: A Review of Mathematical Models of Biophysical Regulation of Cell Responses. *Phys. Life Rev.* **2017**, *22–23*, 88–119. [CrossRef] [PubMed]

Article

Towards a Fully Automated Scanning Probe Microscope for Biomedical Applications

Witold K. Szeremeta [1], Robert L. Harniman [2], Charlotte R. Bermingham [1] and Massimo Antognozzi [1,*]

[1] School of Physics, University of Bristol, Tyndall Avenue, Bristol BS8 1TL, UK; w.szeremeta@bristol.ac.uk (W.K.S.); cbermingham@hotmail.co.uk (C.R.B.)

[2] School of Chemistry, University of Bristol, Cantock's Close, Bristol BS8 1TS, UK; rob.harniman@bristol.ac.uk

* Correspondence: massimo.antognozzi@bristol.ac.uk; Tel.: +44-117-928-8749

Abstract: The increase in capabilities of Scanning Probe Microscopy (SPM) has resulted in a parallel increase in complexity that limits the use of this technique outside of specialised research laboratories. SPM automation could substantially expand its application domain, improve reproducibility and increase throughput. Here, we present a bottom-up design in which the combination of positioning stages, orientation, and detection of the probe produces an SPM design compatible with full automation. The resulting probe microscope achieves sub-femtonewton force sensitivity whilst preserving low mechanical drift (2.0 $\pm$ 0.2 nm/min in-plane and 1.0 $\pm$ 0.1 nm/min vertically). The additional integration of total internal reflection microscopy, and the straightforward operations in liquid, make this instrument configuration particularly attractive to future biomedical applications.

Keywords: SPM; automation; femtonewton resolution; vertical probes; translation stages; inertial drive; piezo actuators; vertical positioning

Citation: Szeremeta, W.K.; Harniman, R.L.; Bermingham, C.R.; Antognozzi, M. Towards a Fully Automated Scanning Probe Microscope for Biomedical Applications. *Sensors* **2021**, *21*, 3027. https://doi.org/10.3390/s21093027

Academic Editor: Bruno Tiribilli

Received: 23 February 2021
Accepted: 22 April 2021
Published: 26 April 2021

Publisher's Note: MDPI stays neutral with regard to jurisdictional claims in published maps and institutional affiliations.

1. Introduction

Over the past few decades, the discovery of new nanoscale phenomena has produced scientific and technological breakthroughs across various disciplines from natural sciences to engineering [1] and medicine [2–4]. The increased interdisciplinarity of these discoveries, and their comparable length scale, reveals a convergence at the nanoscale between these different disciplines [5,6] with the promise for even stronger future integration. In medical science, the transition from microscale to nanoscale observations has enabled faster, more accurate and cheaper diagnostic tools [7]. Whether it is the relation between cell wall mechanics and cancer cells [8–14] or the detection of cellular nanoscale fluctuations [15–17], it seems clear that the exploitation of nanoscale effects is crucial to the future of biomedical diagnostics.

Scanning Probe Microscopy (SPM), one of the most diverse techniques in this area, has become an essential research tool in cellular and molecular biology [18]. When operating in force mode, Atomic Force Microscopy (AFM) [19] extends the intuitive sense of touch directly from the user's frame of reference down to the nanoscale. As a result, AFM data add a unique mechanical assessment of the specimen that is lacking in other characterisation techniques, such as Dynamic Light Scattering [20] (DLS), Raman [21], or Nuclear Magnetic Resonance [22,23] (NMR) spectroscopy. It is precisely the combination of extreme spatial resolution and the ability to manipulate the sample at the nanoscale, that gives AFM the potential to become an ideal tool for biomedical applications [24–26]. Despite these capabilities, it has been challenging to transition SPM technology from research laboratories to clinical or medical applications [27]. The main difficulties can be summarised in three points: (1) the technology requires a highly skilled specialised operator to be present at all times, (2) the most advanced SPM applications are difficult to reproduce [14], and (3) the statistical basis underpinning SPM measurements can be considered weak from a biomedical perspective due to the low throughput [28].

Here, we suggest that a fully automated SPM with femtonewton sensitivity and the ability to perform measurements in physiological conditions could solve most of the above problems and enable widespread adoption of this technology in the biomedical area. It is essential to notice that the requirements for automation and a bio-compatible environment have to preserve the unique SPM sensitivity and versatility to ensure the full impact of this technology.

This work presents a bottom-up design approach to the challenges described above and its initial implementation on an actual Lateral Molecular Force Microscope (LMFM) [29]. The LMFM uses vertically oriented probes (VOP) with a much higher force sensitivity than conventional horizontal AFM cantilevers due to their orientation [30]. The sensitivity of the LMFM has already been successfully demonstrated [31] and is essential when investigating weak biomolecular processes [32,33] or imaging soft nanostructures [34]. Presented below are a series of design solutions, which significantly increase the stability, precision and usability of the modified LMFM. Finally, these changes align with the constraints imposed by future automation levels and retain the versatility required for biomedical applications.

2. Materials and Methods

2.1. Sensitivity and Versatility of Vertically Oriented Probes

LMFM [29] is part of the SPM family with a vertically oriented probe and can use commercially available ultra-compliant silicon nitride cantilevers (NuNano, Ltd, UK) [35]. Standard AFM probes are mounted horizontally and are at risk of jumping into contact with the surface due to attractive interaction forces between the tip and the substrate [36]. This effect limits the minimum cantilever stiffness necessary to counteract the force gradient above the surface [37]. A vertically oriented cantilever benefits from the fact that the minimum stiffness needed to prevent a jump-to-contact has an angular dependence given by $cos^2(90° - \theta)$ [38], where θ is the angle of the cantilever from the vertical, as shown in Figure 1. A vertical orientation enables more compliant cantilevers resulting in greater in plane force sensitivity and a much greater tip-sample separation control. The silicon nitride probes routinely used in this work have a minimum stiffness of the order of 10^{-6} N/m–10^{-5} N/m, compared with 10^{-3} N/m–10 N/m for a standard AFM cantilever. LMFM cantilever stiffness is comparable to optical tweezers stiffness with the advantage of a vertical positional control with sub-nanometre resolution. Additionally, LMFM maintains the ability to image the surface of very soft samples in a liquid environment [34].

2.2. Scattered Evanescent Wave Detection System

The small size of LMFM cantilevers [35], and their vertical orientation, makes it difficult to use conventional AFM optical detection systems [18,19], so LMFM detects the probe's tip position using the Scattered Evanescent Wave (SEW) detection system [29]. The SEW detection is based on the fact that objects entering the evanescent field scatter the light, transforming it from near-field to far-field. The scatterer's three-dimensional position is subsequently measured using a four-sector photodetector. The SEW system has been successfully tested in ambient and liquid environments [35]. The evanescent wave is created upon total internal reflection of a laser beam at the glass-air (or glass-water) interface. The LMFM setup described in Figure 1 makes use of two fibre-coupled lasers; laser #1 is a OBIS FP 660LX (Coherent, Inc., Santa Clara, CA, USA) and laser #2 is a Stradus VersaLase (Vortran Laser Technology, Roseville, CA, USA) with multiple wavelengths (488 nm, 561 nm, 642 nm). The two perpendicular laser beams produce two concentric evanescent fields with perpendicular wave vectors. The SEW detection system uses one wavelength (642 nm-laser #2), while the other laser lines can be used to study optical forces in evanescent fields [31] or combine SPM with Total Internal Reflection Fluorescence (TIRF) microscopy. Two adjustable mirrors are located underneath the TIRF objective lens (Nikon, 100x TIRF objective $NA = 1.49$) and direct the laser beams into the lens. Changing their position affects how far from the axis of the objective lens the beam propagates, which, in turn, affects the angle (α) of incidence of the beam onto the glass surface. Two further mirrors are positioned underneath the objective to stop the exit beams from reaching the detectors. When the tip of the probe scatters the evanescent field, the TIRF objective lens collects the light and produces a high-magnification image in the detection plane. Before reaching the four-sectors photodiode (Hamamatsu S5990-01), the scattered light goes through a dichroic mirror which separates the detection wavelength from the other wavelengths. A second objective lens (Olympus, Plan Fluorite Oil Immersion Objective $NA = 1.30$) is positioned directly in front of the photodetector for further angular magnification. An sCMOS camera (Hamamatsu Orca Flash 4.0) is used for conventional evanescence light scattering microscopy or TIRF microscopy. An automatic routine uses the vertical probe image from the camera with the signal from the photodiode to position the cantilever on the optical axis of the TIRF lens. This operation uses the motorised stages described in the following section.

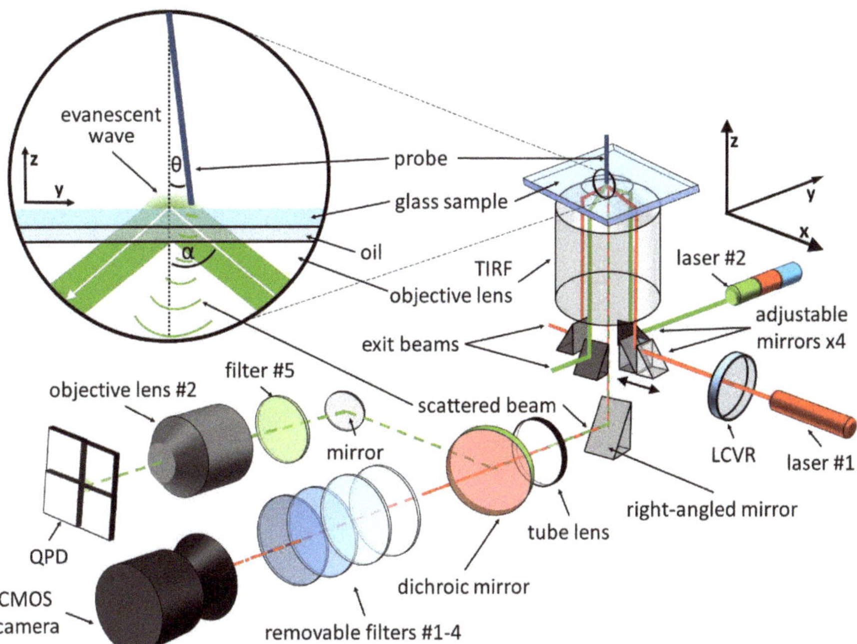

Figure 1. Diagram of the LMFM set up, including the SEW detection system. A closeup diagram shows the vertically mounted cantilever forming an angle (θ) with the normal. A totally internally reflected laser beam, forming an angle (α) with the normal, creates an evanescent wave on the sample side.

2.3. Design of Position Control Stages Compatible with Automation

Remote control of sample and probe position is the initial step towards autonomous operations, and it demands reliable, precise, and stable positioning mechanisms. Furthermore, designing a fully autonomous SPM system requires consideration of how the positioning system interfaces with an automatic sample and tip exchange.

After the invention of the first dynamic piezoelectric translation stage by D.W. Pohl in 1986 [39], the stick-slip design has become one of the most commonly used techniques in SPM for remote operations [40–42].

The upgrade of the LMFM described here incorporates seven motorised degrees of freedom (DoF). The probe Vertical Positioning System (VPS) uses a new shuttle-and-tube design with stick-slip actuation (Figure 2). The Horizontal Positioning System (HPS) for the sample and the probe uses a stick-slip translation stage (Figure 3) inspired by

Drevniok et al. [43]. The sample and the probe positioning systems are both connected to the microscope focusing plate. This plate moves in the vertical direction to adjust the microscope's focus. Three DC-motors, with magnetic encoders, are embedded in the microscope's base and provide the focusing movement.

To produce stick-slip motion in the vertical direction, the acceleration $a(t)$ of the actuator (i.e., piezoelectric crystal) needs to satisfy the following relation:

$$|a(t)| > \left| \frac{F_s}{m} + g \right|,$$ (1)

where g is the acceleration due to gravity, F_s is the static friction, and m is the mass of the moving part. For horizontal movements, we can disregard the acceleration due to gravity.

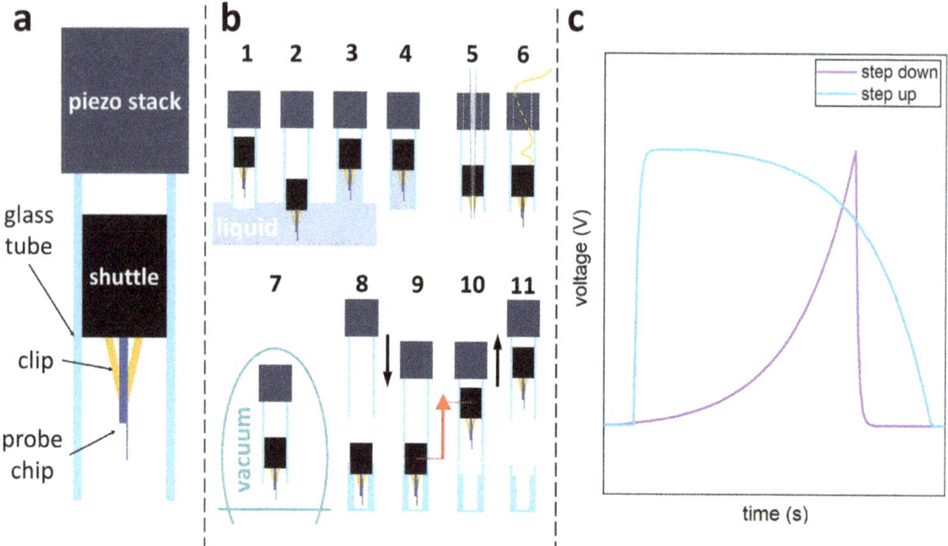

Figure 2. (**a**) Schematic drawing of the Vertical Positioning System (side view) (**b**) schematic drawings of different applications of the VPS. (1–4) The VPS has been tested in liquid environment, and the glass tube can act as a functionalisation chamber. (5) The VPS can be used for other SPM techniques, e.g., NSOM, which uses tapered waveguide probes or (6) STM by establishing an electrical connection to the conductive probe. (7) The VPS has been tested and is compatible with working in a vacuum of 10^{-6} mbar. (8–11) The VPS can expel and regain the plug under simple stick-slip operations. This simple probe exchange can be automated. (**c**) Asymmetric sawtooth waveforms with exponential deceleration used to drive the VPS.

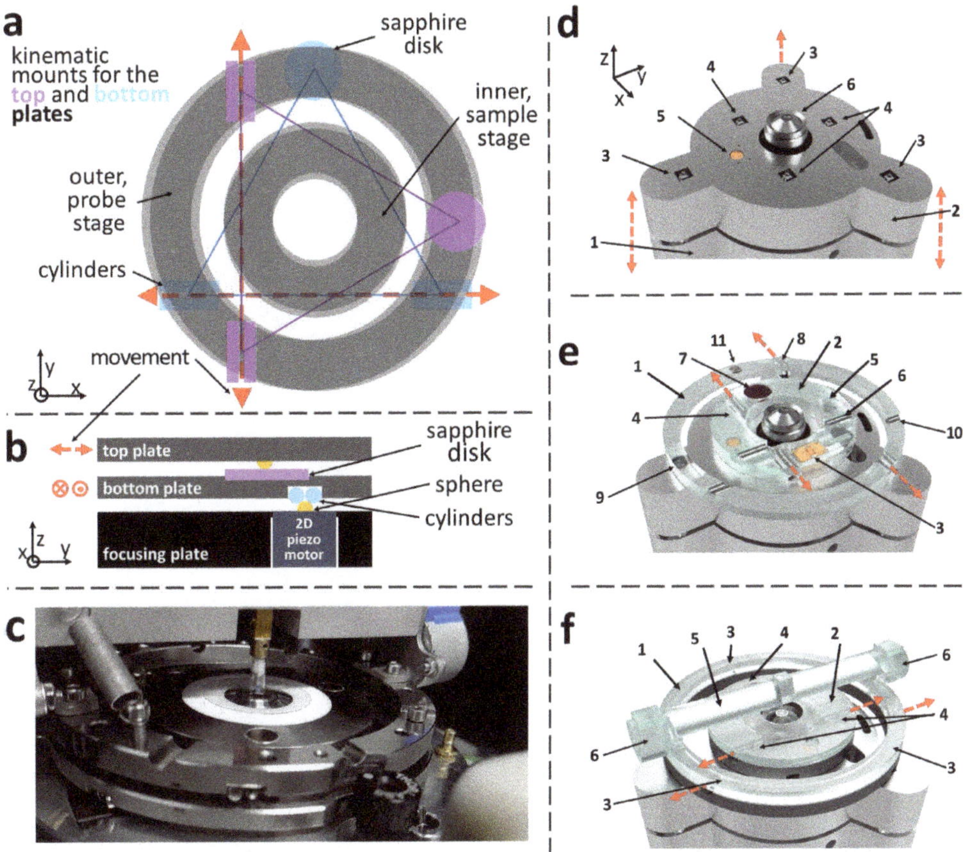

Figure 3. (**a**,**b**) Top view of the probe and sample HPS and kinematic mount details. Only the outer ring is described. Two sets of three ball bearings are positioned at the vertices of two orthogonal equilateral triangles. The first set is fixed on the piezo motors, while the second set is mounted on the top plate's lower part. The bottom plate can only move in the x-direction under stick-slip action, while the top plate's movement is confined to the y-axis. (**c**) Photograph of the current implementation of the HPS and the VPS. (**d**–**f**) CAD drawings, including the microscope's focusing plate and the sample and probe HPSs. (**d**) (1) Microscope's base where the objective lens (6) is mounted. (2) Focusing plate which can move in z-direction using three DC-motors (not shown). The two sets of three XY piezo motors, (3) and (4), are mounted on the focusing plate (2). (5) Otical encoder that measures the movement in the x-direction of the sample stage. (**e**) HPSs' bottom plates for the probe (1) and sample (2). The two pairs of lower cylinders (8) and sapphire disk (9) ensure ring (1) can only move in the x-direction. Similarly, ring (2) can only move in the x-direction thanks to the pairs of lower cylinders (4) and sapphire disk (5). The same type of kinematic mount is present on the two rings' top surface to ensure both top plates can only move in the y-direction (cylinders (6) and (10) and sapphire disks (9) and (7), respectively). The optical encoder (3) measures the movement of the top sample plate in the y-direction. (**f**) Top probe's (1) and sample's plate (2). The mechanical coupling between the top and bottom plates is ensured by two sets of three ball bearings (3) and (4) mounted on the lower surface of the probe and sample plate, respectively. The Vertical Positioning System is partially visible at the centre of the bridge (5). The tilt of the VPS is controlled by rotating the two stepper motors (6).

The VPS design consists of 3 main elements: (1) a moving shuttle (a stainless steel cylindrical plug, 4 mm in diameter and 7 mm long), (2) a smooth glass tube (an NMR glass tube with 4 mm inner diameter and a length of 14 mm), and (3) a piezoelectric stack actuator (PK4FA2H3P2 from Thorlabs, Inc., Newton, NJ, USA). Figure 2a shows a diagram of the VPS. A micro-fabricated cantilever is mounted on the stainless steel plug, which fits into the glass tube. A small amount of Teflon tape is wrapped around the plug to adjust the static friction between the tube and the plug. The glass tube is directly connected (i.e. glued) to the piezoelectric stack actuator and acts as a guide rail for the cylindrical shuttle. The static friction between the shuttle and the glass tube is sufficient to counteract the gravitational pull and keeps the shuttle stationary. At the same time, the friction coefficient is sufficiently low to allow for the slip to happen when the actuator's acceleration satisfies Equation (1). In other words, a sufficiently rapid movement of the glass cylinder in the upwards (downwards) direction produces a slippage of the plug downwards (upwards), moving the plug towards the bottom (top) end of the tube. When an asymmetric sawtooth waveform is applied to the actuator, the plug starts moving [44]. A modified sawtooth waveform, with exponential deceleration, (Figure 2c) produced the most consistent results in terms of step size and plug speed. The continuous (non-stick-slip) vertical extension of the piezoelectric actuator is used for fine vertical positioning of the probe (with 0.1 nm resolution).

Figure 2b describes some of the modes in which the VPS can work. The VPS can operate in a liquid medium and the tube around the probe acts as a container for the liquid (steps 1–4). Different cylindrical probes can be mounted to perform Near-field Scanning Optical Microscopy (NSOM) or Scanning Tunnelling Microscopy (STM) (5 and 6). The VPS can operate reliably in a vacuum of 10^{-6} mbar (7). The VPS can be used to deposit (regain) the plug to (from) a second cylinder. This capability can lead to an automatic probe exchange as described in the discussion section (8–11).

The lateral positioning of the probe and sample is achieved using two concentric Horizontal Positioning Systems (HPSs), as shown in Figure 3a. The inner HPS supports the sample, while the outer stage houses a tilting bridge carrying at its centre the Vertical Positioning System for the probe (Figure 3c,f).

Each HPS comprises two stacked plates, a bottom B-plate and a top T-plate and sits on three shear-piezoelectric motors (PN5FC2 from Thorlabs, Inc., Newton, NJ, USA) with ball bearing ends (Figure 3b). The three piezoelectric actuators can move in the x and y directions (Figure 3d) and they couple with the B-plate of the HPS via a 3-point kinematic mounting. The B-plate lower side contains a sapphire disk and two aligned pairs of parallel cylinders that contacts the three ball bearings and constrain the plate movement along the x-axis. The B-plate top side has the same configuration as the lower side but rotated by 90° (Figure 3e). Three ball bearings on the lower side of the T-plate couple with the B-plate constraining the T-plate movement in the y-axis. All six piezo actuators (three for each HPS) are fixed on the microscope's focusing plate (component 2 in Figure 3d) in the same orientation, ensuring a mechanical link between the sample and probe HPSs. Consequently, focus adjustments do not change the tip-sample separation—a critical requirement when the probe is a few tens of nanometres away from the sample.

A similar sawtooth waveform to the one shown in Figure 2c is used in the HPS to produce the stepping action for long-range movements (millimetre range), while the conventional piezoelectric extension (7 μm range) is used for fine positioning and scanning (with 0.1 nm resolution). The sample's position is measured with 1 μm resolution using two optical encoders (Atom^{TM} from Renishaw plc, UK). The system shown in Figure 3f has a footprint of 400 cm^2 and a range of $\pm$ 1cm for the sample stage. It is essential to notice that the top plate in both HPSs is kept in position only by gravity allowing it to be easily lifted and replaced.

Due to its compact size, the VPS can be easily integrated into a bridge design with motorised tilt, as can be seen in Figure 4a. Two opposing stepper motors are fixed on the top plate of the probe's HPS via the central shafts. When powered, the outer casing of the motors rotates rather than the shafts. The stepper motors are integrated into the two piers of the bridge, causing the bridge to tilt with an angular resolution of 0.17°. The axis of rotation, determined by the shafts' position, is at the level of the sample surface. In this way, when the cantilever's tip is close to the sample, tilting the bridge does not produce any lateral movement of the tip. In other words, the cantilever pivots around its tip.

To obtain consistent results in LMFM, it is essential to determine the cantilever's vertical tilt within a fraction of a degree. An acoustic actuator is used to oscillate the cantilever out of resonance as this ensures stable phase measurements. When the cantilever is not vertical, this oscillation causes the tip to move laterally and up and down with respect to the surface. The vertical oscillation can be observed as a modulation in the sum signal detected using the SEW method (see Figure 4b,c). A lock-in amplifier can measure this oscillation amplitude and its relative phase. A large cantilever tilt angle corresponds to a large oscillation of the sum signal. The cantilever tilt is adjusted by minimising the amplitude measured by the lock-in amplifier. As illustrated in Figure 4b, when the probe is tilted in one direction, the sum signal has a specific phase relation with the acoustic signal. If the tilt direction changes, the phase shifts by approximately 180°. The vertical position of the cantilever can be estimated within 0.17° using an automatic routine that finds the minimum amplitude with an associated sudden change in the phase signal (see Figure 4c).

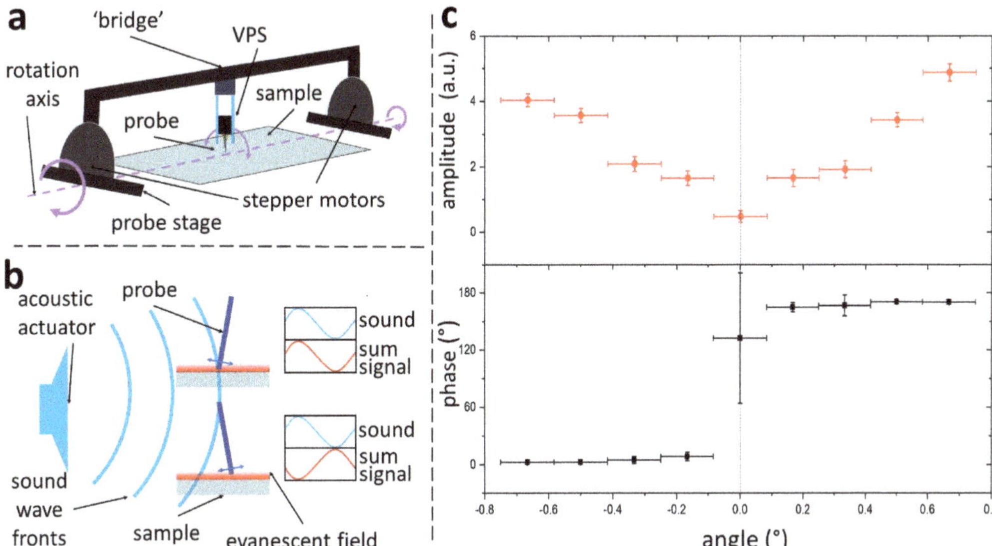

Figure 4. (**a**) Diagram of the probe's angular adjustment method. Two opposing stepper motors are embedded in the bridge carrying the VPS. A rotation of the motors causes a tilt in the bridge changing the vertical orientation of the cantilever. The tilt axis (dashed line) is level with the sample surface, so the cantilever pivots around its tip when changing the tilt. (**b**) Acoustic actuation of the probe can be used to determine its exact vertical position as described in the text. (**c**) Experimental results showing how the amplitude of vertical oscillation and relative phase change as a function of the probe's angle. Upon going through the vertical position (tilt angle $\theta = 0°$), a sharp change in phase is observed. This position also corresponds to the minimum amplitude of vertical oscillations.

2.4. Force Measurements in Intermittent Mode

The LMFM low-compliant cantilevers can be used to measure extremely small in-plane forces. In this mode, the applied force is periodically switched on and off, while the correspondent distribution of the cantilever's positions is recorded. The force is then calculated by multiplying the difference between the two distributions' means by the cantilever's spring constant. The precision of these measurements can be increased by accumulating several on/off cycles, while the accuracy is determined by the error in the spring constant k. The spring constant is determined experimentally from the power spectrum density (PSD) of the thermally fluctuating probe (not actively driven) which can be described by a Lorentzian function [45]

$$S_x(f) = \frac{k_B T}{\gamma \pi^2 (f_c^2 + f^2)},$$

(2)

where k_B is Boltzmann's constant, T is the temperature of the surrounding environment, f_c is the corner frequency, and γ is the drag coefficient. The cantilever's spring constant can be found by fitting the experimental PSD with Equation (2) and knowing that $k = 2f_c\pi\gamma$. By repeating this process multiple times, one can obtain a distribution of k values, the mean of which is used to calculate the force experienced by the probe.

Optical forces caused by an evanescent field [31] were measured using ultra-compliant silicon nitride cantilevers with spring constants in the range between 10^{-6} N/m and 10^{-5} N/m. The position of the tip was recorded at a sampling rate of 48 kHz. The data were subsequently decimated by a factor of 6 to account for an auto-correlation time of 0.125 ms [46–48]. Laser #2 (561 nm) in Figure 1 was switched on and off using a TTL signal with a frequency of 1 Hz.

3. Results and Discussion

3.1. Microscope Positioning Resolution

The VPS combines a stick-slip mechanism (coarse positioning) and a continuous voltage actuation (fine positioning). In stick-slip mode, it is possible to vary the step size allowing for millimetre-long movements and nanoscale positioning. For finer vertical positioning (e.g., less than 10 nm), the continuous voltage mode is used. Figure 5a shows the smallest steps obtained using the VPS when positioning the probe in the vertical direction. The smallest steps towards the surface ('steps down') were recorded to be $S_d = 2.79 \pm 0.09$ nm with a standard deviation of $\sigma_d = 0.52$ nm. The smallest steps away from the surface ('steps up') had a value $S_d = 2.7 \pm 0.1$ nm with a standard deviation of $\sigma_d = 0.7$ nm. Similar values are obtained when the system operates in water. The smallest step size is almost two orders of magnitude smaller than the decay length of the SEW detection system's evanescent field, ensuring a safe approach of the tip to the surface using the stick-slip mode exclusively.

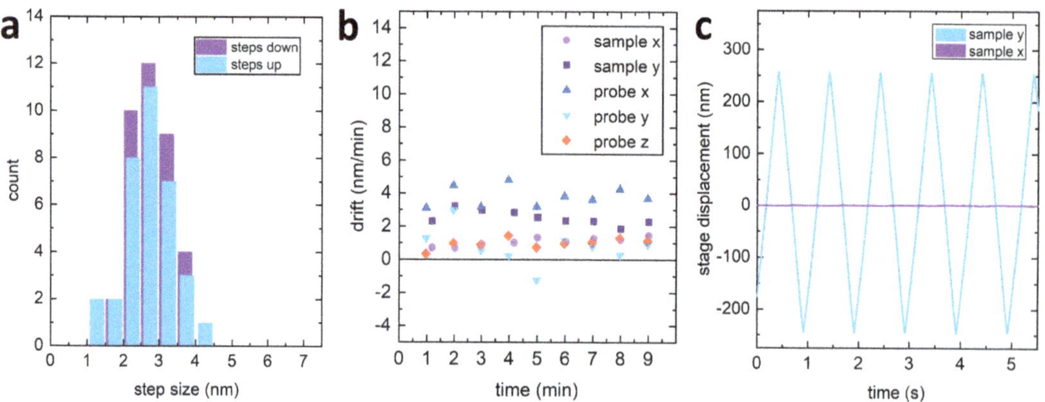

Figure 5. (**a**) Histogram of the average step size of the VPS in both upwards and downwards directions. (**b**) Measurements of the sample and probe's drift in the xy-plane and of the probe's drift in the vertical direction. (**c**) The cross-talk between displacements in orthogonal directions of the sample's HPS is around 0.3%.

3.2. Microscope Stability

The microscope's overall stability can significantly improve performance for measurements with low signal-to-noise ratio, and where long integration time is required. The use of kinematic mounts and the symmetric arrangement of the constraints result in drift values comparable to commercial AFM systems [49]. The measured mechanical drift of the probe and sample positions (see Figure 5b) had an in-plane velocity of 2.0 ± 0.2 nm/min for both the probe and the sample. A smaller drift velocity of 1.0 ± 0.1 nm/min was recorded in the vertical direction for the VPS.

The piezo actuators position in the microscope focusing plate further increases the overall stability of the HPS. The sample top plate does not have any electrical connection and relies on gravity to connect to the kinematic mount. This design allows for easy sample exchange that can potentially be automated. A conveyor belt solution, similar to the ones used for silicon wafer handling, could be implemented to automate this process (see Supplementary Video_S1).

Using the sample stage as an SPM scanning unit, the cross-talk between the fast and slow scan direction was minimised by applying a linear correction in the orthogonal axis. Figure 5c shows that the cross-talk between the two scan-axes after compensation is approximately 0.3%.

3.3. Microscope Force Resolution

Combining cantilever's stiffness between 10^{-6} N/m and 10^{-5} N/m with sub-nanometre sensitivity of the probe's position, enables force measurement with sub-femtonewton resolution. Figure 6 shows the Gaussian distributions of positions of the cantilever with and without an applied optical force. The measurements are obtained by illuminating the cantilever tip for 16 min with an intermittent evanescent field at a rate of 1 Hz. The shift between the two distributions is equal to 0.52 nm ± 0.03 nm that corresponds to an optical force of 2.7 fN ± 0.2 fN.

A time accumulation longer than 1 min results consistently in a sub-femtonewton standard error. On the opposite side of the spectrum, the system's stability allowed a 3-hour-long measurement of an evanescent optical force of 0.3 fN ± 0.1 fN. The force resolution at these accumulation times reaches a limit due to laser intensity fluctuations.

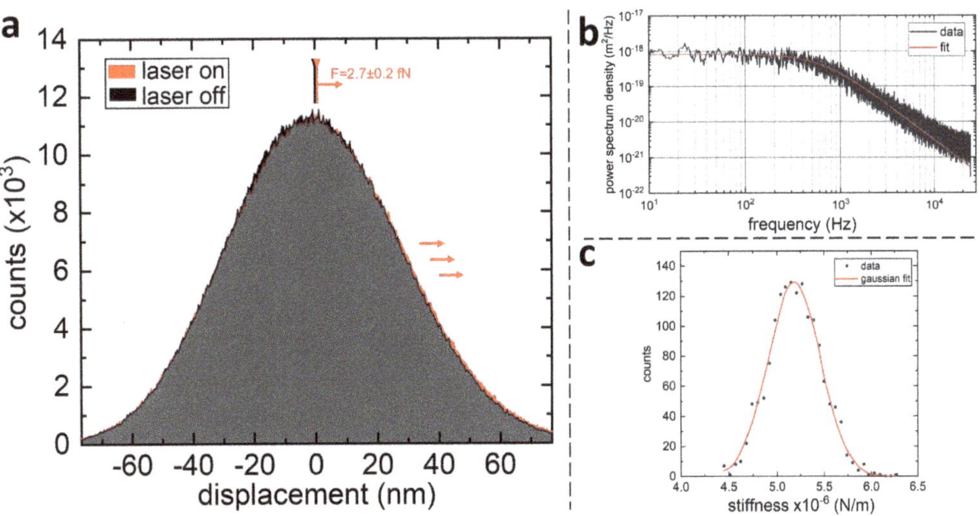

Figure 6. Force measurement with sub-fN precision. (**a**) A shift in the positive direction of the position distribution upon laser illumination (red) is visible to the right of the grey area which corresponds to the overlap of the two distributions. (**b**) A PSD of the probe's position used to evaluate the probe's stiffness. (**c**) A distribution of stiffness values for the cantilever used to calculate the force in (**a**).

So far, we have shown how an original SPM bottom-up design can combine remote control of the instrument with extreme force resolution and stability. This approach preserves various SPM modes (e.g., STM and NSOM) and can use commercially available ultra-compliant, micro-cantilevers. As with any design solution, this particular solution has intrinsic constraints; the SEW detection method requires a transparent surface, whilst the decay length of the evanescent field confines the tip's position to within 150 nanometres from the optical interface. On the other hand, many biomedical applications may benefit from a transparent substrate and integration with fluorescence microscopy. Furthermore, simple and effective operations in a liquid environment may have priority over other aspects.

Usually, AFM operating in a liquid environment is deemed unsuitable for the femtonewton and sub-femtonewton force regime [50]. This limitation is only valid when using standard micro-cantilevers in a horizontal orientation. Several examples in the literature have demonstrated that ultra-compliant custom-made cantilevers in a vertical orientation can operate in a much lower force regime [51–55] and almost any environment [32,56]. For the use of this technology in the biomedical area and based on the cantilevers' unique properties, we envisage three main types of operation. Firstly, the imaging of soft nanostructures or biomolecules in a liquid environment using a constant-height dynamic mode. This method has shown clear advantages over conventional AFM in observing self-assembling peptide cages [34] and provides molecular resolution with simpler operations [57]. Secondly, the in-situ detection of conformation changes in cell membrane proteins triggered by ligand binding to a specific receptor domain. This approach has already given a unique insight into the binding mechanism of Moraxella catarrhalis, unveiling adhesin UspA1 conformational change upon binding to fibronectin and CEACAM1 [33]. Thirdly, an in-vitro force spectroscopy mode in which conformational changes in proteins tethered between the cantilever and the surface are directly observed. The sub-nanometre control of the tip-sample separation, when using vertically oriented cantilevers, provides a unique advantage amplified by the possibility of using the large set of AFM force spectroscopy tools. This mode of operation has successfully detected the stepping action of single kinesin molecules processing on microtubules [32]. In this experiment, motor proteins are first bound to the cantilever and then moved closer to an immobilised microtubule. As soon as one protein starts processing, the microscope records the changes in the cantilever position, revealing the individual steps produced by the kinesin protein. Recently, this technique has been updated using small cylindrical glass cells around the cantilever that facilitate the cantilever incubation with kinesin molecules and its handling [58]. The VPS described here builds on this improvement and radically simplifies this type of experiments by providing an automatic approach routine with integrated vertical angle correction. The HPS with optical encoder further simplifies the positioning of the cantilever directly above immobilised microtubules.

It becomes apparent how the bottom-up design here presented builds on the already demonstrated capabilities of LMFM introducing significant advantages to the various experimental protocols, while simultaneously improving the force sensitivity and stability of the overall instrument. The following section will show how automation and machine learning are interweaved into this bottom-up design, enabling a new generation of autonomous SPMs with an ultralow-force regime.

3.4. Implementing Full-Automation and Artificial Intelligence

Automation of the SPM experiments is key to reduce the operator's bias, improve reproducibility of results and enable widespread adoption of this technology in the biomedical area [59].

One can divide the SPM workflow into different levels of automation. The first level is related to sample preparation and cantilever functionalisation for biomedical applications. Here, the technology supporting robotic-assisted assays used in high-throughput screening can provide the necessary solutions to reach this level of automation. The second level of automation includes sample and cantilever exchange. This level is essential to enable 24/7 continuous operations, and future integration with artificial intelligence (AI) [60]. Incidentally, this level is the most challenging in terms of instrument design. Most, if not all, the commercial SPMs available today are not compatible with this type of automation. The best examples of SPM with this level of automation are currently tailored for vacuum environment or semiconductor industry [61], but they would be unsuitable for biological samples.

Figure 7 describes a possible workflow in which the specific characteristics of the VPS and HPS presented here are used to ensure the first two levels of automation. More specifically, The HPS can use a conveyor belt system to enable sample exchange. First, the HPS is temporarily lowered, connecting the top (sample) plate with the conveyor belt that moves it away from the microscope. Then, a new sample plate can be loaded, performing the same actions in reverse (see Supplementary Video_S1). For automated probe exchange, the unique VPS ability of expelling and loading a shuttle carrying a probe means that we can use two VPSs to split the functionalisation step from the measurement step. A first VPS (i.e., functionalisation VPS) can be used to pick up a new plug with cantilever, perform various functionalisation steps and deliver the plug on a carrier plate. The plate is then transported to the microscope HPS as if it was a new sample. At this point, the microscope VPS can load the new plug and the carrier plate can be removed from the microscope (see Supplementary Video_S2).

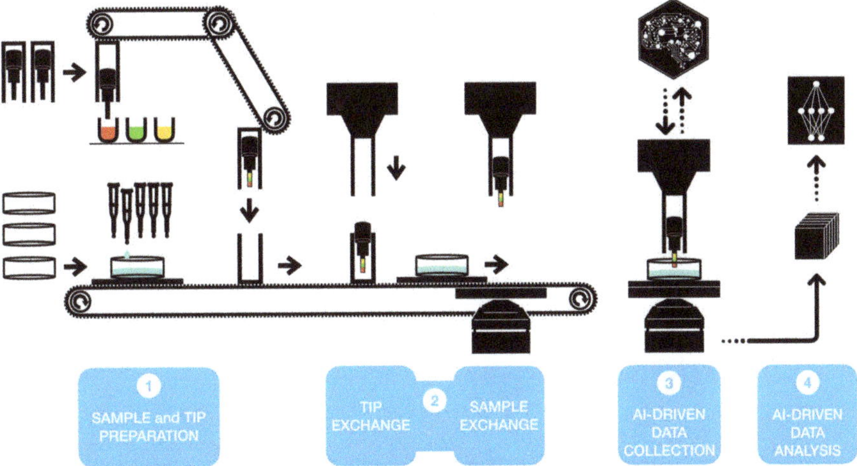

Figure 7. How the VPS and HPS could unlock the four automation levels described in the text. A conventional automated micro-pipetting combined with a conveyor system can directly prepare and deliver samples to the microscope (level 1 and 2). Similarly, the VPS can be used for probe functionalisation, while the same conveyor belt design can deliver the probe to the microscope head (see Supplementary Video_S2). AI can control the full motorised microscope for data acquisition (3) and perform the data analysis (4).

The fully motorised microscope, as described in the text, allows the implementation of the third level of automation through AI-driven operations. Recently, significant progress has been made in the automation of the AFM data collection process [59], including integration with AI-based solutions [60,62]. The significant similarities in imaging mode between conventional non-contact AFM and LMFM [34,57] imply that the AI-based solutions for AFM could be adapted for LMFM image acquisition modes. We predict that a fully motorised SPM will benefit considerably from rapid advances in AI-assisted data collection. The last level of automation is related to SPM data analysis. Application of AI to this level is already giving promising results in biomedical applications [14,63]. Rapid progress should be expected in this area, considering the vast and growing library of machine learning algorithms dedicated to interpreting images.

In conclusion, we have shown how an implementation of a bottom-up design for a particular type of SPM can combine femtonewton sensitivity with automation and become the first step towards a new class of instruments that operate fully autonomously in the force regime and environment ideal for biomedical use. Moreover, leveraging high throughput in SPM data collection and analysis will enable progressively more effective AI algorithms to transform medical research and diagnostics.

Supplementary Materials: The following are available online at https://www.mdpi.com/article/10.3390/s21093027/s1, Video_S1 and Video_S2.

Author Contributions: W.K.S. performed all the measurements (results in Figure 5c were taken with C.R.B.). W.K.S., C.R.B., R.L.H., and M.A. worked on the technical development. M.A. supervised the work. W.K.S. and M.A. wrote the manuscript. All authors have read and agreed to the published version of the manuscript.

Funding: This research was funded by EPSRC grant number EP/N509619/1 and the APC was covered by Sensors.

Informed Consent Statement: Not applicable.

Data Availability Statement: Data presented in this study is available on request from the corresponding author.

Acknowledgments: The authors would like to acknowledge Adrian Crimp for his technical assistance and contribution to building of the microscope components.

Conflicts of Interest: The authors declare no conflict of interest.

Abbreviations

The following abbreviations are used in this manuscript:

AFM	Atomic Force Microscope
DLS	Dynamic Light Scattering
DoF	Degrees of Freedom
HPS	Horizontal Positioning System
LMFM	Lateral Molecular Force Microscope
NMR	Nuclear Magnetic Resonance
NSOM	Near-field Scanning Optical Microscope
PSD	Power Spectrum Density
SEW	Scattered Evanescent Wave
SPM	Scanning Probe Microscope
STM	Scanning Tunneling Microscope
TIRF	Total Internal Reflection Fluorescence
VOP	Vertically Oriented Probe
VPS	Vertical Positioning System

References

1. Son, I.H.; Park, J.H.; Park, S.; Park, K.; Han, S.; Shin, J.; Doo, S.G.; Hwang, Y.; Chang, H.; Choi, J.W. Graphene Balls for Lithium Rechargeable Batteries with Fast Charging and High Volumetric Energy Densities. *Nat. Commun.* **2017**, *8*, 1561. [CrossRef]
2. Flores, A.M.; Hosseini-Nassab, N.; Jarr, K.U.; Ye, J.; Zhu, X.; Wirka, R.; Koh, A.L.; Tsantilas, P.; Wang, Y.; Nanda, V.; et al. Pro-Efferocytic Nanoparticles Are Specifically Taken up by Lesional Macrophages and Prevent Atherosclerosis. *Nat. Nanotechnol.* **2020**, *15*, 154–161. [CrossRef] [PubMed]
3. He, X.; Hwang, H.M. Nanotechnology in Food Science: Functionality, Applicability, and Safety Assessment. *J. Food Drug Anal.* **2016**, *24*, 671–681. [CrossRef]
4. Comparetti, E.J.; Pedrosa, V.d.A.; Kaneno, R. Carbon Nanotube as a Tool for Fighting Cancer. *Bioconjugate Chem.* **2018**, *29*, 709–718. [CrossRef]
5. Roco, M.C. Nanotechnology: Convergence with Modern Biology and Medicine. *Curr. Opin. Biotechnol.* **2003**, *14*, 337–346. [CrossRef]

.   Tegart, G. Nanotechnology: The Technology for the 21st Century. *The Second International Conference on Technology Foresight.* 2003. Available online: https://www.nistep.go.jp/IC/ic030227/pdf/p2-3.pdf (accessed on 23 February 2021).

.   Dinarelli, S.; Girasole, M.; Kasas, S.; Longo, G. Nanotools and Molecular Techniques to Rapidly Identify and Fight Bacterial Infections. *J. Microbiol. Methods* **2017**, *138*, 72–81. [CrossRef] [PubMed]

.   Cross, S.E.; Jin, Y.S.; Rao, J.; Gimzewski, J.K. Nanomechanical Analysis of Cells from Cancer Patients. *Nat. Nanotechnol.* **2007**, *2*, 780–783. [CrossRef]

.   Cross, S.E.; Jin, Y.S.; Tondre, J.; Wong, R.; Rao, J.; Gimzewski, J.K. AFM-Based Analysis of Human Metastatic Cancer Cells. *Nanotechnology* **2008**, *19*, 384003. [CrossRef] [PubMed]

0.   Plodinec, M.; Loparic, M.; Monnier, C.A.; Obermann, E.C.; Zanetti-Dallenbach, R.; Oertle, P.; Hyotyla, J.T.; Aebi, U.; Bentires-Alj, M.; Lim, R.Y.H.; et al. The Nanomechanical Signature of Breast Cancer. *Nat. Nanotechnol.* **2012**, *7*, 757–765. [CrossRef]

1.   Xu, W.; Mezencev, R.; Kim, B.; Wang, L.; McDonald, J.; Sulchek, T. Cell Stiffness Is a Biomarker of the Metastatic Potential of Ovarian Cancer Cells. *PLoS ONE* **2012**, *7*, e46609. [CrossRef]

2.   Sharma, S.; Santiskulvong, C.; Rao, J.; Gimzewski, J.K.; Dorigo, O. The Role of Rho GTPase in Cell Stiffness and Cisplatin Resistance in Ovarian Cancer Cells. *Integr. Biol.* **2014**, *6*, 611–617. [CrossRef]

3.   Yu, W.; Sharma, S.; Gimzewski, J.K.; Rao, J. Nanocytology as a Potential Biomarker for Cancer. *Biomark. Med.* **2017**, *11*, 213–216. [CrossRef]

4.   Ciasca, G.; Mazzini, A.; Sassun, T.E.; Nardini, M.; Minelli, E.; Papi, M.; Palmieri, V.; de Spirito, M. Efficient Spatial Sampling for AFM-Based Cancer Diagnostics: A Comparison between Neural Networks and Conventional Data Analysis. *Condens. Matter* **2019**, *4*, 58. [CrossRef]

5.   Longo, G.; Alonso-Sarduy, L.; Rio, L.M.; Bizzini, A.; Trampuz, A.; Notz, J.; Dietler, G.; Kasas, S. Rapid Detection of Bacterial Resistance to Antibiotics Using AFM Cantilevers as Nanomechanical Sensors. *Nat. Nanotechnol.* **2013**, *8*, 522–526. [CrossRef]

6.   Parry, B.R.; Surovtsev, I.V.; Cabeen, M.T.; O'Hern, C.S.; Dufresne, E.R.; Jacobs-Wagner, C. The Bacterial Cytoplasm Has Glass-like Properties and Is Fluidized by Metabolic Activity. *Cell* **2014**, *156*, 183–194. [CrossRef] [PubMed]

7.   Syal, K.; Iriya, R.; Yang, Y.; Yu, H.; Wang, S.; Haydel, S.E.; Chen, H.Y.; Tao, N. Antimicrobial Susceptibility Test with Plasmonic Imaging and Tracking of Single Bacterial Motions on Nanometer Scale. *ACS Nano* **2016**, *10*, 845–852. [CrossRef]

8.   Dufrêne, Y.F.; Ando, T.; Garcia, R.; Alsteens, D.; Martinez-Martin, D.; Engel, A.; Gerber, C.; Müller, D.J. Imaging Modes of Atomic Force Microscopy for Application in Molecular and Cell Biology. *Nat. Nanotechnol.* **2017**, *12*, 295–307. [CrossRef] [PubMed]

9.   Binnig, G.; Quate, C.F.; Gerber, C. Atomic Force Microscope. *Phys. Rev. Lett.* **1986**, *56*, 930–933. [CrossRef] [PubMed]

0.   Stetefeld, J.; McKenna, S.A.; Patel, T.R. Dynamic Light Scattering: A Practical Guide and Applications in Biomedical Sciences. *Biophys. Rev.* **2016**, *8*, 409–427. [CrossRef] [PubMed]

1.   Jones, R.R.; Hooper, D.C.; Zhang, L.; Wolverson, D.; Valev, V.K. Raman Techniques: Fundamentals and Frontiers. *Nanoscale Res. Lett.* **2019**, *14*, 231. [CrossRef]

2.   Passe, T.J.; Charles, H.; Rajagopalan, P.; Krishnan, K. Nuclear Magnetic Resonance Spectroscopy: A Review of Neuropsychiatric Applications. *Prog. Neuro-Psychopharmacol. Biol. Psychiatry* **1995**, *19*, 541–563. [CrossRef]

3.   Zia, K.; Siddiqui, T.; Ali, S.; Farooq, I.; Zafar, M.S.; Khurshid, Z. Nuclear Magnetic Resonance Spectroscopy for Medical and Dental Applications: A Comprehensive Review. *Eur. J. Dent.* **2019**, *13*, 124–128. [CrossRef]

4.   Maver, U.; Velnar, T.; Gaberšček, M.; Planinšek, O.; Finšgar, M. Recent Progressive Use of Atomic Force Microscopy in Biomedical Applications. *TrAC Trends Anal. Chem.* **2016**, *80*, 96–111. [CrossRef]

5.   Sharma, S.; Gimzewski, J.K. Application of AFM to the Nanomechanics of Cancer. *MRS Adv.* **2016**, *1*, 1817–1827. [CrossRef]

6.   Sharma, S.; LeClaire, M.; Gimzewski, J.K. Ascent of Atomic Force Microscopy as a Nanoanalytical Tool for Exosomes and Other Extracellular Vesicles. *Nanotechnology* **2018**, *29*, 132001. [CrossRef]

7.   Dinu, C.Z.; Dong, C.; Hu, X. Current Status and Perspectives in Atomic Force Microscopy-Based Identification of Cellular Transformation. *Int. J. Nanomed.* **2016**, 2107. [CrossRef]

8.   Kuznetsova, T.G.; Starodubtseva, M.N.; Yegorenkov, N.I.; Chizhik, S.A.; Zhdanov, R.I. Atomic Force Microscopy Probing of Cell Elasticity. *Micron* **2007**, *38*, 824–833. [CrossRef] [PubMed]

9.   Antognozzi, M.; Ulcinas, A.; Picco, L.; Simpson, S.H.; Heard, P.J.; Szczelkun, M.D.; Brenner, B.; Miles, M.J. A New Detection System for Extremely Small Vertically Mounted Cantilevers. *Nanotechnology* **2008**, *19*, 384002. [CrossRef] [PubMed]

0.   Rugar, D.; Budakian, R.; Mamin, H.J.; Chui, B.W. Single Spin Detection by Magnetic Resonance Force Microscopy. *Nature* **2004**, *430*, 4. [CrossRef]

1.   Antognozzi, M.; Bermingham, C.R.; Harniman, R.L.; Simpson, S.; Senior, J.; Hayward, R.; Hoerber, H.; Dennis, M.R.; Bekshaev, A.Y.; Bliokh, K.Y.; et al. Direct Measurements of the Extraordinary Optical Momentum and Transverse Spin-Dependent Force Using a Nano-Cantilever. *Nat. Phys.* **2016**, *12*, 731–735. [CrossRef]

2.   Scholz, T.; Vicary, J.A.; Jeppesen, G.M.; Ulcinas, A.; Hörber, J.K.H.; Antognozzi, M. Processive Behaviour of Kinesin Observed Using Micro-Fabricated Cantilevers. *Nanotechnology* **2011**, *22*, 095707. [CrossRef] [PubMed]

3.   Agnew, C.; Borodina, E.; Zaccai, N.R.; Conners, R.; Burton, N.M.; Vicary, J.A.; Cole, D.K.; Antognozzi, M.; Virji, M.; Brady, R.L. Correlation of in Situ Mechanosensitive Responses of the Moraxella Catarrhalis Adhesin UspA1 with Fibronectin and Receptor CEACAM1 Binding. *Proc. Natl. Acad. Sci. USA* **2011**, *108*, 15174–15178. [CrossRef] [PubMed]

4.   Fletcher, J.M.; Harniman, R.L.; Barnes, F.R.H.; Boyle, A.L.; Collins, A.; Mantell, J.; Sharp, T.H.; Antognozzi, M.; Booth, P.J.; Linden, N.; et al. Self-Assembling Cages from Coiled-Coil Peptide Modules. *Science* **2013**, *340*, 595–599. [CrossRef]

35. Vicary, J.; Ulcinas, A.; Hörber, J.; Antognozzi, M. Micro-Fabricated Mechanical Sensors for Lateral Molecular-Force Microscopy. *Ultramicroscopy* **2011**, *111*, 1547–1552. [CrossRef] [PubMed]
36. Burnham, N.A.; Colton, R.J. Measuring the Nanomechanical Properties and Surface Forces of Materials Using an Atomic Force Microscope. *J. Vac. Sci. Technol. Vacuum Surf. Film.* **1989**, *7*, 2906–2913. [CrossRef]
37. Smith, D.P.E. Limits of Force Microscopy. *Rev. Sci. Instrum.* **1995**, *66*, 3191–3195. [CrossRef]
38. Attard, P.; Carambassis, A.; Rutland, M.W. Dynamic Surface Force Measurement. 2. Friction and the Atomic Force Microscope. *Langmuir* **1999**, *15*, 553–563. [CrossRef]
39. Pohl, D.W. Dynamic Piezoelectric Translation Devices. *Rev. Sci. Instrum.* **1987**, *58*, 54–57. [CrossRef]
40. Han, W.; Mou, J.; Sheng, J.; Yang, J.; Shao, Z. Cryo Atomic Force Microscopy: A New Approach for Biological Imaging at High Resolution. *Biochemistry* **1995**, *34*, 8215–8220. [CrossRef]
41. Shao, Z.; Zhang, Y. Biological Cryo Atomic Force Microscopy: A Brief Review. *Ultramicroscopy* **1996**, *66*, 141–152. [CrossRef]
42. Qin, L.; Zhang, J.; Sun, J.; Czajkowsky, D.M.; Shao, Z. Development of a Low-Noise Cryogenic Atomic Force Microscope for High Resolution Imaging of Large Biological Complexes. *Acta Biochim. Biophys. Sin.* **2016**, *48*, 859–861. [CrossRef]
43. Drevniok, B.; Paul, W.M.P.; Hairsine, K.R.; McLean, A.B. Methods and Instrumentation for Piezoelectric Motors. *Rev. Sci. Instrum.* **2012**, *83*, 033706. [CrossRef] [PubMed]
44. Edeler, C.; Meyer, I.; Fatikow, S. Modeling of Stick-Slip Micro-Drives. *J. -Micro-Nano Mechatron.* **2011**, *6*, 65–87. [CrossRef]
45. Gittes, F.; Schmidt, C.F. Thermal Noise Limitations on Micromechanical Experiments. *Eur. Biophys. J.* **1998**, *27*, 75–81. [CrossRef]
46. Pralle, A.; Florin, E.L.; Stelzer, E.; Hörber, J. Local Viscosity Probed by Photonic Force Microscopy. *Appl. Phys. Mater. Sci. Process* **1998**, *66*, S71–S73. [CrossRef]
47. Tischer, C.; Pralle, A.; Florin, E.L. Determination and Correction of Position Detection Nonlinearity in Single Particle Tracking and Three-Dimensional Scanning Probe Microscopy. *Microsc. Microanal.* **2004**, *10*, 425–434. [CrossRef] [PubMed]
48. Tassieri, M.; Giudice, F.D.; Robertson, E.J.; Jain, N.; Fries, B.; Wilson, R.; Glidle, A.; Greco, F.; Netti, P.A.; Maffettone, P.L.; et al. Microrheology with Optical Tweezers: Measuring the Relative Viscosity of Solutions 'at a Glance'. *Sci. Rep.* **2015**, *5*, 8831 [CrossRef]
49. Marinello, F.; Balcon, M.; Schiavuta, P.; Carmignato, S.; Savio, E. Thermal Drift Study on Different Commercial Scanning Probe Microscopes during the Initial Warming-up Phase. *Meas. Sci. Technol.* **2011**, *22*, 094016. [CrossRef]
50. Neuman, K.C.; Nagy, A. Single-Molecule Force Spectroscopy: Optical Tweezers, Magnetic Tweezers and Atomic Force Microscopy. *Nat. Methods* **2008**, *5*, 491–505. [CrossRef] [PubMed]
51. Mueller, F.; Heugel, S.; Wang, L.J. Femto-Newton Light Force Measurement at the Thermal Noise Limit. *Opt. Lett.* **2008**, *33*, 539 [CrossRef] [PubMed]
52. Longenecker, J.G.; Moore, E.W.; Marohn, J.A. Rapid Serial Prototyping of Magnet-Tipped Attonewton-Sensitivity Cantilevers by Focused Ion Beam Manipulation. *J. Vac. Sci. Technol. Nanotechnol. Microelectron.* **2011**, *29*, 032001. [CrossRef] [PubMed]
53. Kim, Y.W.; Choi, H.H.; Choi, J.H.; Lee, S.G. Fabrication and Characterization of an Attonewton-Sensitivity Si Cantilever with an Nb Micro-Ring. *J. Korean Phys. Soc.* **2012**, *60*, 973–977. [CrossRef]
54. Hickman, S.A.; Moore, E.W.; Lee, S.; Longenecker, J.G.; Wright, S.J.; Harrell, L.E.; Marohn, J.A. Batch-Fabrication of Cantilevered Magnets on Attonewton-Sensitivity Mechanical Oscillators for Scanned-Probe Nanoscale Magnetic Resonance Imaging. *ACS Nano* **2010**, *4*, 7141–7150. [CrossRef]
55. Mamin, H.J.; Rugar, D. Sub-Attonewton Force Detection at Millikelvin Temperatures. *Appl. Phys. Lett.* **2001**, *79*, 3358–3360 [CrossRef]
56. de Lépinay, L.M.; Pigeau, B.; Besga, B.; Vincent, P.; Poncharal, P.; Arcizet, O. A Universal and Ultrasensitive Vectorial Nanomechanical Sensor for Imaging 2D Force Fields. *Nat. Nanotechnol.* **2016**, *12*, 156–162. [CrossRef] [PubMed]
57. Harniman, R.L.; Vicary, J.A.; Hörber, J.K.H.; Picco, L.M.; Miles, M.J.; Antognozzi, M. Methods for Imaging DNA in Liquid with Lateral Molecular-Force Microscopy. *Nanotechnology* **2012**, *23*, 085703. [CrossRef] [PubMed]
58. Bermingham, C.R. Measurement of Pico/Femto-Newton Scale Forces Using the Lateral Molecular Force Microscope. Ph.D. Thesis, University of Bristol, Bristol, UK, 2016.
59. Dujardin, A.; De Wolf, P.; Lafont, F.; Dupres, V. Automated Multi-Sample Acquisition and Analysis Using Atomic Force Microscopy for Biomedical Applications. *PLoS ONE* **2019**, *14*, e0213853. [CrossRef]
60. Huang, B.; Li, Z.; Li, J. An Artificial Intelligence Atomic Force Microscope Enabled by Machine Learning. *Nanoscale* **2018**, *10*, 21320–21326. [CrossRef]
61. Hasumura, S.; Wakiyama, S.; Iyoki, M.; Ando, K. Measurement of Microscopic Three-Dimensional Profiles with High Accuracy and Simple Operation. *Hitachi Rev.* **2016**, *65*, 5.
62. Krull, A.; Hirsch, P.; Rother, C.; Schiffrin, A.; Krull, C. Artificial-Intelligence-Driven Scanning Probe Microscopy. *Commun. Phys.* **2020**, *3*, 54. [CrossRef]
63. Sokolov, I.; Dokukin, M.E.; Kalaparthi, V.; Miljkovic, M.; Wang, A.; Seigne, J.D.; Grivas, P.; Demidenko, E. Noninvasive Diagnostic Imaging Using Machine-Learning Analysis of Nanoresolution Images of Cell Surfaces: Detection of Bladder Cancer. *Proc. Natl Acad. Sci. USA* **2018**, *115*, 12920–12925. [CrossRef] [PubMed]

MDPI
St. Alban-Anlage 66
4052 Basel
Switzerland
Tel. +41 61 683 77 34
Fax +41 61 302 89 18
www.mdpi.com

Sensors Editorial Office
E-mail: sensors@mdpi.com
www.mdpi.com/journal/sensors